ENGINEER-IN-TRAINING EXAMINATION REVIEW

THIRD EDITION

Donald G. Newnan, Ph.D., P.E.

Bruce E. Larock, Ph.D., P.E.
University of California, Davis

A WILEY-INTERSCIENCE PUBLICATION

JOHN WILEY & SONS, INC.

New York / Chichester / Brisbane / Toronto

Library of Congress Cataloging-in-Publication Data:

Newnan, Donald G.
    Engineer-in-training examination review / Donald G. Newnan, Bruce
E. Larock.--3rd ed.
        p.   cm.
    Includes bibliographical references.
    1. Engineering--Examinations, questions, etc.   2. Engineering-
-Problems, exercises, etc.   I. Larock, Bruce E., 1940–
II. Title.
TA159.N416   1990
620'.0076--dc20
    ISBN 0-471-50827-6   90-12329
                                                                          CIP
Printed in the United States of America

10 9 8 7 6 5 4 3

Printed and bound by Courier Companies, Inc.

CONTENTS

Appendices   569

Index   593

PREFACE

One of the first steps an engineer takes in his or her transition from college student to practicing engineer is to begin the process of becoming a registered Professional Engineer. The starting point is taking and passing the Engineer-In-Training examination. (It may also be called the EIT or Engineering Fundamentals examination.) To assist in this process, we have assembled in this book relevant information about the examination and its content.

The first chapter, Introductory Comments, outlines the examination, its structure and technical content. In the eleven chapters that follow, each subject area in the examination is methodically reviewed. There is an orderly discussion of the important concepts and principles in a subject followed by typical problems, each with a detailed step-by-step solution. We have carefully tailored each chapter with just one purpose—to help the reader prepare successfully and efficiently for the Engineer-In-Training examination.

This third edition continues our interest and involvement in, and commitment to, the preparation of engineers and other technical staff members for careers as productive professionals. In updating and modernizing the previous edition we have added two new chapters and made more than minor editorial changes in four other chapters. In preparing this edition we have had the assistance of Lincoln D. Jones, Professor of Electrical Engineering, and K. S. Sree Harsha, Professor of Materials Engineering, San Jose State University. We have also been helped by Michael Ames, Director of the Computer Center, and Professor Jan K. Wolski, both of the New Mexico Institute of Mining and Technology. Of course, the final responsibility for the correctness of all material in this edition resides with the authors. If you note any errors in this book, we would appreciate being told about them. To write to us, send a letter to Prof. Bruce E. Larock, Civil Engineering Department, University of California, Davis, California 95616. Finally, we wish you success with the exam!

DONALD G. NEWNAN
BRUCE E. LAROCK

1

INTRODUCTORY
COMMENTS

BECOMING A PROFESSIONAL ENGINEER

In each of the 50 states, laws regulate the practice of engineering and the right of an individual to designate himself or herself as an engineer. These laws were enacted to protect the public from people who might call themselves engineers but who in reality are not qualified to perform competent engineering work. Thus registration as a professional engineer (PE) is a desirable and often mandatory goal for most engineers.

There are four steps to becoming a registered professional engineer.

1. Education. The usual goal is the completion of a B.S. degree in engineering from an accredited college or university. While this is not an absolute requirement, it is much more difficult for people to become registered if they either lack the college degree or have a degree from a non-accredited institution.

2. Engineer-In-Training (Engineering Fundamentals) Examination. Most people are required to take and pass this eight-hour examination. The content covers the broad common body of knowledge that all engineers receive in their undergraduate education. The requirements to be eligible to take this examination vary from state to state. In some states one may only need good moral character and the ability to pay the application fee. Other states may also require up to four years of engineering education and/or experience.

3. Experience. After passing the Engineering Fundamentals examination there is a further experience requirement before one is permitted to take the Professional Engineer examination. The typical requirement is from two to four years of experience, following the Engineering Fundamentals examination, along with engineering references. It may, of course, take more than two years to acquire two years of acceptable engineering experience.

4. *Professional Engineer (PE) Examination.* The second exam is called the *Principles and Practice of Engineering* by the people who prepare it, but probably everyone else calls it the *Professional Engineer* or *PE* exam. This examination presents current professional situations in the applicant's particular branch of engineering. The problems require an understanding of engineering fundamentals along with additional knowledge, judgment, and insight. Passing this eight-hour examination is the final hurdle to becoming a Registered Professional Engineer.

THE ENGINEER-IN-TRAINING (EIT) EXAMINATION

The examination on the fundamentals of engineering is prepared by the National Council of Examiners for Engineering and Surveying (NCEES) and is known by various names. In most states it is called the Engineer-In-Training or EIT examination. Elsewhere it may be called the Engineering Fundamentals examination or the Intern Engineer examination. Engineers in all 50 states (and the District of Columbia, Puerto Rico, and Guam) take the *same* nationally prepared examination at the same time. Engineers from all branches of engineering (CE, ME, EE, and so on) take this common engineering fundamentals examination. The examination is given about the third week in both April and October of each year.

The eight-hour examination is split into two four-hour sessions. The morning session consists of 140 multiple-choice questions, each with five possible answers. Each examinee is expected to answer all of the questions. The topics and the approximate number of questions are as follows:

	Questions
Mathematics	20
Statics	14
Dynamics	14
Mechanics of Materials	11
Fluid Mechanics	14
Thermodynamics	14
Chemistry	14
Materials Science/Structure of Matter	14
Electrical Circuits	14
Engineering Economics	11
	140

The afternoon session consists of 70 required questions:

	Questions
Engineering Mechanics (Statics, Dynamics, and Mechanics of Materials)	20
Applied Mathematics	20
Electrical Circuits	10
Thermodynamics/Fluid Mechanics	10
Engineering Economics	10
	70

EXAMINATION TOPICS

The National Council of Examiners for Engineering and Surveying provides a detailed listing of the typical content of each topic.

Morning Session

Mathematics

Analytic geometry
Differential calculus
Integral calculus
Differential equations
Laplace transforms
Probability and statistics
Difference equations
Linear algebra
Roots of equations
Vector analysis

This topic may also include some number systems and computer algorithm problems.

Statics

Vector forces
Two-dimensional equilibrium
Three-dimensional equilibrium
Concurrent force systems

Centroid of area
Moment of inertia
Friction

Dynamics

Friction
Kinematics
Force, mass, and acceleration (kinetics)
Impulse and momentum
Work and energy
Response of first- and second-order systems including vibration

Mechanics of Materials

Stress and strain
Tension and compression
Shear
Beams
Columns
Torsion
Bending

Fluid Mechanics

External flows
Fluid properties
Fluid statics
Pipe and other internal flows
Flow measurement
Impulse and momentum
Similitude and dimensional analysis

Thermodynamics

Properties: enthalpy, entropy, free energy
Thermodynamic processes
Energy, heat, and work
Phase changes
First law
Second law
Availability-reversibility
Cycles

Ideal gases
Mixture of gases
Heat transfer

Chemistry

Nomenclature
Equations
Stoichiometry
Periodicity
States of matter
Metals and nonmetals
Oxidation and reduction
Acids and bases
Solutions
Kinetics
Equilibrium
Electrochemistry
Organic chemistry

Materials Science and Structure of Matter

Atomic structure
Properties
Materials
Crystallography
Phase diagrams
Processing and testing
Diffusion
Corrosion

Electrical Circuits

DC circuits
AC circuits
Three-phase circuits
Capacitance and inductance
Electric and magnetic fields
Operational amplifiers (ideal)

Engineering Economics

Annual cost

Present worth
Future worth or value
Breakeven analysis
Valuation and depreciation

Afternoon Session

Engineering Mechanics

Statics
    Resultants of force systems
    Equilibria of rigid bodies
    Analysis of internal forces
    Friction
Dynamics
    Kinematics
    Friction
    Inertia and force
    Particles and solid bodies
    Impulse and momentum
    Work and energy
Mechanics of materials
    Stress and strain
    Tension and compression
    Shear
    Simple beams
    Torsion
    Bending
    Columns

Applied Mathematics

Analytic geometry
Differential equations
Differential calculus
Integral calculus
Linear algebra and vector analysis
Probability and statistics

Electrical Circuits

DC circuits

AC circuits
Diode applications
Three-phase circuits
Capacitance and inductance
Electric and magnetic fields
Operational amplifiers (ideal)
Transients

This topic may also include some computer organization problems and problem solving with FORTRAN and BASIC.

Engineering Economics

Annual cost
Benefit-cost analysis
Future worth or value
Present worth
Breakeven analysis
Rate of return analysis
Risk analysis
Tax considerations
Valuation and depreciation

Thermodynamics/Fluid Mechanics

Thermodynamics
   Chemical reactions
   Cycles
   Refrigeration
   Air conditioning
   Combustion
   Gas mixtures
   Flow processes
   Availability and reversibility
   Heat transfer
Fluid mechanics
   Compressible flow
   External flow
   Flow measurement
   Fluid statics
   Hydraulics and fluid machines

Impulse and momentum

Pipe and channel flow

Similitude and dimensional analysis

EXAMINATION ORGANIZATION

The problems for each topic (like mathematics) are grouped together in the examination. Thus one might first encounter 14 mathematics problems, followed by 14 thermodynamics problems. To reduce the possibility of copying in the examination, it is prepared in two or three forms so that adjacent examinees take the same test but encounter the groups of problems in a different sequence.

TAKING THE EXAMINATION

The EIT examination covers the topics that are normally taken no later than the junior year of an engineer's undergraduate education. Thus one should take the examination as soon as he or she has completed the courses covered by the examination and meets the examination eligibility requirements of the particular state. The longer one delays, the harder it will be to recall the material and pass the examination.

There are lots of questions and very little time. In the morning one has 240 minutes to answer 140 problems, or one minute and forty seconds each. Time must be allocated carefully. A practical strategy would be to go through the examination working and answering all questions that are familiar and which can be handled promptly (and hopefully correctly). Problems that are passed up in this first round could be marked so they can quickly be found later and either worked on or guessed at, as time runs down. The afternoon session is similar to the morning, but with twice as much time (3 minutes 20 seconds) per problem.

If time permits, it just makes sense to check your work. You should make sure that all questions have been answered and no duplicate marks are recorded, and so on. You may want to look up some equations, or check some units. If one worries over a multiple-choice problem too long, after a while nothing looks right. We recommend therefore that you do *not* change a multiple choice answer unless there is a clear reason for doing so.

WHAT TO TAKE TO THE EXAMINATION

The EIT exam is open-book. One may bring to the exam:

Textbooks

Handbooks

Bound reference materials
Battery operated silent non-printing calculators

One may *not* bring:

Writing tablets (there is space for scratch work in the examination book)
Unbound tables or notes

Since the examination allows little time to look up anything, it may be unrealistic to bring very much to the examination. If you have retained your college textbooks, you may want to bring the appropriate ones. They will be like old friends, in comparison with unfamiliar new reference materials. Apparently there is no regulation that prevents you from preparing reference notes and tables and putting them into three-ring binders. This seems to represent "bound" reference materials. It would be prudent to check with your own state board of registration if you plan to bring materials in three-ring binders. A listing of the state boards is given in Appendix D.

Since so much depends on your hand-held calculator, you should bring an extra set of fresh batteries, and maybe even a second calculator. Also, it is imperative that you have a time plan for the examination and bring a watch.

If the examination site is not too far away, it would be a good idea to make an advance visit. That way you can plan where to park, where to eat lunch (or to bring a lunch), where the restrooms are, the water fountain, and so on. Possibly the more important reason for the advance visit is to give one confidence about the situation and how to cope with the various potential problems on examination day.

EXAMINATION SCORING

You must answer 140 questions in the morning and another 70 in the afternoon. The afternoon problems are more likely to require data manipulation, and they each have twice the grading weight of the morning problems. The EIT examination is administered by the state engineering boards, but it is prepared by NCEES with the assistance of the Educational Testing Service. Thus this is a professionally prepared examination. The completed examination books are sent to NCEES/ETS for scoring.

The examination is scored on the basis of correct answers only. No penalty is subtracted for incorrect answers. Thus it is important that answers be provided for all the 140 morning and 70 afternoon problems. Guessing is clearly better than no answer.

The score on the 140 morning plus 70 double-weighted afternoon problems provides the examinee's raw score (280 possible). Using test scaling procedures, a raw score of about 140 is set as 70 on a 100-point scale. Since 70 is the passing grade usually required in the various state laws, it is important to note that this does *not* mean one must answer 70% of the examination correctly. *The actual passing grade of 70 is obtained by answering about half the raw score points correctly.* The exact raw score that is set equal to 70 on the 100-point scale will vary slightly from examination to examination to adjust for slightly differing examination difficulty. About two-thirds of all the people who take the EIT examination pass.

THIS BOOK

This book has two purposes: to improve the reader's knowledge of the content of the engineering fundamentals/engineer-in-training examination, and to sharpen the reader's skill and confidence in dealing with multiple-choice questions at this difficulty level in a timed situation.

Each chapter begins with a short review of the fundamental principles that are conceptually important and practically useful in the topic. The goal is to present basic ideas compactly and directly; lengthy, detailed derivations are avoided. The result is a selective overview of the discipline rather than an encyclopedic coverage. The remainder of each chapter consists of multiple-choice problems. An important feature of the book is the presentation of a complete solution for each problem. The International System of Units, SI, is used in some of the problems. Conversion factors and other information on SI units are provided in an Appendix.

By reviewing the technical content of the engineer-in-training examination and working the more than 550 problems in the book, you should be able to assess where additional work and study is required. In this manner both your technical competency and test-taking skills can be improved.

2

MATHEMATICS

According to some, mathematics is *the* most fundamental branch of all science. Indeed, the goal of much scientific and technical work is to express in precise mathematical terms the behavior of our universe and its smaller component parts. In varying degrees mathematics is used in all the disciplines that together make up engineering fundamentals. For this reason it is highly desirable to have a working knowledge of some basic relations in algebra and linear algebra, trigonometry, plane and analytic geometry, calculus, ordinary differential equations and some elements of probability and statistics. We present here a *brief* review of fundamental principles; a thorough review, including proofs, is outside the scope of this volume.

ALGEBRA AND LINEAR ALGEBRA

The basic rules of algebra apply equally well to real and complex numbers, that is, numbers expressible in the form $a_1 + ia_2$, where a_1 and a_2 are real numbers, zero or nonzero, and $i^2 = -1$. The basic rules, given in additive and multiplicative form, are three:

Commutative: $\quad a + b = b + a \qquad\qquad ab = ba$

Distributive: $\quad a(b + c) = ab + ac$

Associative: $\quad a + (b + c) = (a + b) + c \quad a(bc) = (ab)c$

All but one of these rules also hold when a, b and c are matrices. The important exception is that matrix multiplication is in general *not* commutative; that is, $ab \neq ba$.

The laws of exponents and logarithms are intimately related. For positive numbers a and b and any positive or negative exponents x and y, the rules for exponents are as follows:

$$b^{-x} = \frac{1}{b^x} \qquad b^x b^y = b^{x+y}$$

$$(ab)^x = a^x b^x \qquad b^{xy} = (b^x)^y$$

If $b^y = x$ for positive b and x, then $y = \log_b x$ is the definition of the logarithm of x to the base b. The logarithm is therefore a kind of exponent. The most commonly used base numbers are $b = 10$ for common logarithms and $b = e = 2.718\ldots$ for natural logarithms. (When $b = 10$ it is often not written down; when $b = e$ often $\log_e = \ln$ is written.) Regardless of the value of b these laws hold for logarithms:

$$b^{\log_b x} = x \qquad \log_b b^x = x$$

$$\log_b (xy^n) = \log_b x + n \log_b y \qquad \text{for any value of } n$$

To change, for example, the base of a logarithm from any base b to the base e,

$$\log_b x = \frac{\log_e x}{\log_e b} = \frac{\ln x}{\ln b} = \log_b e \times \ln x$$

since $(\log_b e)(\log_e b) = \log_e (b^{\log_b e}) = \log_e e = 1$.

An entire branch of mathematics, linear algebra, has grown out of an interest in solving sets of linear, simultaneous algebraic equations. The field is a generalization of solving the equation $ax = b$, which is linear in the one unknown x and has the obvious solution $x = b/a$. For two simultaneous equations in the unknowns x and y, the equations are often solved by eliminating y and solving for x:

$$a_{11}x + a_{12}y = b_1$$

$$a_{21}x + a_{22}y = b_2$$

From the first equation $y = (1/a_{12})(b_1 - a_{11}x)$. Insertion of this expression for y into the second equation yields an equation of the form of the single linear equation, which is easily solved.

The foregoing problem and also larger sets of simultaneous linear equations can be solved by using determinants. The determinant of the coefficients in the problem is

$$D = \begin{vmatrix} a_{11} & a_{12} \\ a_{21} & a_{22} \end{vmatrix} = a_{11}a_{22} - a_{12}a_{21}$$

If D is nonzero, then Cramer's rule gives the solution for x and y as

$$x = \frac{D_1}{D} \qquad y = \frac{D_2}{D}$$

D_1 is formed from D by replacing a_{11} and a_{21} by b_1 and b_2, respectively. To find D_2, replace the second column of a's by the b's. The same procedure is followed for three or more unknown variables. A 3×3 determinant can be reduced to 2×2 determinants by expanding it in terms of minors along one column or row. For example, expanding along the first row we have

$$\begin{vmatrix} a_{11} & a_{12} & a_{13} \\ a_{21} & a_{22} & a_{23} \\ a_{31} & a_{32} & a_{33} \end{vmatrix} = a_{11}\begin{vmatrix} a_{22} & a_{23} \\ a_{32} & a_{33} \end{vmatrix} - a_{12}\begin{vmatrix} a_{21} & a_{23} \\ a_{31} & a_{33} \end{vmatrix} + a_{13}\begin{vmatrix} a_{21} & a_{22} \\ a_{31} & a_{32} \end{vmatrix}$$

Systems of three linear equations in three unknowns can still be solved by successive elimination, but the use of Cramer's rule and determinants is often more efficient. For four or more unknowns the required bookkeeping becomes formidable by either method, but it is then preferable to use Cramer's rule because it is more systematic.

Quadratic equations are always solvable by algebra. If $ax^2 + bx + c = 0$, then the two solutions are

$$x = \frac{1}{2a} \left[-b \pm (b^2 - 4ac)^{1/2} \right]$$

If $b^2 < 4ac$, the roots of the equation are complex numbers. Formulas also exist that give the solutions to third- and fourth-order equations, but it is usually easier to try solving the equation by (a) attempting to factor the equation algebraically (often not successful), (b) graphing the equation and noting the points of intersection with the x-axis, or (c) substituting numerically by trial and error.

Another useful formula in algebra is the binomial theorem, which is a special form of the Taylor's series of calculus:

$$(a + b)^n = a^n + \frac{n}{1!} a^{n-1}b + \frac{n(n-1)}{2!} a^{n-2}b^2 + \cdots$$

$$+ \frac{n(n-1)\ldots(n-r+1)}{r!} a^{n-r}b^r + \cdots + b^n$$

For a positive integer n this expansion has $(n + 1)$ terms. In the formula the convenient "factorial" notation $n! = n(n-1)(n-2)\ldots(3)(2)(1)$ has been used.

TRIGONOMETRY

Trigonometry deals with the relations between the angles and the sides of triangles. The periodic functions defined by these relations, however, have vastly wider applications. In using these functions we often deal with angle measurement, of which there are two kinds. One system divides one revolution into 360° (degrees). Each degree is further divisible into 60′ (minutes), and each minute into 60″ (seconds), although often a fraction of a degree is written as a decimal (e.g., $30' = 0.5°$). The unit of measurement in the second system is the radian (rad); 2π rad equals one revolution, or $180° = \pi$ rad $= 3.14159\ldots$ rad.

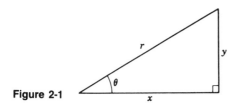

Figure 2-1

The two most basic trigonometric functions are the sine and cosine (Fig. 2-1), which are defined as follows:

$$\sin \theta = \frac{y}{r} \qquad \cos \theta = \frac{x}{r}$$

Here x and y may assume any value, but r is always positive. The other four basic functions are

$$\tan \theta = \frac{\sin \theta}{\cos \theta} = \frac{y}{x} \qquad \cot \theta = \frac{1}{\tan \theta} = \frac{x}{y}$$

$$\sec \theta = \frac{1}{\cos \theta} = \frac{r}{x} \qquad \csc \theta = \frac{1}{\sin \theta} = \frac{r}{y}$$

The sine and cosine are odd and even periodic functions, respectively, with periods of 2π (Fig. 2-2). By learning the variations of these two functions, one can easily deduce the variation of the other functions.

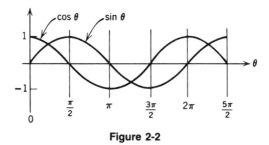

Figure 2-2

Much can be done in trigonometry by remembering a few fundamental identities. Among them are:

$$\sin^2 \theta + \cos^2 \theta = 1 \qquad 1 + \tan^2 \theta = \sec^2 \theta \qquad 1 + \cot^2 \theta = \csc^2 \theta$$

$$\sin (\theta \pm \phi) = \sin \theta \cos \phi \pm \cos \theta \sin \phi$$

$$\cos (\theta \pm \phi) = \cos \theta \cos \phi \mp \sin \theta \sin \phi$$

From these last two identities the double-angle ($\sin 2\theta$, $\cos 2\theta$) formulas can be derived by letting $\theta = \phi$. The half-angle ($\sin \theta/2$, $\cos \theta/2$) formulas can also be derived by replacing θ and ϕ by $\theta/2$ and rearranging the resulting expressions.

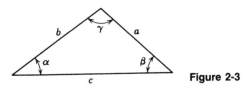

Figure 2-3

In solving for the unknown parts of a plane triangle (Fig. 2-3), two or three basic formulas are often useful. These are

Sum of angles: $\qquad \alpha + \beta + \gamma = 180°$

Law of sines: $\qquad \dfrac{a}{\sin \alpha} = \dfrac{b}{\sin \beta} = \dfrac{c}{\sin \gamma}$

Law of cosines: $\qquad a^2 = b^2 + c^2 - 2bc \cos \alpha$

Note that if $\alpha = 90°$ the triangle is a right triangle, and the law of cosines then becomes a statement of the Pythagorean formula.

PLANE AND ANALYTIC GEOMETRY

Here we group together some elements of elementary plane geometry, which describes some spatial properties of objects of various shapes, and analytic geometry, which employs algebraic notation in its more detailed description of some of these same objects.

The triangle and rectangle are basic geometric figures; they may also be considered as special cases of the trapezoid. From Fig. 2-4 we can consider

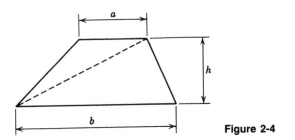

Figure 2-4

the trapezoid as the sum of two triangles. The area A of each shape is as follows:

$$\text{Trapezoid:} \quad A = \frac{h}{2}(a + b)$$

$$\text{Rectangle:} \quad A = hb \quad (a = b)$$

$$\text{Triangle:} \quad A = \frac{hb}{2} \quad (a = 0)$$

Another important shape is the regular polygon having n sides. The central angle subtended by one side is the vertex angle; its value is $2\pi/n$. The included angle between two successive sides of the polygon is $(n-2)\pi/n$.

The most important nonpolygonal geometric shape is the circle. For a circle of radius r and diameter $d = 2r$, the circumference is $c = \pi d$ and the enclosed area $A = \pi r^2$. In three dimensions its counterpart is the sphere which has a surface area $S = 4\pi r^2$ and an enclosed volume $V = \frac{4}{3}\pi r^3$.

Plane analytic geometry describes algebraically the properties of one- and two-dimensional geometric forms in an (x, y) plane.

The general equation of a straight line is $Ax + By + C = 0$. This equation is often more usefully written in one of the three following forms:

$$\text{Point-slope:} \quad y - y_1 = m(x - x_1)$$

$$\text{Slope-intercept:} \quad y = mx + b$$

$$\text{Two-intercept:} \quad \frac{x}{a} + \frac{y}{b} = 1$$

For a straight line passing through the points $P_1 = (x_1, y_1)$ and $P_2 = (x_2, y_2)$, the slope $m = (y_2 - y_1)/(x_2 - x_1)$; the intercepts a and b are the coordinate values occurring where the line intersects the x- and y-axes, respectively. The distance between points P_1 and P_2 is $D = [(x_2 - x_1)^2 + (y_2 - y_1)^2]^{1/2}$. Also, parallel lines have equal slopes, whereas perpendicular lines have negative reciprocal slopes.

Included in the general equation of second degree $Ax^2 + Bxy + Cy^2 + Dx + Ey + F = 0$ are a set of geometric shapes called the conic sections. The different conic sections can be recognized by investigating $B^2 - 4AC$:

If $B^2 - 4AC > 0$, the section is a hyperbola

If $B^2 - 4AC = 0$, the section is a parabola

If $B^2 - 4AC < 0$, the section is an ellipse

If in the last case $A = C$ and they are not zero, the section is a circle. If A, B, and C are all zero, the straight line again results.

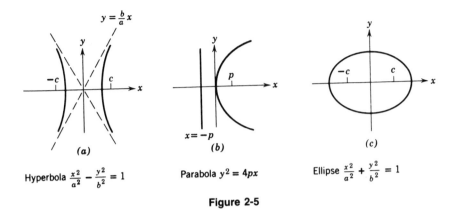

Hyperbola $\frac{x^2}{a^2} - \frac{y^2}{b^2} = 1$ Parabola $y^2 = 4px$ Ellipse $\frac{x^2}{a^2} + \frac{y^2}{b^2} = 1$

Figure 2-5

The basic equation for the hyperbola can be found from the general second-degree equation. For a hyperbola centered at the coordinate origin with limbs opening left and right (Fig. 2-5a), the equation is

$$\frac{x^2}{a^2} - \frac{y^2}{b^2} = 1$$

The difference of distances from the two focuses $(\pm c, 0)$ to a point on the hyperbola is always constant, where $c^2 = a^2 + b^2$. The limbs are asymptotic to the straight lines $y = \pm (b/a)x$. For a hyperbola centered at the point (h, k), replace x by $(x - h)$ and y by $(y - k)$. (This procedure for shifting the location of figures applies generally for all conic sections.)

The parabola, which geometrically is the locus of points equidistant from a point and a line (Fig. 2-5b), may be written in type form as

$$y^2 = 4px$$

when the vertex of the parabola is at the coordinate origin and the parabola opens to the right. Here the parabola is equidistant from the focus point $(p, 0)$ and the directrix line $x = -p$. Change the sign of p to obtain a parabola opening to the left; interchange the roles of y and x to obtain parabolas opening upward or downward.

The type equation for an ellipse, centered on the coordinate origin (Fig. 2-5c), is

$$\frac{x^2}{a^2} + \frac{y^2}{b^2} = 1$$

The semimajor and semiminor axes are a and b. The focuses of the ellipse are at $(\pm c, 0)$, where $c^2 = a^2 - b^2$. Any point on the ellipse is such that the sum of the distances to that point from the two focuses is a constant. If $a = b = r$, the ellipse then becomes a circle of radius r.

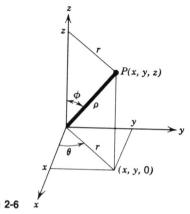

Figure 2-6

Sometimes it is more convenient to use the polar (r, θ), cylindrical (r, θ, z), or spherical (ρ, θ, ϕ) coordinate system in place of the two- or three-dimensional Cartesian (x, y, z) coordinate system. By reference to Fig. 2-6, these coordinate systems can be related to one another. The relations between the polar, cylindrical, and Cartesian coordinate systems are

$$x = r \cos \theta \qquad y = r \sin \theta \qquad z = z$$

Since we also have

$$z = \rho \cos \phi \qquad r = \rho \sin \phi$$

the relations between the spherical and Cartesian coordinate systems are

$$x = \rho \sin \phi \cos \theta \qquad y = \rho \sin \phi \sin \theta \qquad z = \rho \cos \phi$$

CALCULUS

At a point on the curve $y = f(x)$ the slope of the curve is the ratio of the change in $f(x)$ to the change in x when the change in x approaches zero in the limit; mathematically,

$$\text{Slope} = \frac{dy}{dx} = \frac{df(x)}{dx} = f'(x)$$

This is also called the rate of change of y with respect to x or the first derivative of $f(x)$. Second and higher derivatives are in turn defined as the rate of change of the next lower ordered derivative.

Derivatives of basic functions of x are given now. Let f and g be functions of x; c, m, and n are constants.

$$\frac{d}{dx}(c) = 0$$

$$\frac{d}{dx}(cx^n) = cnx^{n-1} \qquad \text{for } n \neq 0$$

$$\frac{d}{dx}(f \pm g) = \frac{df}{dx} \pm \frac{dg}{dx}$$

$$\frac{d}{dx}(f^m) = mf^{m-1}\frac{df}{dx}$$

$$\frac{d}{dx}(f^m g^n) = f^m \frac{d}{dx}(g^n) + g^n \frac{d}{dx}(f^m)$$

Here m and n assume any value. If $m = 1$, $n = 1$ this rule treats the simple product of two functions; if $m = 1$, $n = -1$ the rule governs the differentiation of the quotient of two functions. If $x = x(t)$,

$$\frac{df}{dt} = \frac{df}{dx}\frac{dx}{dt}$$

This is the chain rule of differentiation.

$$\frac{d}{dx}(\sin f) = \cos f \frac{df}{dx}$$

$$\frac{d}{dx}(\cos f) = -\sin f \frac{df}{dx}$$

At a maximum or a minimum of the function $f(x)$, the rate of change of f is zero; that is, $f'(x) = 0$. At that point f is a maximum if $f'' < 0$; it is a minimum if $f'' > 0$. Often, however, other physical considerations indicate whether the function has a maximum or minimum, and the second derivative test is not actually needed. If $f'' = 0$, the point is usually (but not always) a point of inflection, a point where the curvature of the function changes from concave upward to concave downward or vice versa.

If the function $f(x)$ approaches the value c as x approaches the value x_0, then we say c is the limiting value of $f(x)$ at the point x_0 and express this mathematically as

$$\lim_{x \to x_0} f(x) = c$$

The algebra of limits is no different from ordinary algebra:

$$\lim (f + g) = \lim f + \lim g$$

$$\lim (fg) = (\lim f)(\lim g)$$

$$\lim \frac{f}{g} = \frac{\lim f}{\lim g} \qquad \text{if } \lim g \neq 0$$

In the use of this last equation, however, the indeterminate forms $0/0$ or ∞/∞ may be encountered. In this case L'Hospital's rule is useful. Let f and g be functions having continuous derivatives with $g'(x_0) \neq 0$. If on approaching the limit point $x = x_0$ we have

$$\lim_{x \to x_0} f(x) = \lim_{x \to x_0} g(x) = 0$$

or

$$\lim_{x \to x_0} f(x) = \lim_{x \to x_0} g(x) = \pm\infty$$

then

$$\lim_{x \to x_0} \left[\frac{f(x)}{g(x)} \right] = \lim_{x \to x_0} \left[\frac{f'(x)}{g'(x)} \right]$$

An example is

$$\lim_{x \to 0} \frac{\sin x}{x} = \lim_{x \to 0} \frac{\cos x}{1} = 1$$

Continuous functions may be expressed as power series expansions around a point $x = c$ by use of the Taylor's series

$$f(x) = f(c) + f'(c) \frac{(x - c)}{1!} + f''(c) \frac{(x - c)^2}{2!} + \cdots + f^{(n)}(c) \frac{(x - c)^n}{n!} + \cdots$$

This series is particularly useful when a polynomial representation for a function is desired for x near c. The series can often then be truncated after only a few terms with little loss in accuracy.

Integration is the inverse of the process of differentiation. It may also be defined as the limit of a sequence; by this process the integral may be used to evaluate the exact area under a curve (Fig. 2-7). The area under this curve between a and b is $A = \int_a^b f(x)\, dx$.

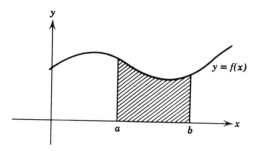

Figure 2-7

Some basic integration formulas follow, using the same notation as was used for derivatives.

$$\int \frac{df(x)}{dx}\,dx = f(x) + C \qquad (C = \text{constant of integration})$$

$$\int 0 \times dx = C$$

$$\int cf(x)\,dx = c \int f(x)\,dx$$

$$\int (f \pm g)\,dx = \int f\,dx \pm \int g\,dx$$

$$\int x^n\,dx = \frac{x^{n+1}}{n+1} + C \qquad (n \neq -1)$$

$$\int \frac{dx}{x} = \ln x + C$$

$$\int e^x\,dx = e^x + C$$

$$\int \sin x\,dx = -\cos x + C$$

$$\int \cos x\,dx = \sin x + C$$

$$\int f\,dg = fg - \int g\,df$$

This last formula, involving the two functions $f(x)$ and $g(x)$, is called integration by parts and is a powerful tool of integration. Evaluating any integral between definite limits will eliminate the constant of integration. We should also note that, for $a \leq b \leq c$,

$$\int_a^c f(x)\,dx = \int_a^b f(x)\,dx + \int_b^c f(x)\,dx$$

and

$$\int_a^b f(x)\,dx = -\int_b^a f(x)\,dx$$

DIFFERENTIAL EQUATIONS

Only the barest introduction to ordinary differential equations can be given here. A few problems in this chapter will introduce the topic further, but for a comprehensive review of the subject the reader should consult an appropriate mathematics text.

First-order ordinary differential equations are separable if they can be put in the form

$$M(x) \, dx + N(y) \, dy = 0$$

The equation is solved by direct integration; as in all differential equations a boundary condition must be specified before the constant of integration can be evaluated. If $M = M(x, y)$ and $N = N(x, y)$, a solution $F(x, y)$ can be found if the differential equation is exact, that is, if $\partial M / \partial y = \partial N / \partial x$. Then the solution $F(x, y)$ must satisfy the requirements $M = \partial F / \partial x$, $N = \partial F / \partial y$.

Many additional terms and types of basic ordinary differential equations are important to engineering but cannot be reviewed here. Just one representative important differential equation will be mentioned, however, to illustrate some concepts. The equation is

$$\frac{d^2 y}{dx^2} + p^2 y = f(x) \qquad (p = \text{constant})$$

plus two boundary conditions since this equation involves a second derivative. The equation is linear (y and/or derivatives of y are not multiplied together to form quadratic or higher order terms) and nonhomogeneous (the right term does not involve y; if it were not present, the equation would be homogeneous). The general solution is the sum of a complementary solution [solution of the equation with $f(x) = 0$] plus one particular solution [solution with $f(x)$ present]. If, for example, $f(x) = x$, a particular solution is $y = x/p^2$. Since $p^2 > 0$, the general complementary solution is $y = A \sin px + B \cos px$. If the plus sign in the equation were a minus sign, the solution would involve terms of the form $e^{\pm px}$.

STATISTICS AND PROBABILITY

Statistics is used as an aid in drawing conclusions from masses of data. The computation of certain properties of the data is useful in answering the questions "How big is it?" and "How much variation in size is there?"

The arithmetic mean, the median, and the mode are valuable tools in ascertaining the answer to the first question. The arithmetic mean $\bar{x}$ is the average value of N individual values x_i:

$$\bar{x} = \frac{1}{N} \sum_{i=1}^{N} x_i$$

The median value is the middle value when all data are arranged in order by magnitude; half the values are larger than the median value, half are smaller. The mode, on the other hand, is the value that occurs most frequently. The standard deviation σ is a statistic that helps to answer the second question; it gives the rms deviation from the mean value $\bar{x}$:

$$\sigma = \left[\frac{1}{N-1} \sum_{i=1}^{N} (x_i - \bar{x})^2 \right]^{1/2}$$

This expression gives a conservative and unbiased value for σ, but it is not unusual to find the divisor $N-1$ replaced by N. The square of the standard deviation σ^2 is called the *variance*.

Much of basic probability theory deals with mutually exclusive events or independent events. If only one of a set of possible events can occur, the events are mutually exclusive. Events are independent if the occurrence or nonoccurrence of one event does not affect the probability of occurrence of the other events. In this connection it is sometimes necessary to compute the number of permutations or set arrangements of n things taken r at a time, which is $n!/(n-r)!$, and the number of combinations (no set arrangement) of n things taken r at a time, which can be expressed in the equivalent forms

$$\binom{n}{r} = \frac{n!}{r!(n-r)!}$$

The probability of success of an event plus the probability of failure of an event is obviously unity, a fact of great simplicity and use. The probability of success itself is the ratio of the number of ways of achieving success to the total number of possible events (successes and failures). For independent events the following two rules are also useful:

1. The probability of A *or* B occurring equals the *sum* of the probability of occurrence of A and the probability of occurrence of B.
2. The probability of A *and* B occurring equals the *product* of the two individual probabilities.

PROBLEM 2-1

The number below that has four significant figures is

(A) 1414.0
(B) 1.4140
(C) 0.141
(D) 0.01414
(E) 0.0014

Solution

A significant figure is any of the digits 1 through 9 as well as 0 except when 0 is used to fix the decimal point or to fill the places of unknown or discarded digits.

Part	No. of Significant Figures
A	5
B	5
C	3
D	4
E	2

Answer is (D)   ok

PROBLEM 2-2

If $i = \sqrt{-1}$, the quantity i^{27} is equal to

(A)   0
(B) i
(C) $-i$
(D)   1
(E) -1

Solution

$(i)^n$ is a periodic function; that is,

$$i^1 = i \qquad i^2 = -1 \qquad i^3 = -i \qquad i^4 = +1 \qquad i^5 = i \ \ldots$$

More generally,

$$i^{4n+1} = i \qquad i^{4n+2} = -1 \qquad i^{4n+3} = -i \qquad i^{4n+4} = +1$$

for any integer n.
  Here $27 = 4n + 3$ and $i^{27} = -i$.

Answer is (C)

PROBLEM 2-3

If $x = +6$ and -4, the equation satisfying both these values would be

(A) $2x^2 + 3x - 24 = 0$
(B) $x^2 + 10x - 24 = 0$
(C) $x^2 - 2x - 24 = 0$
(D) $x^2 - 4x - 32 = 0$
(E) $x^2 - 3x + 18 = 0$

Solution

If a quadratic equation has solutions of $+6$, -4, it must have the factored form

$$(x - 6)(x + 4) = 0$$

Expanding this, we obtain

$$x^2 - 6x + 4x - 24 = 0$$

$$x^2 - 2x - 24 = 0$$

Answer is (C) ✔

PROBLEM 2-4

$\dfrac{7! \times 6!}{8! \times 0!}$ is equal to

(A) ∞
(B) 0
(C) 720
(D) 5760
(E) 90 ✔

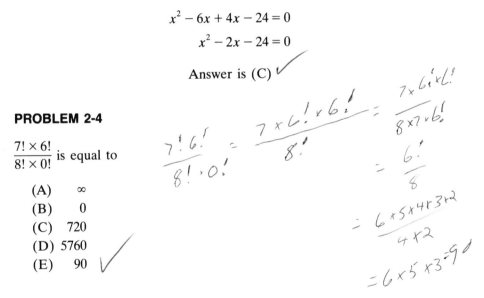

Solution

In this problem it is necessary to know that zero factorial is equal to one.

$$\frac{7! \times 6!}{8! \times 0!} = \frac{7! \times 6 \times 5 \times 4 \times 3 \times 2 \times 1}{8 \times 7! \times 1} = 90$$

Answer is (E)

PROBLEM 2-5

The fourth term in the expansion of $(1 - 2x^{-1/2})^{-2}$ is

(A) $-6x^{-1}$
(B) $12x^{-1}$
(C) $32x^{-3/2}$
(D) $64x^{-3/2}$
(E) $80x^{-2}$

Solution

Using the binomial expansion

$$(1 - A)^{-n} = 1 + nA + \frac{n(n+1)}{2!} A^2 + \frac{n(n+1)(n+2)}{3!} A^3$$

$$+ \frac{n(n+1)(n+2)(n+3)}{4!} A^4 + \cdots$$

in which we have $A = 2x^{-1/2}$ and $n = 2$ here, we obtain

$$(1 - 2x^{-1/2})^{-2} = 1 + 2(2x^{-1/2}) + \frac{(2)(3)}{2} (2x^{-1/2})^2$$

$$+ \frac{(2)(3)(4)}{(3)(2)} (2x^{-1/2})^3 + \frac{(2)(3)(4)(5)}{(4)(3)(2)} (2x^{-1/2})^4 + \cdots$$

$$= 1 + 4x^{-1/2} + 12x^{-1} + 32x^{-3/2} + 80x^{-2} + \cdots$$

Answer is (C)

PROBLEM 2-6

If $x = \log_a N$, then

(A) $x = N^a$
(B) $a = x^N$
(C) $x = a^N$
(D) $N = a^x$
(E) $N = x^a$

Solution

If x is the logarithm of N to the base a $(x = \log_a N)$, then by definition $N = a^x$.

Answer is (D)   ✓

PROBLEM 2-7

If $\log_a 10 = 0.250$, $\log_{10} a$ equals

(A) 4 ✓

(B) 0.50

(C) 2

(D) 0.25

(E) 1000

Handwritten:
$0.250 = \log_q 10$
$10 = q^{.250} = q^{\frac{1}{4}}$
$q = 10^4 = 10,000$
$\log_{10} 10^4 = 4$

Solution

$\log_a 10 = 0.250$ can be written as $10 = a^{0.250}$. Taking $\log_{10}$,

$$\log_{10} 10 = \log_{10} a^{0.250}$$
$$1 = 0.250 \log_{10} a$$

Since $1 = 0.250 \log_{10} a$,

$$\log_{10} a = \frac{1}{0.250} = 4$$

Answer is (A)   ✓

PROBLEM 2-8

In the equation

$$\log_{10} (X - 1) + \log_{10} X = 1$$

Handwritten: $= \text{antilog}_{10} (X)(X-1)$

the value of X is most nearly

(A) −2.7

(B) 1.6

(C) 3.7

(D) −0.6

(E) 5.5

Handwritten:
$\log_{10} (X)(X-1) = 1$
$\log X^2 - X = 1$
$X^2 - X = 10$
$X^2 - X - 10 = 0$

Solution

Since the sum of logarithms of numbers is their product and $\log 10 = 1$, then

$$(X - 1)(X) = 10 \quad \text{or} \quad X^2 - X = 10$$

$$X^2 - X - 10 = 0$$

$$X = \frac{1 \pm \sqrt{1 + 40}}{2} = \frac{1 \pm 6.4}{2} = 3.7$$

Note that there is only one real root. The other answer from the quadratic equation (-2.7) is not real, for the log of a negative number has no meaning.

Answer is (C) OK

PROBLEM 2-9

If $x = \frac{1}{2} \ln \dfrac{1 + u}{1 - u}$ ($\ln$ = natural logarithm), then

(A) $u = e^x$

(B) $u = \tanh x$

(C) $u = (e^x - 1)/(e^x + 1)$

(D) $u = (2x - 1)/(2x + 1)$

(E) $u = \ln(\sin x)$

Solution

$$2x = \ln \frac{1 + u}{1 - u}$$

$$e^{2x} = \exp\left(\ln \frac{1 + u}{1 - u}\right) = \frac{1 + u}{1 - u}$$

$$(1 - u)e^{2x} = 1 + u$$

$$u(e^{2x} + 1) = e^{2x} - 1$$

$$u = \frac{e^{2x} - 1}{e^{2x} + 1} = \frac{e^x - e^{-x}}{e^x + e^{-x}} = \tanh x$$

Answer is (B)

PROBLEM 2-10

A certain job can be performed by group X in 100 hr. Group Y can perform the same job in 25 hr, and group Z requires 20 hr. If the three groups, X, Y, and Z, work together, the number of hours required to complete the job is nearest

(A) 8
(B) 10
(C) 12
(D) 14
(E) 16

$R_x = Rate_x = \frac{1}{100}$ $R_y = \frac{1}{25}$ $R_z = \frac{1}{20}$

$\frac{1}{100} + \frac{1}{25} + \frac{1}{20} = \frac{1}{5\times20} + \frac{1}{5\times5} + \frac{1}{5\times4}$

$= \left(\frac{1}{20} + \frac{1}{5} + \frac{1}{4}\right)\frac{1}{5}$

$= \left(\frac{400}{20} + \frac{400}{5} + \frac{400}{4}\right)\left(\frac{1}{2000}\right)$

$= (20 + 80 + 100)\left(\frac{1}{2000}\right)$

Solution

Let N = number of hours to complete the job. The hourly progress is

$$\text{Group } X = \frac{N}{100} \qquad \text{Group } Y = \frac{N}{25} \qquad \text{Group } Z = \frac{N}{20}$$

$= \frac{200}{2000}$

$= \frac{1}{10}$

The combined effort is

$$\frac{N}{100} + \frac{N}{25} + \frac{N}{20} = 1 \qquad (0.01 + 0.04 + 0.05)N = 1$$

$\frac{1}{10}$ job/HR

JOB requires 10 HRS

$$N = \frac{1}{0.10} = 10 \text{ hr}$$

Answer is (B)

PROBLEM 2-11

Assuming that the sizes of the matrices **R**, **S** and **T** are such that the following operations can be done, which one of these matrix operations is not valid?

(A) $(\mathbf{S} + \mathbf{T})\mathbf{R} = \mathbf{SR} + \mathbf{TR}$
(B) $\mathbf{R}(\mathbf{ST}) = (\mathbf{RS})\mathbf{T}$
(C) $\mathbf{RS} = \mathbf{SR}$
(D) $3(\mathbf{S} - \mathbf{T}) = 3\mathbf{S} - 3\mathbf{T}$
(E) $\mathbf{R} + (\mathbf{S} + \mathbf{T}) = (\mathbf{R} + \mathbf{S}) + \mathbf{T}$

Solution

The five operations listed here are, in turn, called the distributive law and the associative law for matrix multiplication, the commutative law for multiplication, multiplication by a scalar (in this case, 3), and the commutative law for addition.

The commutative law for multiplication of scalars is rs = sr, but this law does *not* also apply to matrices; in fact **RS** and **SR** are in general unequal.

Answer is (C)

PROBLEM 2-12

If **A** is an $n \times n$ matrix, then the only *in*correct statement about **A** is

(A) The determinant of **A** is nonzero.
(B) The row vectors of **A** are linearly dependent.
(C) The matrix **A** has an inverse.
(D) The rank of **A** is n.
(E) $Ax = 0$ has only the solution $x = 0$.

Solution

Except for statement (B), all the statements are equivalent; the row vectors of **A** are actually linearly independent in this case.

Answer is (B)

PROBLEM 2-13

When the determinant D is

$$D = \begin{vmatrix} 1 & 1 & 1 \\ 2 & -1 & 1 \\ 1 & 2 & -1 \end{vmatrix}$$

its value is

(A) -5
(B) -3
(C) $+1$
(D) $+3$
(E) $+7$

Solution

Expanding by minors along the top row,

$$D = \begin{vmatrix} 1 & 1 & 1 \\ 2 & -1 & 1 \\ 1 & 2 & -1 \end{vmatrix} = 1 \begin{vmatrix} -1 & 1 \\ 2 & -1 \end{vmatrix} - 1 \begin{vmatrix} 2 & 1 \\ 1 & -1 \end{vmatrix} + 1 \begin{vmatrix} 2 & -1 \\ 1 & 2 \end{vmatrix}$$

$$= 1(1-2) - 1(-2-1) + 1(4+1)$$

$$= 1(-1) - 1(-3) + 1(+5)$$

$$= -1 + 3 + 5$$

$$= +7$$

Answer is (E)

PROBLEM 2-14

We are given the following three equations:

$$\frac{L}{2} + \frac{m}{3} + \frac{n}{4} = 62$$

$$\frac{L}{4} + \frac{m}{5} + \frac{n}{6} = 38$$

$$\frac{L}{3} + \frac{m}{4} + \frac{n}{5} = 47$$

The value of L is

(A) 12
(B) 24
(C) 60
(D) 72
(E) 120

Solution

The problem could also be solved by successive elimination, but we choose to use Cramer's rule and determinants here:

$$D = \begin{vmatrix} \frac{1}{2} & \frac{1}{3} & \frac{1}{4} \\ \frac{1}{4} & \frac{1}{5} & \frac{1}{6} \\ \frac{1}{3} & \frac{1}{4} & \frac{1}{5} \end{vmatrix} = \frac{1}{2} \begin{vmatrix} \frac{1}{5} & \frac{1}{6} \\ \frac{1}{4} & \frac{1}{5} \end{vmatrix} - \frac{1}{3} \begin{vmatrix} \frac{1}{4} & \frac{1}{6} \\ \frac{1}{3} & \frac{1}{5} \end{vmatrix} + \frac{1}{4} \begin{vmatrix} \frac{1}{4} & \frac{1}{5} \\ \frac{1}{3} & \frac{1}{4} \end{vmatrix}$$

$$D = -\frac{1}{1200} + \frac{1}{540} - \frac{1}{960} = -\frac{1}{43,200} \qquad \frac{1}{D} = -43,200 \neq 0$$

$$D_L = \begin{vmatrix} 62 & \frac{1}{3} & \frac{1}{4} \\ 38 & \frac{1}{5} & \frac{1}{6} \\ 47 & \frac{1}{4} & \frac{1}{5} \end{vmatrix} = -\frac{62}{600} - \frac{38}{240} + \frac{47}{180} = -\frac{1}{1800}$$

$$L = \frac{D_L}{D} = \frac{43,200}{1800} = 24$$

Answer is (B)

Additional calculations would show that $m = 60$ and $n = 120$.

PROBLEM 2-15

We are given a set of three simultaneous equations:

$$5X + 2Y + 4Z = 4 \tag{1}$$
$$3X - Y + 2Z = -11 \tag{2}$$
$$7X - 3Y - 3Z = 8 \tag{3}$$

The correct value Y is nearest to

(A) -5
(B) -1
(C) 2
(D) 7
(E) 10

Solution

Although determinants could be used here, we use the successive elimination approach.

$$
\begin{array}{rl}
(1) & 5X + 2Y + 4Z = 4 \\
-2 \times (2) & \underline{-6X + 2Y - 4Z = 22} \\
(4) & -X + 4Y = 26
\end{array}
$$

$$
\begin{array}{rl}
3 \times (2) & 9X - 3Y + 6Z = -33 \\
2 \times (3) & \underline{14X - 6Y - 6Z = 16} \\
(5) & 23X - 9Y = -17
\end{array}
$$

$$
\begin{array}{rl}
23 \times (4) & -23X + 92Y = 598 \\
(5) & \underline{23X - 9Y = -17} \\
& 83Y = 581 \qquad Y = 7
\end{array}
$$

Answer is (D)

Additional calculations show that $X = 2$ and $Z = -5$.

PROBLEM 2-16

In the matrix multiplication $\mathbf{AB} = \mathbf{C}$, where

$$\mathbf{A} = [A] = \begin{bmatrix} 2 & 1 & -3 \\ 0 & -4 & 6 \end{bmatrix}, \qquad \mathbf{B} = [B] = \begin{bmatrix} 0 & -2 \\ 4 & 2 \\ 1 & 3 \end{bmatrix}$$

the element c_{12} is closest to

(A) -11
(B) -10
(C) 1
(D) 10
(E) 11

Solution

The element c_{12} is a single entry in the 2×2 matrix $[C]$ which is determined by multiplying together, in pairs, elements of row 1 of $[A]$ and column 2 of $[B]$ and summing the results; thus

$$c_{12} = 2(-2) + 1(2) + (-3)(3) = -11$$

Answer is (A)

PROBLEM 2-17

If the sine of angle α is given as K, the tangent of angle α is equal to

(A) $1 - K$
(B) $1/K$
(C) $\sqrt{1 - K^2}$

(D) $\dfrac{1}{\sqrt{1 - K^2}}$

(E) $\dfrac{K}{\sqrt{1 - K^2}}$

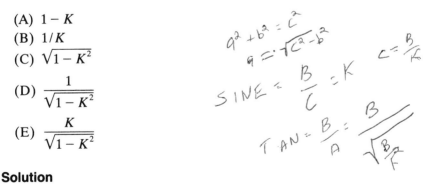

Solution

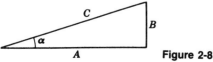

Figure 2-8

Sin $\alpha = B/C = K$. If we let $C = 1$, then $B = K$. Since $A^2 + B^2 = C^2$, $A^2 + K^2 = 1^2$ or $A = \sqrt{1 - K^2}$. Then the triangle appears as

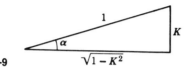

Figure 2-9

$$\tan \alpha = \frac{K}{\sqrt{1 - K^2}}$$

Answer is (E)

PROBLEM 2-18

Which of the following is incorrect?

(A) $\cos^2 A = \tan A \cot A$
(B) $\sin A = \cos A \tan A$
(C) $\cos 2A = \cos^2 A - \sin^2 A$
(D) $2 \sin^2 A = 1 - \cos 2A$
(E) $\sin (A + B) = \sin A \cos B + \cos A \sin B$

Solution

An identity is an equality that is valid for all values of the variable(s) in the equation. Considering the first expression,

$$\tan A \cot A = 1 \quad \text{or} \quad \cos^2 A = 1$$

which is not true for all values of A.
The four other expressions are correct.

Answer is (A)

PROBLEM 2-19

It is 3.8 km from point A to the north end of the lake and 5.3 km from A to the south end of the lake. The lake subtends an angle of 110° at A. The length of the lake from north to south is nearest to

(A) 5.4 km
(B) 6.5 km
(C) 7.5 km
(D) 8.1 km
(E) 9.1 km

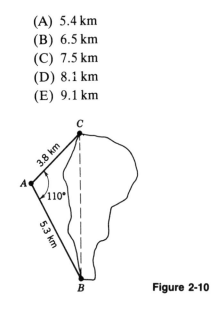

Figure 2-10

Solution

The law of cosines applies directly:

$$a^2 = b^2 + c^2 - 2bc \cos \alpha$$
$$a^2 = (3.8)^2 + (5.3)^2 - 2(3.8)(5.3) \cos 110°$$

Since $\cos 110° = -0.342$,

$$a^2 = 14.4 + 28.1 - (-13.8) = 56.3$$
$$a = 7.5 \text{ km}$$
$$\text{Answer is (C)}$$

PROBLEM 2-20

Two points lie on a horizontal line directly south of a tower 100 ft high. The angles of depression to the points are 28°10′ and 42°50′. The distance between the points is closest to

(A)   39.7 ft
(B)   64.8 ft
(C)   70.4 ft
(D)   78.9 ft
(E) 104.0 ft

Solution

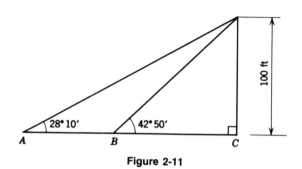

Figure 2-11

The distance $\overline{AB} = \overline{AC} - \overline{BC}$

$$\tan(28°10') = \frac{100\,\text{ft}}{\overline{AC}} \qquad \tan(42°50') = \frac{100\,\text{ft}}{\overline{BC}}$$

$$\overline{AB} = \frac{100\,\text{ft}}{\tan(28°10')} - \frac{100\,\text{ft}}{\tan(42°50')}$$

$$= 100\,\text{ft}[\cot(28°10') - \cot(42°50')]$$

$$= 100\,\text{ft}(1.8676 - 1.0786)$$

$$= 100\,\text{ft}(0.7890) = 78.9\,\text{ft}$$

Answer is (D)

PROBLEM 2-21

Given the following relations:

$$r \cos \phi \cos \theta = 4 \qquad (1)$$

$$r \cos \phi \sin \theta = 3 \qquad (2)$$

$$r \sin \phi \qquad\quad = 5 \qquad (3)$$

The value of r is most nearly

(A)   6.00
(B)   7.07
(C)   7.50
(D)   8.66
(E)   10.00

Solution

Divide Eq. (2) by Eq. (1):

$$\frac{r \cos \phi \sin \theta}{r \cos \phi \cos \theta} = \frac{3}{4} = \tan \theta$$

Therefore

$$\theta = \tan^{-1}\left(\tfrac{3}{4}\right) = 36.9°$$

Divide Eq. (3) by Eq. (1):

$$\frac{r \sin \phi}{r \cos \phi \cos \theta} = \frac{5}{4} \qquad \tan \phi \frac{1}{\cos \theta} = \frac{5}{4}$$

but $(1/\cos \theta) = (1/\cos 36.9°) = (1/0.8) = 1.25$, $1.25 \tan \phi = \tfrac{5}{4}$, $\tan \phi = 1$, $\phi = 45°$, $r \sin \phi = 5$, and $r \sin 45° = 5$. Therefore

$$r = \frac{5}{0.707} = 7.07$$

This result is easily checked by squaring and adding the three original equations and recalling that $\cos^2 \theta + \sin^2 \theta = 1$ for both θ and ϕ; $r^2[\cos^2 \phi(\cos^2 \theta + \sin^2 \theta) + \sin^2 \phi] = r^2 = 50$.

Answer is (B)

PROBLEM 2-22

Each interior angle of a regular polygon with eight sides is nearest to

(A) 100°
(B) 80°
(C) 150°
(D) 125°
(E) 135°

$(n-2) \times 180°$
$= 6 \times 180°$
$= 1080°$
$\frac{1080}{8} = 540$ $\frac{540 \cdot 2}{4}$ $\frac{270}{2}$ $= 135$

Solution

The sum of the interior angles of a polygon is equal to $(n-2) \times 180°$. In a regular polygon all sides are equal, hence all angles are equal.

$$\frac{(n-2) \times 180°}{8} = \frac{6}{8}(180) = 135°$$

Answer is (E)

PROBLEM 2-23

As shown, a circle can be circumscribed around a triangle ABC. AC is a diameter of the circle.

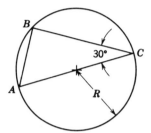

Figure 2-12

In terms of the radius R, the area of the triangle is most nearly

(A) $0.750R^2$
(B) $0.785R^2$
(C) $0.866R^2$
(D) $1.000R^2$
(E) $1.414R^2$

Solution

A geometric theorem states that a triangle inscribed in a semicircle is a right triangle, so angle $ABC = 90°$.

$\text{Sin } 30° = \dfrac{AB}{2R} = \dfrac{1}{2}$. Therefore $AB = R$. $\text{Cos } 30° = \dfrac{BC}{2R} = \dfrac{\sqrt{3}}{2}$. Therefore $BC = \sqrt{3}R$.

$$\text{Area} = \tfrac{1}{2}(AB)(BC) = \tfrac{1}{2}R\sqrt{3}R$$
$$= 0.866R^2$$

Answer is (C)

PROBLEM 2-24

The surface area of a tetrahedron is described by

(A) 4 equilateral triangles
(B) 6 squares
(C) 12 pentagons
(D) 3 trapeziums
(E) 8 pentagons

Solution

A tetrahedron (triangular pyramid) is bounded by four equilateral triangles.

Answer is (A)

PROBLEM 2-25

Two ships leave the same port at the same time, one sailing due northeast at the rate of 6 miles/hr and the other sailing due north at the rate of 10 miles/hr. The distance between the two ships after 3 hr of sailing is nearest to

(A) 12 miles
(B) 21 miles
(C) 24 miles
(D) 28 miles
(E) 45 miles

Solution

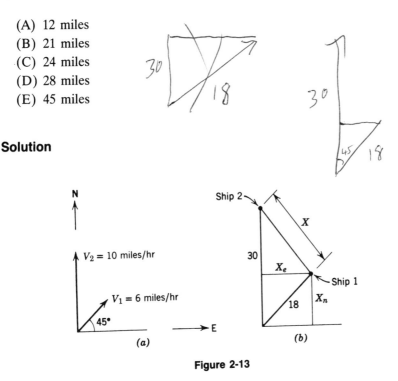

Figure 2-13

After 3 hr

ship 1 will be $3 \times 6 = 18$ miles from the port
ship 2 will be $3 \times 10 = 30$ miles from the port

Ship 1 is X_e miles east of ship 2: $X_e = 18 \sin 45°$
Ship 1 is X_n miles north of the port: $X_n = 18 \sin 45°$
Solving for X,

$$X^2 = X_e^2 + (30 - X_n)^2$$
$$X^2 = (18 \sin 45°)^2 + (30 - 18 \sin 45°)^2$$
$$X^2 = 162 + 298 = 460$$
$$X = 21.4 \text{ miles}$$

Answer is (B)

PROBLEM 2-26

The equation of a straight line that has a slope of $+2$ and passes through a point with x and y coordinates of 4 and 5, respectively, is

(A) $x + 2y = 14$
(B) $xy = 20$
(C) $2x + y = 13$
(D) $4y = 5x$
(E) $y = 2x - 3$

Solution

The point-slope equation for a straight line is $y - y_1 = m(x - x_1)$, where m is the slope and x_1 and y_1 are the coordinates of a point on the line.

$$y - 5 = 2(x - 4) \qquad y - 5 = 2x - 8 \qquad y = 2x - 3$$

Answer is (E)

PROBLEM 2-27

The equation (in rectangular coordinates) of the plane passing through the three points $(1, 3, 5)$, $(2, 4, 4)$, and $(3, 4, 2)$ is

(A) $xyz = -2$
(B) $y - z = 2(x + 2)$
(C) $2x - y + z - 4 = 0$
(D) $x - 2 = yz$
(E) $x + 2(y + z) = 17$

Solution

The general equation of a plane may be written

$$Ax + By + Cz + D = 0$$

Substituting the three points:

$$A + 3B + 5C + D = 0 \qquad (1)$$
$$2A + 4B + 4C + D = 0 \qquad (2)$$
$$3A + 4B + 2C + D = 0 \qquad (3)$$

Solving for A, B, and D in terms of C, we have

Equation $(3) - (2)$,

$$A - 2C = 0 \qquad A = 2C \qquad (4)$$

Equation $(3) - (1)$,

$$2A + B - 3C = 0 \qquad (5)$$

Substituting (4) into (5),

$$2(2C) + B - 3C = 0 \qquad B = -C \qquad (6)$$

Substituting (4) and (6) back into (1) gives

$$2C + 3(-C) + 5C + D = 0 \qquad D = -4C$$

From the general equation

$$2Cx - Cy + Cz - 4C = 0$$
$$C(2x - y + z - 4) = 0$$

The equation of the plane is $2x - y + z - 4 = 0$.

Answer is (C)

PROBLEM 2-28

The curve represented by the equation $\dfrac{x^2}{a^2} - \dfrac{y^2}{b^2} = 1$ is a

(A) straight line
(B) circle
(C) ellipse
(D) parabola
(E) hyperbola

Solution

The simple equations for the various curves are

$$\begin{array}{ll}
\text{straight line} & y = mx + b \\[6pt]
\text{circle} & x^2 + y^2 = a^2 \\[6pt]
\text{ellipse} & \dfrac{x^2}{a^2} + \dfrac{y^2}{b^2} = 1 \\[6pt]
\text{parabola} & y^2 = ax \\[6pt]
\text{hyperbola} & \dfrac{x^2}{a^2} - \dfrac{y^2}{b^2} = 1
\end{array}$$

Answer is (E)

PROBLEM 2-29

The equation of the largest circle that is tangent to both coordinate axes and has its center on the line $2X + Y - 6 = 0$ is

(A) $X^2 + Y^2 = 36$
(B) $(X - 2)^2 + (Y - 2)^2 = 4$
(C) $(X + 6)^2 + (Y - 6)^2 = 36$
(D) $(X - 6)^2 + (Y + 6)^2 = 36$
(E) $X^2 + Y^2 = 4(X + Y - 1)$

Solution

The center of the circle must be at the intersection of two straight lines: $2X + Y - 6 = 0$, and $X - Y = 0$ or $X + Y = 0$. Solving the two cases, we obtain

$$\begin{array}{rl}
2X + Y - 6 = 0 & \\
\underline{X - Y \qquad = 0} & \\
3X \qquad - 6 = 0 & \\
X = 2 & \\
Y = 2 &
\end{array}$$

The circle has its center at $(2, 2)$ with radius $= 2$.

$$2X + Y - 6 = 0$$
$$-X - Y \quad = 0$$
$$\overline{X \quad -6 = 0}$$
$$X = 6$$
$$Y = -6$$

The circle has its center at $(6, -6)$ with radius $= 6$.

The second circle is the correct one, since the largest circle is desired. The general equation for a circle is $(X - a)^2 + (Y - b)^2 = R^2$. Substituting, we obtain $(X - 6)^2 + (Y + 6)^2 = 6^2$.

Answer is (D)

PROBLEM 2-30

The cable of a suspension bridge hangs in the shape of an arc of a parabola AB. The supporting towers are 70 ft high and 200 ft apart, and the lowest point on the cable is 20 ft above the roadway. The length of the supporting rod L 50 ft from the middle of the bridge is nearest

(A) 30.0 ft
(B) 32.5 ft
(C) 36.7 ft
(D) 38.2 ft
(E) 45.0 ft

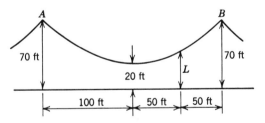

Figure 2-14

Solution

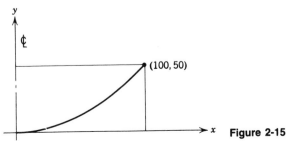

Figure 2-15

When $x = \pm 100$, $y = 70 - 20 = 50$

$$y = kx^2$$
$$50 = k(100)^2$$

Therefore $k = 0.005$ and $y = 0.005x^2$.

When $x = 50$ ft, $y = 0.005(50)^2 = 0.005(2500) = 12.5$ ft. The length of the supporting rod is $L = 12.5 + 20 = 32.5$ ft.

Answer is (B)

PROBLEM 2-31

In the equation $x = \dfrac{t^2 + t}{2t^2 + 1}$, the limit of x as t approaches infinity is

(A) ∞
(B) 2
(C) 1
(D) $\frac{1}{2}$
(E) 0

Solution

$$\lim_{t \to \infty} \frac{t^2 + t}{2t^2 + 1} = ?$$

The usual rule is to divide both the numerator and denominator by the highest power of the variable occurring in either. In this case, divide by t^2. This step is valid since $t \neq 0$.

$$\lim_{t \to \infty} \frac{t^2 + t}{2t^2 + 1} = \lim_{t \to \infty} \frac{1 + (1/t)}{2 + (1/t^2)} = \frac{1}{2}$$

The limit of each term in the numerator and denominator containing t is zero. $(1/\infty = 0.)$

Answer is (D)

PROBLEM 2-32

In the equation $y = \dfrac{-x^3 + 3x + 2}{x^2 + 2x + 1}$, the limit of y as x approaches a value of -1 is

(A) 0
(B) 1
(C) 2
(D) 3
(E) ∞

Solution

$$\lim_{x \to -1} \frac{-x^3 + 3x + 2}{x^2 + 2x + 1} = \frac{-(-1)^3 + 3(-1) + 2}{(-1)^2 + 2(-1) + 1} = \frac{+1 - 3 + 2}{+1 - 2 + 1} = \frac{0}{0}$$

which is indeterminate.

Not obtaining a solution above, we try factoring the numerator and denominator:

$$\lim_{x \to -1} \frac{-x^3 + 3x + 2}{x^2 + 2x + 1} = \lim_{x \to -1} \frac{(x + 1)(x + 1)(-x + 2)}{(x + 1)(x + 1)}$$

$$= \lim_{x \to -1} (-x + 2) = +3$$

We find that y is not continuous at $x = -1$ but does approach a limit of $+3$ as x approaches -1.

Answer is (D)

PROBLEM 2-33

In the equation of $y = \dfrac{\ln (1 - z)}{z}$, the limit of y as z approaches a value of zero is

(A) ∞
(B)   3
(C)   1
(D)   0
(E) -1

Solution

$$\lim_{z \to 0} y = \frac{\ln (1 - z)}{z} = \frac{\ln (1 - 0)}{0} = \frac{0}{0}$$

The form is indeterminate. Applying L'Hospital's rule,

$$\lim_{z \to 0} \frac{\ln(1-z)}{z} = \lim_{z \to 0} \frac{\frac{d}{dz}[\ln(1-z)]}{\frac{d}{dz}(z)}$$

$$= \lim_{z \to 0} \frac{-1}{1-z} = -1$$

Answer is (E)

PROBLEM 2-34

The slope of the curve $y = x^3 - 4x$ as it passes through the origin ($x = 0$, $y = 0$) is equal to

(A) +4
(B) +2
(C) 0
(D) −2
(E) −4

Solution

$$\frac{dy}{dx}\bigg|_{x=0} = [3x^2 - 4]_{x=0} = -4$$

Therefore the slope of the curve at $x = 0$ is -4.

Answer is (E)

PROBLEM 2-35

For the position-time function $x = 3t^2 + 2t$, the velocity in the x direction at $t = 1$ is

(A) 9
(B) 8
(C) 7
(D) 6
(E) 5

Solution

$$\text{Velocity} = \frac{dx}{dt} = [6t + 2]_{t=1} = 6 + 2 = 8$$

Answer is (B)

PROBLEM 2-36

The stiffness of a rectangular timber is proportional to the width and the cube of the depth. The width of the stiffest beam that can be made of a circular log whose diameter is 20 in. is closest to

(A) 10 in.
(B) 12 in.
(C) 14 in.
(D) 16 in.
(E) 17 in.

Solution

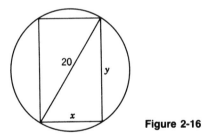

Figure 2-16

Here $x =$ beam width, $y =$ beam depth, and log diameter $= 20$ in.
    To maximize stiffness, the function xy^3 must be maximized. From the figure, $y = (20^2 - x^2)^{1/2}$. Writing xy^3 as a function of one variable,

$$xy^3 = x(20^2 - x^2)^{3/2}$$
$$f(x) = x(20^2 - x^2)^{3/2} \qquad \text{for } 0 \le x \le 20$$

Since $f(0) = f(20) = 0$ and $f(x) > 0$ for intermediate values of x, it is clear that $f(x)$ must attain a maximum value. To find the maximum, we set $f'(x) = 0$ and solve for x. [Alternatively, we could check that $f(x)$ is indeed maximized by verifying that $f''(x) < 0$ at that point.] Using the chain rule for differentiation of a product,

$$f'(x) = x\tfrac{3}{2}(20^2 - x^2)^{1/2} \times (-2x) + (20^2 - x^2)^{3/2} \times (1)$$
$$= -3x^2(20^2 - x^2)^{1/2} + (20^2 - x^2)^{3/2} = 0$$

Dividing by $(20^2 - x^2)^{1/2}$, we obtain $-3x^2 + 20^2 - x^2 = 0$, $-4x^2 + 20^2 = 0$, and $x^2 = 400/4 = 100$. Then

$$x = 10 \text{ in.}$$

Answer is (A)

PROBLEM 2-37

Circular cylindrical cans of volume V_0 are to be manufactured with both ends closed. The ratio between the diameter and height that will require the minimum amount of metal to make each can is nearest to

(A) $d/h = 0.6$
(B) $d/h = 0.8$
(C) $d/h = 1.0$
(D) $d/h = 1.2$
(E) $d/h = 1.4$

Solution

Total surface area = 2 end areas + side surface area

$$= 2\left(\frac{\pi}{4} d^2\right) + \pi \, dh$$

The surface area is to be a minimum. To find extreme values we must find the first derivative of the function. But the function contains two variables (d and h), so one must be defined in terms of the other to eliminate one variable.

$$V_0 = \frac{\pi}{4} d^2 h$$

Therefore

$$h = \frac{4V_0}{\pi d^2}$$

Hence

$$\text{Surface area} = 2\left(\frac{\pi}{4}d^2\right) + \pi d\left(\frac{4V_0}{\pi d^2}\right) = \frac{\pi}{2}d^2 + \frac{4V_0}{d}$$

Taking the first derivative and equating it to zero,

$$f'(d) = \frac{\pi}{2}(2d) + 4V_0\left(\frac{-1}{d^2}\right) = 0$$

$$\pi d = \frac{4V_0}{d^2} \quad \text{and} \quad d^3 = \frac{4V_0}{\pi}$$

[Check: $f''(d) = \pi + 8V_0/d^3 > 0$ and therefore is a minimum.]
Since $V_0 = (\pi/4)d^2 h$, we also have $d^2h = 4V_0/\pi$. Therefore

$$d^2h = d^3 \quad \text{or} \quad h = d$$

$$\text{Ratio } d/h = 1$$

$$\text{Answer is (C)}$$

PROBLEM 2-38

The area under the curve $y = x^2$ between the values $x = +1$ ft and $x = +7$ ft
is nearest to

(A)   96 ft^2
(B)   114 ft^2
(C)   147 ft^2
(D)   171 ft^2
(E)   342 ft^2

$y = x^2$

Solution

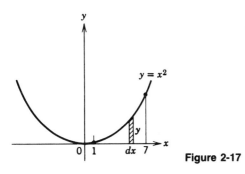

$$\int_1^7 x^2 \, dx$$

$$= \left.\frac{x^3}{3}\right|_1^7$$

$$= \frac{7^3 - 1}{3}$$

$$= \frac{343 - 1}{3} = \frac{342}{3} = 114$$

y = x²

y = x2

0 | 1   dx 7

Figure 2-17

$$dA = y \, dx$$

$$A = \int dA = \int y \, dx$$

Substituting for y in terms of x,

$$A = \int_1^7 x^2 \, dx = \left. \frac{x^3}{3} \right|_1^7$$

$$= \tfrac{343}{3} - \tfrac{1}{3} = \tfrac{342}{3}$$

$$= 114 \text{ ft}^2$$

Answer is (B)

PROBLEM 2-39

The area formed by the boundaries $y = 1$, $x = 1$, and $y = e^{-x}$ is closest to

(A) 0.50
(B) 0.46
(C) 0.42
(D) 0.38
(E) 0.34

Solution

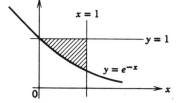

Figure 2-18

$$\text{Area} = 1^2 - \int_0^1 e^{-x} \, dx = 1 - [-e^{-x}] \Big|_0^1 = 1 - [-e^{-1} - (-e^{-0})]$$

$$= 1 + e^{-1} - 1 = e^{-1}$$

$$= \frac{1}{2.718} = 0.368$$

Answer is (D)

PROBLEM 2-40

The value of

$$\int_2^4 \frac{x+1}{x^2-x} \, dx$$

is closest to

(A) 2.0
(B) 1.5
(C) 1.0
(D) 0.5
(E) 0.1

Solution

First we decompose the integrand into partial fractions. Since $x^2 - x = x(x-1)$, we write

$$\frac{x+1}{x^2-x} = \frac{A}{x} + \frac{B}{x-1}$$

or $x + 1 = A(x - 1) + Bx$. Thus we find $A = -1$, $B = 2$, and the integral becomes

$$\int_2^4 \left[\frac{-1}{x} + \frac{2}{x-1} \right] dx = \left[-\ln x + 2 \ln (x-1) \right] \Big|_2^4 = \ln \left[\frac{(x-1)^2}{x} \right] \Big|_2^4$$

$$= \ln \left(\frac{9}{4} \right) - \ln \left(\frac{1}{2} \right) = 0.811 - (-0.693) = 1.504$$

Answer is (B)

PROBLEM 2-41

The value of the integral

$$\int_0^{\pi/2} \sin^2 \theta \, d\theta$$

is nearest to

(A) π
(B) $\pi/2$
(C) 1
(D) $\pi/4$
(E) 0

Solution

One could use the trigonometric identity $\sin^2 \theta = (1 - \cos 2\theta)/2$ to evaluate the integral, but we will take a different approach here. In the interval $(0, \pi/2)$ the plots of the functions $\sin^2 \theta$ and $\cos^2 \theta$ are similar in form, and we also have the relation $\sin^2 \theta + \cos^2 \theta = 1$. Hence

$$\int_0^{\pi/2} (\sin^2 \theta + \cos^2 \theta)\, d\theta = \int_0^{\pi/2} (1)\, d\theta = \pi/2$$

From the symmetry of the two integrands one concludes that

$$\int_0^{\pi/2} \sin^2 \theta\, d\theta = \pi/4$$

Answer is (D)

PROBLEM 2-42

The differential equation

$$\frac{d^2 y}{dx^2} + \int_0^1 \cos (x - t) y(t)\, dt = 3x$$

can be classified as

(A) exact
(B) variables separable
(C) nonlinear
(D) linear and homogeneous
(E) linear but not homogeneous

Solution

The equation is linear in y (no powers of y above 1 appear); since the right side does not contain y, the equation is not homogeneous in y. The variables x and y cannot be separated, and the equation is not an exact differential form.

Answer is (E)

PROBLEM 2-43

The general solution of the ordinary differential equation

$$\frac{dy}{dx} = x - 2xy$$

with $C = $ constant is

(A) $\ln(1 - 2y) = x^2 + C$
(B) $2y = 1 + Ce^{-x^2}$
(C) $-\ln(1 - 2y) = (x^2/2) + C$
(D) $-\frac{1}{2}\ln(1 - 2y) = x^2 + C$
(E) impossible to determine

Solution

This is a linear differential equation of order one which has the standard form

$$dy + P(x)y\, dx = Q(x)\, dx$$

Here we have

$$dy + 2xy\, dx = x\, dx$$

so $P = 2x$ and $Q = x$. The integrating factor for this equation is

$$\exp\left(\int P\, dx\right) = \exp\left(\int 2x\, dx\right) = e^{x^2}$$

The exact equation is

$$e^{x^2}\, dy + 2xye^{x^2}\, dx = xe^{x^2}\, dx$$

and its solution is

$$ye^{x^2} = \frac{1}{2}e^{x^2} + c_1$$

If we let $C = 2c_1$, the solution can be written as

$$2y = 1 + Ce^{-x^2}$$

Answer is (B)

PROBLEM 2-44

The rate of decay of radioactive elements is usually assumed to be proportional to the number of atoms that have not decayed, where λ is the
proportionality constant. If at time $t = 0$ there are X_0 atoms of a given
element, the expression for the number of atoms, X, that have not decayed
(as a function of time t, λ, and X_0) is

   (A) $X_0(1 - \lambda t)$
   (B) $X_0 e^{-\lambda t}$
   (C) $X_0(1 - e^{-\lambda t})$
   (D) $X_0/(1 + \lambda t)$
   (E) $X_0(1 - \lambda t^{1/2})$

Solution

X = number of atoms that have not decayed. The rate of decay of X is
proportional to X, or $dX/dt = -\lambda X$. Rearranging this equation,

$$\frac{dX}{X} = -\lambda \, dt$$

Integrating,

$$\ln X = -\lambda t + A$$
$$X = e^{-\lambda t} e^{A}$$

When $t = 0$, $X = X_0 = e^{A}$. Therefore $X = X_0 e^{-\lambda t}$.

$$\text{Answer is (B)}$$

PROBLEM 2-45

A 1000-ft^3 storage tank is filled with natural gas at 80°F and 1 atm pressure.
The tank is flushed out with nitrogen gas at 80°F and 1 atm pressure, at a
constant rate of 300 cfm. The flushing process is carried out at constant
temperature and pressure, under conditions of perfect mixing in the tank at
all times. The time required to reach a gas composition of 95 vol.% nitrogen
in the tank is nearest to

   (A)   3 min
   (B)   5 min
   (C)   7 min
   (D) 10 min
   (E) 15 min

Solution

Let g = quantity of pure natural gas in the tank at any time, and
$\quad x$ = quantity of nitrogen added to the tank
$\quad\quad$ = quantity of mixture removed from the tank.
$\quad$ The tank volume = 1000 ft³, so 5% of tank volume = $0.05 \times 1000 = 50$ ft³.
Suppose a volume Δx of the mixture is removed from the tank. The amount of natural gas thus removed will be $(g/1000)\,\Delta x$. Hence the change in the amount of natural gas in the tank is given by $\Delta g = -(g/1000)\,\Delta x$. Then the ratio of the quantity of natural gas removed to the volume of nitrogen added is

$$\frac{\Delta g}{\Delta x} = -\frac{g}{1000}$$

When $\Delta x \to 0$ we obtain the instantaneous rate of change of g with respect to x:

$$\frac{dg}{dx} = \frac{-g}{1000}$$

Separating the variables of the differential equation,

$$\frac{dg}{g} + \frac{dx}{1000} = 0$$

Integrating,

$$\int \frac{dg}{g} + \int \frac{dx}{1000} = C \qquad \ln g + \frac{x}{1000} = C$$

When $g = 1000$, $x = 0$. Therefore $C = \ln 1000$.

$$\frac{x}{1000} = \ln 1000 - \ln g = \ln \frac{1000}{g}$$

We want to find x when g is 5% by volume or 50 ft³. Hence

$$\frac{x}{1000} = \ln \frac{1000}{50} = \ln 20 = 3.0$$
$$x = 3000 \text{ ft}^3 \text{ nitrogen}$$

Since nitrogen flows in at 300 cfm, the time required to reach 5% by volume natural gas = 3000/300 = 10 min.

$$\text{Answer is (D)}$$

PROBLEM 2-46

The governing differential equation for a body of mass m suspended from a spring (spring constant is k) and subjected to periodic forcing $F = F_m \sin \omega t$ is

$$m \frac{d^2 x}{dt^2} + kx = F_m \sin \omega t$$

Here F_m is the maximum applied force, and the forcing frequency is ω.

Question 1. If the initial equilibrium displacement is $x = 0$, the maximum amplitude x_m to be expected is

(A) $x_m = F_m/(2k)$
(B) $x_m = F_m/k$
(C) $x_m = F_m/(k - m\omega^2)$
(D) $x_m = F_m/(k + m\omega^2)$
(E) none of these values

Solution

If we assume a particular solution of the differential equation is $x_p = x_m \sin \omega t$, substitution of this expression into the differential equation gives

$$-m\omega^2 x_m \sin \omega t + kx_m \sin \omega t = F_m \sin \omega t$$

and

$$x_m = F_m/(k - m\omega^2)$$

Answer is (C)

Question 2. The complete solution to the differential equation when the initial conditions at $t = 0$ are $x = x_0$, $dx/dt = 0$ is

(A) $x_0 \cos pt + x_m \sin \omega t$
(B) $x_0 \cos pt$
(C) $x_m[1 - \omega/p] \sin \omega t$
(D) $x_m[1 + \omega/p] \sin \omega t$
(E) $x_0 \cos pt + x_m[1 - \omega/p] \sin \omega t$

Solution

The complete solution is the sum of the particular and complementary solutions; the particular solution may be written as $x_m \sin \omega t$. The complementary solution x_c will satisfy the homogeneous differential equation (i.e., the case when $F_m = 0$) and will be of the form

$$x_c = A \sin pt + B \cos pt$$

Substitution of x_c into the homogeneous differential equation yields

$$m(-Ap^2 \sin pt - Bp^2 \cos pt) + k(A \sin pt + B \cos pt) = 0$$

Hence $-mp^2 + k = 0$ or $p^2 = k/m$. When $t = 0$, $\sin \omega t = \sin pt = 0$ and $x_0 = B$. In general $dx/dt = Ap \cos pt - Bp \sin pt + x_m \omega \cos \omega t$. When $t = 0$, $dx/dt = 0 = Ap + x_m \omega$, leading to $A = -x_m \omega/p$ and

$$x = x_0 \cos pt + x_m[1 - \omega/p] \sin \omega t$$

Answer is (E)

PROBLEM 2-47

A factory has measured the diameter of 100 random samples of its product. The results, arranged in ascending order, were:

(45 results between 0.859 and 0.900, inclusive) . . .
0.901 0.902 0.902 0.902 0.903 0.903
0.904 0.904 0.904 0.904 . . .
(45 more different results from 0.905 to 0.958, inclusive)

The sum of all 100 observations is 91.170. No observed value among those not numerically shown above occurred more than twice. The smallest value observed was 0.859; the largest, 0.958. From the information given, the median for the data is

(A) 0.901
(B) 0.902
(C) 0.903
(D) 0.904
(E) 0.912

Solution

Mean: The arithmetic mean is what is commonly called the *average*, that is, the sum of the values divided by the number of values.

$$\bar{X} = \frac{1}{N} \sum_{i=1}^{N} X_i = \frac{91.170}{100} = 0.9117 = 0.912$$

Median: The median of a set of data is the middle value in order of size if N is odd or the value midway between the two middle items if N is even. Here the median is halfway between 0.903 and 0.903, or 0.903.

Mode: This is the most frequent value. In this case it is 0.904, which occurred four times.

Answer is (C)

PROBLEM 2-48

The geometric mean of the numbers 4 and 49 is

(A) 14.0
(B) 22.0
(C) 26.5
(D) 33.0
(E) 49.3

Solution

The geometric mean (G) may be defined as the Nth root of the product of N values:

$$G = \sqrt[N]{x_1 x_2 \ldots x_n}$$

In this case

$$G = \sqrt[2]{(4)(49)} = \sqrt{4} \times \sqrt{49} = 2(7) = 14$$

Answer is (A)

PROBLEM 2-49

The probability of obtaining at least one 6 in three throws of a die is nearest to

(A) 0.17
(B) 0.42
(C) 0.50
(D) 0.58
(E) 0.83

Solution

Let P_1 be the probability of getting one or more 6's and P_2 be the probability of getting no 6's. For any situation the sum of probabilities for all possibilities is 1:

$$P_1 + P_2 = 1$$

The probability of *not* getting a 6 on any single roll is $\frac{5}{6}$; thus the probability that a 6 will not turn up in three rolls is

$$P_2 = (\tfrac{5}{6})(\tfrac{5}{6})(\tfrac{5}{6}) = \tfrac{125}{216} = 0.58$$
$$P_1 = 1 - P_2 = 1 - 0.58 = 0.42$$

Therefore the probability of obtaining at least one 6 in three throws of a die is 0.42.

$$\text{Answer is (B)}$$

PROBLEM 2-50

An elementary game is played by rolling a die and drawing a ball from a bag containing three white and seven black balls. The player wins whenever he rolls a number less than 4 and draws a black ball. What is the probability of winning in the first attempt?

(A) 7/20
(B) 12/10
(C) 1/2
(D) 7/10
(E) 13/20

Solution

The probability of rolling a number less than 4 with a die $= \frac{3}{6} = \frac{1}{2}$. The probability of drawing a black ball $= 7/(7+3) = 7/10$. The probability of a series of independent events equals the product of probabilities of the individual events:

$$\frac{1}{2} \times \frac{7}{10} = \frac{7}{20}$$

$$\text{Answer is (A)}$$

PROBLEM 2-51

The table gives the values of x and the frequency f with which they occur.

x	f
2	4
4	6
7	6
12	4

Question 1. The arithmetic mean $\bar{x}$ is nearest to

(A) 6.55
(B) 6.40
(C) 6.25
(D) 6.10
(E) 5.95

Solution

When individual values recur, it is convenient to compute the mean as the sum of each distinct value times its frequency of occurrence f, divided by the total number of values.

x	f	fx
2	4	8
4	6	24
7	6	42
12	4	48
	$N = 20$	$122 = \Sigma\,(fx)$

$$\bar{x} = \frac{\Sigma\,(fx)}{N} \qquad \bar{x} = \frac{122}{20} = 6.1$$

This procedure is equivalent to summing 20 individual values of x and dividing by 20.

Answer is (D)

Question 2. The standard deviation σ is nearest to

(A) 4.7
(B) 4.4
(C) 4.1
(D) 3.8
(E) 3.5

Solution

The standard deviation σ is a measure of the dispersion or scatter of a set of values. It is sometimes called the rms deviation, as this describes its method of calculation. First, square the deviations of individual values from the arithmetic mean. Then take the mean of these squares and extract the square root.

x	f	$x - \bar{x}$	$(x - \bar{x})^2$	$f(x - \bar{x})^2$
2	4	−4.1	16.81	67.24
4	6	−2.1	4.41	26.46
7	6	0.9	0.81	4.86
12	4	5.9	34.81	139.24
	$N = 20$			$\Sigma = 237.80$

$$\sigma = \left[\frac{\Sigma \, f(x - \bar{x})^2}{N - 1} \right]^{1/2} = \left(\frac{237.80}{19} \right)^{1/2} = (12.52)^{1/2} = 3.54$$

Answer is (E)

(If N rather than $N - 1$ were used in the denominator of σ, the result would be $\sigma = 3.45$.)

PROBLEM 2-52

Assume a group of nine people consists of four men and five women. The probability that a committee of three, selected at random, would consist of two men and one woman is nearest to

(A) 0.30
(B) 0.35
(C) 0.40
(D) 0.45
(E) 0.50

Solution

Here we use the notation

$$\binom{n}{r} = \frac{n!}{r!(n - r)!}$$

which gives the total number of ways r objects can be chosen from n objects.

There are nine people (four men and five women). The total number of committees of three people would be equal to the total number of committees consisting of

$$
\begin{array}{llll}
0 \text{ men} \quad 3 \text{ women} \quad \text{or} & \dbinom{4}{0} \times \dbinom{5}{3} = 1 & \times \dfrac{5!}{3!2!} = 10 \\[4mm]
1 \text{ man} \quad 2 \text{ women} & \dbinom{4}{1} \times \dbinom{5}{2} = \dfrac{4!}{3!} & \times \dfrac{5!}{2!3!} = 40 \\[4mm]
2 \text{ men} \quad 1 \text{ woman} & \dbinom{4}{2} \times \dbinom{5}{1} = \dfrac{4!}{2!2!} & \times \dfrac{5!}{4!} = 30 \\[4mm]
3 \text{ men} \quad 0 \text{ women} & \dbinom{4}{3} \times \dbinom{5}{0} = \dfrac{4!}{3!} & \times 1 = 4
\end{array}
$$

Total   84

The number of committees consisting of two men and one woman is 30 out of 84 possible combinations. Hence the probability of this happening is

$$
P = \frac{30}{84} = 0.357
$$

Answer is (B)

PROBLEM 2-53

The sum of all whole numbers from 1 to 100 inclusive is nearest to

(A) 6500
(B) 6000
(C) 5500
(D) 5050
(E) 5005

Solution

We want to determine the sum S of an arithmetic progression. Let a be the first term, d the difference between successive terms, l the last term, and n the number of terms. Then

$$
S = a + (a + d) + (a + 2d) + \cdots + [a + (n - 1)d]
$$

or, in reverse order,

$$S = [a + (n - 1)d] + [a + (n - 2)d] + \cdots + a$$

Adding these two equations,

$$2S = n[2a + (n - 1)d]$$

or, since $[a + (n - 1)d] = l$,

$$S = \frac{n}{2}(a + l)$$

In this problem $a = 1$, $n = l = 100$, and $S = \frac{100}{2}(101)$.

$$S = 5050$$

$$\text{Answer is (D)}$$

PROBLEM 2-54

Given a universe = (1, 2, 3, 4, 5, 6, 7), set $M = (1, 3, 6)$, and set $N = (1, 2, 6, 7)$, the set $M \cap N$ is

(A) (2, 4, 5, 7)
(B) (3, 4, 5)
(C) (1, 6)
(D) (3)
(E) (1, 2, 4, 5, 6, 7)

Solution

$\overline{M}$ represents non-M, or all elements of the universe that are not in set M.

$$\text{(A) } \overline{M} = (2, 4, 5, 7)$$

$\overline{N}$ represents non-N.

$$\text{(B) } \overline{N} = (3, 4, 5)$$

$M \cap N$ denotes the intersection of sets M and N, which includes all elements in *both* sets M and N.

$$\text{(C) } M \cap N = (1, 6)$$

$M \cap \overline{N}$ is the intersection of sets M and $\overline{N}$.

$$(D) \ M \cap \overline{N} = (3)$$

$\overline{M} \cup N$ represents the union of sets $\overline{M}$ and N; it includes all elements that are members of *either* $\overline{M}$ *or* N.

$$(E) \ \overline{M} \cup N = (1, 2, 4, 5, 6, 7)$$

Answer is (C)

3

COMPUTER PROGRAMMING

Computer programming is one aspect of a user's interaction with digital computers. Competent computer programming requires knowledge in three areas. First one should have some understanding of computer hardware and its organization. Next one needs knowledge of computer operating systems. Finally one must know the details of one or more computer languages such as BASIC and FORTRAN.

The Engineer-In-Training examination will not contain a separate section on computer programming. Instead one may find some problems on number systems and computer algorithms in the sections on mathematics, and some problems on computer organization and logic in the electrical circuits sections. Some of the problems may include BASIC or FORTRAN. Thus this material is in effect a part of the mathematics and electricity topics rather than a separate topic in the examination.

COMPUTER ORGANIZATION

All computers process, store and/or recall information units, called data, of one kind or other. The efficiency and capacity of each machine to do this task well depends on the organization of the computer. Currently it is conceptually correct to regard any computer as consisting of four interrelated units: a central processing unit (CPU), a control unit, a memory unit, and an input/output (I/O) unit.

Computer operations on data are directed by a clearly defined set of instructions called a program. Programs written by most engineers are written in a high-level language such as BASIC, FORTRAN, PASCAL or similar language. Some computer scientists and engineers may write some

65

parts of a program in assembly language, an intermediate-level language. Before these programs can be executed by the computer, they must be translated, i.e. compiled, into machine language. Programs in higher level languages are called source code, and the machine-language code is called the object code. Every program is the implementation of an algorithm, which is the name given to any finite set of clearly defined rules. Example algorithms are given later in this chapter.

The central processing unit is where most data and instruction manipulation is done. The computational part of this unit is called the Arithmetic Logic Unit. The CPU also has registers for the temporary storage of data. The control unit (CU) is the executive center of the entire system which precisely directs the activities of each part of the system. Sometimes the control unit is considered a part of the CPU. The memory unit holds the information used in the programs; it provides both temporary and permanent storage with each datum having an individual storage location identified with an address. The input/output unit is the main connection or interface between the other computer units and the outside world via devices such as terminals, keyboards and printers.

NUMBER SYSTEMS

The set of *rational numbers* consists of all positive integers $(1, 2, 3, \ldots)$, all negative integers $(-1, -2, -3, \ldots)$, the number 0, and fractions constructed from these integers. All numbers in this system are of the form a/b where a is any positive or negative integer or 0, and b is any positive or negative integer. Numbers that cannot be represented by a/b are called *irrational numbers*. Examples of irrational numbers are $\sqrt{5}$, $\sqrt[3]{10}$ and $\pi = 3.14159 \ldots$.

The most common method of representing real numbers is constructed on the base 10:

10^5	10^4	10^3	10^2	10^1	10^0
100,000	10,000	1000	100	10	1

To represent the number 123_{10}, for example, we write

$$1 \times 10^2 + 2 \times 10^1 + 3 \times 10^0$$
$$= 100 \quad + 20 \quad + 3 \quad = 123_{10}$$

We usually simply omit the base 10 subscript as being understood.

With computers it is sometimes necessary to work in number systems with bases other than 10; the most common of these is the binary system with base 2, and a convenient intermediate system is octal, which uses base 8.

The binary number system uses only two digits 0 and 1. Each binary number can be written in the form

$$a_n a_{n-1} a_{n-2} \cdots a_1 a_0 = a_n 2^n + a_{n-1} 2^{n-1} + \cdots + a_1 2^1 + a_0 2^0$$

An example is $101\ 101_2 = 2^5 + 2^3 + 2^2 + 2^0 = 45_{10}$. Octal numbers use base 8. Thus $427_8 = (4 \times 8^2) + (2 \times 8^1) + (7 \times 8^0) = 256 + 16 + 7 = 279_{10}$. Since $8 = 2^3$, each octal digit can quickly be converted to a triple of binary digits. An example of converting from octal to binary is $427_8 = 100\ 010\ 111_2$.

To convert from decimal to octal, divide the decimal number repeatedly by 8, truncating the result and saving the remainder each time; the remainders are the coefficients a_0, a_1 etc. As an example

$$279_{10}/8 = 34 \text{ with remainder } a_0 = 7$$
$$34_{10}/8 = 4 \text{ with remainder } a_1 = 2, \text{ and } a_2 = 4$$

As we found earlier, $279_{10} = 427_8$. Fractions are treated in an analogous way when changing bases but with the roles of multiplication and division reversed. Examples are

$$0.125_{10} = \frac{1}{10} + \frac{2}{10^2} + \frac{5}{10^3} \quad \text{and} \quad 0.42_8 = \frac{4}{8} + \frac{2}{8^2}$$

Example 1

Convert 123_{10} to binary.

2^7	2^6	2^5	2^4	2^3	2^2	2^1	2^0	
128	64	32	16	8	4	2	1	
	64 +	32 +	16 +	8 +	0 +	1 +	1	$= 123_{10}$
	1	1	1	1	0	1	1	

Thus $123_{10} = 1\ 111\ 011_2$.

Example 2

Convert 0.125_{10} to binary. Let

$$0.125_{10} = \frac{b_1}{2} + \frac{b_2}{2^2} + \frac{b_3}{2^3} + \cdots$$

Multiply by 2, $0.250 = b_1 + \frac{b_2}{2} + \cdots, \quad b_1 = 0$

Multiply by 2, $0.500 = b_2 + \frac{b_3}{2} + \cdots, \quad b_2 = 0$

Multiply by 2, $1.000 = b_3 + \dfrac{b_4}{2} + \cdots, \quad b_3 = 1$

We also see that $b_n = 0$ for all $n > 3$ so that $0.125_{10} = 0.001_2$.

BASIC

BASIC (Beginners All-purpose Symbolic Instruction Code) was originally created at Dartmouth for time-sharing computer use. It has since become an important microcomputer language. There are three types of BASIC instructions: statements, functions and commands. These are described in the following sections.

Statements

The bulk of a BASIC program consists of statements. The general format is

Line number   BASIC statement

The line number is one to five digits and gives the order in which the program is stored. This is followed by the BASIC statement itself. A list of selected statements, along with a description and example, appears in Table 3.1.

TABLE 3.1 BASIC Statements

Statement	Description	Example
DATA	Assigns values to the constants that are listed in **READ** statements	10 DATA 2.5,20
DIM	Specifies the maximum values for array variable subscripts and allocates storage	10 DIM X(15)
END	Terminates program execution, closes all files, and returns to the command level	80 END

FOR – NEXT	Establishes a loop	10 FOR N=1 TO 10
		   .
		   .
		50 NEXT N

GOSUB – RETURN	Branches to and returns from a subroutine	10 GOSUB 40
		   .
		   .
		30 GOTO 90
		40 REM SUBROUTINE BEGINS
		   .
		   .
		80 RETURN
		90 REM

GOTO	Unconditionally branches to specified line number	10 GOTO 140

TABLE 3.1 Continued

Statement	Description	Example
IF	Alters program flow based on the results of the evaluation of an expression	10 IF X < 10 THEN Y=20
INPUT	Receives input from the keyboard during program execution; a question mark appears on the screen	10 INPUT A, X
LET	Assigns the value of an expression to a variable; the word LET is unnecessary	10 LET Y=10 10 Y=10
LPRINT	Prints data on the printer	10 LPRINT A, B
LPRINT USING	Prints strings or numbers on the printer using a specified format	10 LPRINT USING "&"; A$
ON – GOSUB	Branches to the specified first line number of a subroutine, depending on the value of the expression	10 ON X-20 GOSUB 300,400
ON – GOTO	Branches to the specified line number, depending on the value of the expression	10 ON X-20 GOTO 50,60
OPEN	Allows input or output to a file or device	10 OPEN "EIT.DAT" AS #1

PRINT	Displays data on the output device	`10 PRINT X, X+5`
PRINT USING	Displays strings or numbers on the output device using a specified format	`10 PRINT USING "##.##";.5`
PRINT#	Writes data sequentially to a file	`10 PRINT#2,A;B`
READ	Reads values from a DATA statement and assigns them to variables	`10 READ X,Y`
REM	Inserts remarks in a program	`10 REM Part Two`
RETURN	Stops a subroutine and returns to the main program; see GOSUB	`80 RETURN`
STOP	Stops execution of a program and returns to the command level; files not closed	`80 STOP`
WRITE	Outputs data to the output device; similar to PRINT, except WRITE inserts commas between items as they are displayed	`10 WRITE A,B`
WRITE#	Writes data sequentially to a file	`10 WRITE#2,A$,B$`

Functions

BASIC has a number of functions to ease numeric computations. A function is used like a variable to perform the operation. Fourteen of the more frequently used functions are given here.

Function	Description
ABS (x)	Absolute value of x
ATN (x)	Arctangent of x
CDBL (x)	Converts x to a double-precision number
CINT (x)	Converts x to an integer by rounding
COS (x)	Cosine x, where x is in radians
EXP (x)	e^x
FIX (x)	Truncates x to an integer
INT (x)	Largest integer $\leq x$
LOG (x)	Natural logarithm (base e) of x for $x > 0$
RND (x)	Random number (0–1)
SGN (x)	Returns the sign of x
SIN (x)	Sine x, where x is in radians
SQR (x)	Square root of x, for $x > 0$
TAN (x)	Tangent x, where x is in radians

Commands

Commands generally operate on programs and typically consist of a word or a word and a file name. Examples are

> RUN   (Runs program currently in memory)
>
> RUN AL   (Loads AL and runs it)

Some commands are given in the following list.

Command	Description
AUTO	Automatically generates next line number each time *Enter* is pressed
CLEAR	Sets all numeric variables to zero and all string variables to null
CONT	Resumes program execution after a break
DELETE	Deletes program lines
EDIT	Displays a line for editing

Command	Description
KILL	Deletes a file from a disk
LIST	Lists the program currently in memory on the screen
LLIST	Lists the program currently in memory on the printer
LOAD	Loads a program from the specified device into memory
NEW	Deletes the program currently in memory and clears all variables
RENUM	Renumbers the program lines
RUN	Causes the current program to be compiled and executed
SAVE	Stores a BASIC program file

ALGORITHMS

An algorithm is a set of well-defined rules for the solution of a problem in a finite number of steps. There are many numerical methods that are utilized in computing. One favorite algorithmic examination question deals with the sorting of numerical or alphabetical lists, or elements of a one-dimensional array. This may be called push-down sorting (largest item moved to the bottom) or bubble sorting (smallest item moved to the top). The sort portion of a BASIC program might take this form:

```
100 REM Sort of Array A(j)
110 REM Set FLIP=1 to force first pass through loop
120 FLIP=1
130 IF FLIP=0 THEN GOTO 240
140 FLIP=0
150 FOR I=1 TO J-1
160 IF A(I)>A(I+1) THEN 170 ELSE 210
170 TEMP=A(I+1)
180 A(I+1)=A(I)
190 A(I)=TEMP
200 FLIP=1
210 NEXT I
220 J=J-1
230 GOTO 130
```

The sort begins in statement 160 by comparing the two values at the beginning of the one-dimensional array. If the larger value appears before the smaller value, they are reversed in the array in lines 170 to 190. The sort then proceeds to the second and third elements in the array and makes the same comparison. As the sort proceeds the larger values are moved downward, and the smaller values are moved upward. In the first full pass through the array the largest value is moved to the bottom of the array. In the next iteration of the loop (lines 130 to 230) the array to be sorted is shortened by one (J=J−1) and the process is repeated. The process continues until the sort array has been shortened to one item. At this point the sort is complete with all items ordered; the smallest value is at the top and the largest value is at the bottom.

FORTRAN

FORTRAN (FORmula TRANslation) is a high-level computer language that has been in use for thirty years or more. The best known modern version is called FORTRAN 77 and is the basis for the discussion here.

Constants

Constants can be of numerical or character type. There are four types of numerical constants.

Integers	Sometimes called fixed point; whole numbers, positive, negative, or zero with no fractional value and no decimal point
Real	Sometimes called floating point; contains a decimal point representation
Double precision	Values are stored with an increased number of significant digits
Complex	In the form $a + ib$ where $i = \sqrt{-1}$

Character constants are enclosed in single apostrophes.

Variable Names

Variable names that begin with the letters I−N are automatically typed as integer variables. Those beginning with A−H and O−Z are automatically considered to be real variables. These assignments may be overriden by the use of the REAL or INTEGER declarations. Also the IMPLICIT statement can alter the variable types.

Nonexecutable Statements

Nonexecutable statements are used by the FORTRAN compiler, but they are not translated into machine language; hence they are nonexecutable. In general, nonexecutable statements are placed at the beginning of the program (or subprogram) ahead of the executable statements. Important nonexecutable statements are given in Table 3.2.

Executable Statements

As in BASIC, the heart of a FORTRAN program is its executable statements. Arithmetic assignment is the most common executable statement. Examples are

$$PI = 3.14159 \quad \text{and} \quad AREA = PI * R * * 2$$

Other important executable statements are given in Table 3.3.

TABLE 3.2 Nonexecutable Statements

Statement	Description	Example
CHARACTER	Defines a character string variable of n characters	CHARACTER*n NAME
COMMON	Sets up a memory block common to the main program and to those subprograms where it is present	COMMON A, B
COMPLEX	For variables in $a + ib$ form	COMPLEX AX, AY
DATA	Provides initial values for variables	DATA A, B, I/2.05,3.1,6/
DIMENSION	Specifies maximum values for array variable subscripts and allocates storage	DIMENSION J(10)
DOUBLE PRECISION	Used to define variables to be of type double precision	DOUBLE PRECISION Y
EQUIVALENCE	Variables of different names share the same data storage	EQUIVALENCE (OLE, OLD) EQUIVALENCE (B(1), C(4))
FORMAT	Indicates the layout of input or output	10 FORMAT (1X, F6.2)

FUNCTION	Called from the main program by its name; the name must be assigned a value before a RETURN statement	FUNCTION FNC (X, Y, N)
IMPLICIT	Overrides the default types of variables	IMPLICIT REAL (A-Z)
INTEGER	Declares variables to be integer type	INTEGER A
LOGICAL	Used to define variables to be of logical type; they can have only a .TRUE. or .FALSE. value	LOGICAL TEST
PARAMETER	Assigns a symbolic name to a constant	PARAMETER (X=5)
PROGRAM	Assigns a name to a program and marks its beginning (optional)	PROGRAM PE
REAL	Declares variables to be real (floating point)	REAL I
SUBROUTINE	Assigns a name to a subroutine, marks its beginning, and provides any argument list	SUBROUTINE EIT (X, Y, N)

TABLE 3.3 Executable Statements

Statement	Description	Example
CALL	To call a subroutine	CALL HOME (IN, Y, OUT)
CLOSE	Closes a file previously open	CLOSE (UNIT=10)
CONTINUE	Generally used to indicate the end of a DO loop; it must be preceded by a label (e.g., 100)	100 CONTINUE
DO	Begins a counting loop	DO 100 I=4, 30, 2
END	The required last statement in every program and subprogram	END
ENTRY	May begin a subprogram execution anywhere except within a DO loop or IF block	ENTRY EITMID (X, Y, N)
GO TO	Unconditional transfer	GO TO 40
GO TO	Computed GO TO	GO TO (30, 50, 20), INDEX
IF	If the condition is true, one statement is executed	IF (A.GT.C) GO TO 60
IF – THEN ENDIF	If the condition is true, statements are executed to ENDIF; if not, control jumps to the statement following ENDIF	IF (A.LE.C) THEN ⋮ ENDIF

| IF – THEN<br>ELSE<br>ENDIF | If the condition is true, statements are executed to ELSE and then control jumps to the statement following ENDIF; if the condition is false, control jumps to the first statement after ELSE | ```
IF (A.EQ.C) THEN
   .
   .
   .
ELSE
   .
   .
   .
ENDIF
``` |
| IF – THEN<br>ELSEIF – THEN<br>ELSE<br>ENDIF | A more elaborate structure typically used to test different cases | ```
IF (A.LT.C) THEN
   .
   .
   .
ELSEIF(A.EQ.C) THEN
   .
   .
   .
ELSE
   .
   .
   .
ENDIF
``` |
| OPEN | Assigns a unit number to a data file and declares whether it already contains data ('OLD') or not ('NEW'). The OPEN statement must precede any READ or WRITE of the file | ```
OPEN (UNIT=10, FILE=
'EIT', STATUS='NEW')
``` |
| PAUSE | Halts program execution; *enter* key resumes the program execution | PAUSE |
```

TABLE 3.3 Continued

Statement	Description	Example
PRINT	Formatted output	PRINT 60, MO, IYR 60 FORMAT(1X, I2, 5X, I4)
PRINT*	Output with format and spacing determined by the compiler	PRINT*, MO, IYR
READ	Input from a data file	READ (10, *) MO, IYR
READ	Formatted input from a data file	READ (10, 50) MO, IYR 50 FORMAT (I2, I4)
READ*	Input from keyboard (or in batch processing, from data lines following the program)	READ*, MO, IYR
RETURN	Used in function and subroutine subprograms to return control to the statement that referenced it	FUNCTION PIX(X) PIX=3.14159 * X RETURN END
STOP	Terminates program execution	STOP
WRITE	Writes information into a data file	WRITE(10, *) MO, IYR
WRITE	Writes information into a data file in a specified format	WRITE(10, 50) MO, IYR 50 FORMAT(1X, I1, I4)

Functions

There are two classes of functions in FORTRAN: functions written by the programmer and library (or intrinsic) functions. An example of a programmer-written function to compute $N!$ is

```
    FUNCTION NFAC(N)
    J=1
    DO 100 I=1, N
    J=J*I
100 CONTINUE
    NFAC=J
    RETURN
    END
```

Library functions are included in the compiler and are available for direct use in programs. The major ones are given in the following list.

Function	Description
AINT(x)	Truncates x to a whole number
ANINT(x)	Round x to the nearest whole number
ABS(x)	Absolute value of x
ALOG10(x)	Logarithm (base 10) of x
ATAN(x)	Arctangent of x
COS(x)	Cosine of x, where x is in radians
DBLE(x)	Converts x to double precision
DIM(x, y)	x set equal to the minimum of x and y
EXP(x)	e^x
INT(x)	Truncates x to an integer
LOG(x)	Natural log (base e) of x
MOD(x, y)	Remainder from integer division x/y
REAL(x)	Converts x to a real value
SIN(x)	Sine of x, where x is in radians
SQRT(x)	Square root of x, $x > 0$
TAN(x)	Tangent of x, where x is in radians

PROBLEM 3-1

The component of a computer system that ties all the hardware components together in an integrated manner is the

(A) Hard disk
(B) Read only memory (ROM)
(C) Operating system
(D) Random access memory (RAM)
(E) FORTRAN

Solution

The operating system is a computer program that brings together the hardware components and provides resource allocation, scheduling, input/output control, and data management. It controls the execution of programs.

Answer is (C)

PROBLEM 3-2

The binary number 110 010 is closest to which of these base 10 numbers?

(A) 25
(B) 50
(C) 75
(D) 100
(E) 125

Solution

2^5	2^4	2^3	2^2	2^1	2^0
32	16	8	4	2	1
1	1	0	0	1	0

$$32 \ + 16 \qquad\qquad +2 \qquad\qquad = 50_{10}$$

Answer is (B)

PROBLEM 3-3

For the base 10 number of 26, the equivalent binary number is

(A)   26
(B) 1 101
(C) 11 010
(D) 00 101
(E) None of the above

Solution

2^5	2^4	2^3	2^2	2^1	2^0	
32	16	8	4	2	1	
	1	1	0	1	0	$= 26_{10}$

Answer is (C)

PROBLEM 3-4

The number 5_8 is written in binary as

(A) 101
(B) 011
(C) 100
(D) 010
(E) 110

Solution

The binary numbers are

$$001_2 = 1_8 \quad 010_2 = 2_8 \quad 011_2 = 3_8 \quad 100_2 = 4_8$$
$$101_2 = 5_8 \quad 110_2 = 6_8 \quad 111_2 = 7_8 \quad 1000_2 = 10_8 \text{ and so on}$$

Answer is (A)

PROBLEM 3-5

The number 100_{10} is written in octal as

(A) 100
(B) 144
(C) 1210
(D) 1250
(E) 8808

Solution

Divide the decimal number repeatedly by 8, truncating the result and saving the remainder each time. The remainders are the coefficients of the octal number.

$$100_{10}/8 = 12 \text{ with remainder } a_0 = 4$$
$$12_{10}/8 = 1 \text{ with remainder } a_1 = 4$$
$$1_{10}/8 = 0 \text{ with remainder } a_2 = 1$$

Thus $100_{10} = 144_8$.

Answer is (B)

PROBLEM 3-6

The number 144_8 is written in binary as

(A)   144
(B)   1 001
(C)   11 001
(D)   10 010 000
(E) 001 100 100

Solution

Each octal digit is converted to a set of three binary digits

$$144_8 = 001\ 100\ 100_2$$

Problems 3-5 and 3-6 demonstrate that

$$100_{10} = 144_8 = 001\ 100\ 100_2$$

Answer is (E)

PROBLEM 3-7

The sum $3155_8 + 634_8$ is

(A) 4011_8
(B) 3701_8
(C) 3789_8
(D) 3802_8
(E) 4002_8

Solution

$$3^1 1^1 5^1 5_8$$
$$+ \quad 6\,3\,4_8$$

$$4\,0\,1\,1_8$$

Answer is (A)

PROBLEM 3-8

The number 77777_8 is closest to the base 10 number

(A) 11,111
(B) 16,807
(C) 32,767
(D) 131,000
(E) 262,143

Solution

If 1 is added to 77777_8 we have 100000_8. Thus $77777_8 = 100000_8 - 1 = 8^5 - 1 = 32,768 - 1 = 32,767$.

Answer is (C)

PROBLEM 3-9

The number $-6,020,000$ may be represented in exponential form by which one of the following?

(A) $-0.602E-7$
(B) $-0.602E7$
(C) $6.02E-6$
(D) $6E-6.02$
(E) $-4E602$

Solution

The exponential form in computing consists of an integer or fixed point mantissa, with or without a sign, followed by the letter E and an exponent with or without a sign. The number $-6,020,000$ may be written in a variety of ways in exponential form, including $-6.02E6$, $-6.02E+6$, and $-0.602E7$.

Answer is (B)

PROBLEM 3-10

Consider five lines in a computer program:

```
11   X=6
12   Y=10
13   Z=8
14   Y=X
15   X=Y
```

After this portion of the program is executed, which one of these outcomes is correct?

(A) $X=Y$ and $X<Z$
(B) $X=Y$ and $X>Z$
(C) $X<Y$ and $X<Z$
(D) $X<Y$ and $X>Z$
(E) $X>Y$ and $X=Z$

Solution

In line 14 Y is set to 6. In line 15 X is set to the current value of Y, that is, 6. Hence $X=Y=6$. Since $Z=8$, $X=Y$ and $X<Z$.

Answer is (A)

PROBLEM 3-11

Which *one* of the following statements is correctly written in BASIC?

(A) IF (X .GT. 6.5) THEN Y=1.5
(B) DIM=HI/1000.
(C) J%=20 ∧ 5
(D) IF A<<B THEN A=0
(E) B$="EIT EXAM"

Solution

In (A) the relational operator for "greater than" is .GT. in FORTRAN; in BASIC it is $>$. In (B) DIM is a reserved word used for dimensioning arrays in BASIC; it must not be used as a variable name. In (C) J% is an integer variable; since $20^5 = 3,200,000$, it is far too large to store as an integer number without overflow. In (D) BASIC does not recognize $<<$. In (E) the dollar sign following B indicates a string type-declaration. The line is correct BASIC.

Answer is (E)

PROBLEM 3-12

To express the algebraic expression $X(-Y)$ in BASIC, one should write which of the following statements?

(A) X*(-Y)
(B) X(-Y)
(C) X-Y
(D) -XY
(E) XY

Solution

An arithmetic operator must be inserted in the algebraic expression.

Answer is (A)

PROBLEM 3-13

```
10 FOR I=1 TO 3
20 FOR J=1 TO 4
30 A=I+J
40 PRINT A, I, J
50 NEXT J, I
60 END
```

The number of lines of printed output is closest to

(A)   0
(B)   3
(C)   4
(D) 12
(E) 14

Solution

The PRINT statement is within both FOR—NEXT loops. Hence it will be printed 12 times.

Answer is (D)

PROBLEM 3-14

```
10 DIM Y(2,2)
20 READ M,N
30 DATA 2,2,3,3,4,5
40 FOR I=1 TO M
50 FOR J=1 TO N
60 READ Y(I,J)
70 NEXT J,I
80 END
```

Consider the BASIC program and then answer the following questions.

Question 1. How many elements are in Y?

(A) 1
(B) 2
(C) 4
(D) 6
(E) 9

Solution

Each storage location in an array is called an element. In BASIC 0,0 is the first element in a two-dimensional array (and 0 is the first element in a one-dimensional array) unless an OPTION BASE 1 statement precedes the DIM statement. Thus the array Y contains the following elements:

```
Y(0,0)    Y(0,1)    Y(0,2)
Y(1,0)    Y(1,1)    Y(1,2)
Y(2,0)    Y(2,1)    Y(2,2)
```

In BASIC storage is by rows, hence $Y(0,2)$ is, for example, the third element.

Answer is (E)

Question 2. What is the value of $Y(2,1)$ after the program has run?

(A) 0
(B) 2
(C) 3
(D) 4
(E) 5

Solution

A DATA statement may be used by more than one READ statement. Unless a RESTORE statement is used, the second READ statement will begin in the DATA statement following where the first READ statement ended. Here $M=2$ and $N=2$. Then four values are read into Y in this sequence:

$$Y(1,1)=3$$
$$Y(1,2)=3$$
$$Y(2,1)=4$$
$$Y(2,2)=5$$

Answer is (D)

Question 3. What is the value of $Y(2,2)$ after the program has run?

(A) 0
(B) 2
(C) 3
(D) 4
(E) 5

Solution

We have seen in the solution to Question 2 that $Y(2,2)=5$.

Answer is (E)

PROBLEM 3-15

Which one of the following lines has a BASIC syntax error:

(A) 15 IF X < 9 THEN 10
(B) 50 FOR I=1 TO 5
(C) 20 DIM YEARS (2,4)
(D) 70 B=B+1
(E) 50 PRINT A USING 80

Solution

In (E) the correct BASIC statement should be

$$50 \ \text{PRINT USING } 80, \ A$$

The other statements are correct.

Answer is (E)

PROBLEM 3-16

With a microcomputer, if one types the BASIC command line

BASIC EEA

which one of the following will occur?

(A) BASIC will be loaded with its Extended Error Allocation table
(B) The EEA table will be displayed on the screen
(C) An advanced version of BASIC will be loaded
(D) BASIC will be loaded, and the program EEA.BAS will be loaded and executed (run)
(E) BASIC will be loaded, and the program EEA.BAS will be loaded but not executed (run)

Solution

There are a variety of options when one types a BASIC command line. The one in the problem statement is the same as typing

BASIC
LOAD "EEA.BAS"
RUN

Answer is (D)

PROBLEM 3-17

Consider the following BASIC program.

```
10 READ X, N
20 C=0
30 READ A
40 C=C+A
50 PRINT "C= "; C
60 IF C<N THEN 30
70 DATA 2, 4, 1
80 DATA 1, 3, 1, 5
90 END
```

The number of lines of output is closest to

 (A) No output
 (B) 1 line
 (C) 2 lines
 (D) 3 lines
 (E) 4 lines

Solution

The key to this problem is identifying the values of A read in line 30. The first value is read as 1 from the line 70 DATA statement. Subsequent values are read successively from the line 80 DATA statement.

Answer is (D)

PROBLEM 3-18

Here is a short BASIC program.

```
10 READ J
20 K=1
30 FOR L=2 TO J
40 K=K*L
50 NEXT L
60 WRITE K
70 DATA 5
80 END
```

The relation between the input J and the output K is

(A) $K=J^2$
(B) $K=J^J$
(C) $K=J^K$
(D) $K=J!$
(E) None of the above

Solution

$$J = 5$$

$$K = 1 \times 2 \times 3 \times 4 \times 5 = 5!$$

Answer is (D)

PROBLEM 3-19

```
10 K=2
20 L=3.5
30 M=1
40 D=K*L ∧ K+M
```

When these BASIC statements are evaluated, the value of D is closest to

(A)   8
(B)   25
(C)   50
(D)   87
(E)   343

Solution

In BASIC the order of execution of operations is

1. Functions.
2. Arithmetic operations in this order: exponentiation, negation (assigning a negative sign to an expression), multiplication and floating point division, integer division (the operator is the backslash), modulo arithmetic (MOD), followed by addition and subtraction. Multiplication and floating point division are at the same level. This is also true for addition and subtraction. Operations at the same level are performed in left-to-right order.
3. Relational operations (=, <, >).
4. Logical operations in this order: NOT, AND, and OR.

Operations within parentheses are performed first, in the usual order of execution. In this problem 3.5^2 is performed first, then multiplied by 2, and finally 1 is added. $3.5^2 \times 2 + 1 = 25.50$

Answer is (B)

PROBLEM 3-20

```
100 R=4
110 LET S=2
120 FOR I=3 TO 5 STEP S
130 W=I+S+R
140 IF W>=10 THEN 170
150 NEXT I
160 GOTO 180
170 W=W∧2
180 PRINT W
190 END
```

In this short BASIC program the value of the output is closest to

(A)   9
(B)   11
(C)   64
(D)   81
(E)   121

Solution

The FOR−NEXT loop creates values of W of 9 and 11. When $W = 9$ control does not branch to line 170. At $W = 11$ it does. Thus the printed value of W is $11^2 = 121$.

Answer is (E)

PROBLEM 3-21

For a 60° angle (A = 60.), which FORTRAN statement correctly computes the cosine of 60°?

(A) `COSINE=COS(A)`
(B) `COSINE=COS A`
(C) `COSINE=COS(A/360.)`
(D) `COSINE=COS(A*3.14159/180.)`
(E) `COSINE=COS(A*3.14159/360.)`

Solution

The argument for the $COS(x)$ must be in radians. Since there are 2π radians in 360°, for 60° the correct argument is `(A*3.14159/180.)`

Answer is (D)

PROBLEM 3-22

Which one of the following is not an executable FORTRAN statement?

(A) REAL
(B) CALL
(C) DO
(D) STOP
(E) END

Solution

REAL is a nonexecutable declaration that indicates to the compiler the type of data to be stored in the variable that follows this word. It is not translated into machine language.

Answer is (A)

PROBLEM 3-23

Consider the following FORTRAN program.

```
PROGRAM CALC
NUM=64.
ROOT=NUM**(1/2)
END
```

The value of ROOT is closest to

(A)   0
(B)   1
(C)   8
(D)   64
(E) 4096

Solution

Arithmetic operations between two integers yield an integer. The $(1/2)$ is integer division which truncates to zero. Thus $ROOT = 64^0 = 1$.

Answer is (B)

PROBLEM 3-24

For the following FORTRAN arithmetic statement, answer the three questions.

$$T=(3*4)**2+16./4-3$$

Question 1. The first operation performed is

(A) Addition
(B) Subtraction
(C) Multiplication
(D) Division
(E) Exponentiation

Solution

In FORTRAN the order of execution of operations is

1. Functions.
2. Arithmetic operations in this order: exponentiation, multiplication and division, followed by addition and subtraction. Multiplication and division are at the same level. This is also true for addition and subtraction. Operations at the same level are performed in left-to-right order except multiple exponentiation (for example, $X**Y**Z$) is performed right to left.
3. Relational operations (.EQ., .NE., .LT., .LE., .GT., and .GE.).

4. Logical operations in this order: .NOT., .AND., .OR., .EQV. (equivalent), and .NEQV. (not equivalent).

Operations within parentheses are performed first, in the usual order of execution. In this situation the multiplication within parentheses is performed first.

Answer is (C)

Question 2. The last operation performed is

(A) Addition
(B) Subtraction
(C) Multiplication
(D) Division
(E) Exponentiation

Solution

The last two arithmetic operations are addition and subtraction. The left to right rule makes the subtraction last.

Answer is (B)

Question 3. The computed value of T is closest to

(A) 49
(B) 64
(C) 139
(D) 145
(E) 160

Solution

$T = 12^2 + (16/4) - 3 = 145$

Answer is (D)

PROBLEM 3-25

Which *one* of the following statements is correctly written in FORTRAN?

(A) A=32,700.
(B) READ*, (I(J), I=1,10)
(C) 2X=X+X
(D) SQRT(X)=X**0.5
(E) DO 50 TIME=2,20,0.5

Solution

In (A) a comma must not be used within a number. (B) is an implied DO loop to read ten values into the I array; the index in the implied DO should be J = 1,10 and not I = 1,10. In (C) the variable name 2X must begin with an alphabetic character; X2 would be acceptable, but not 2X. In (D) SQRT(X) is an intrinsic (library) function; it may only appear on the *right side* of an assignment statement.

In a DO loop the index may be either real or integer; thus (E) is correctly written.

Answer is (E)

PROBLEM 3-26

Given X = 5, Y = 10, Z = 15, which one of the following FORTRAN logical expressions is FALSE?

(A) X*Y.EQ.Z*3.+5.0
(B) .NOT.(Y.GE.10.).AND.Z.LT.20.
(C) .NOT.(Y.EQ.5.)
(D) Z.LE.15.0.AND.Y.EQ.2.*X
(E) .NOT.(X.GT.5.0).OR.X.NE.5.0

Solution

The relational operators are

.EQ.	Equal to
.NE.	Not equal to
.LT.	Less than
.GT.	Greater than
.GE.	Greater than or equal to

The result of a logical expression may be either true or false.

The logical operators are

.NOT. *expression*	True if *expression* is false
expression(1) .AND. *expression*(2)	True if both *expression*(1) and *expression*(2) are true
expression(1) .OR. *expression*(2)	True if either *expression*(1) or expression(2) is true

The relational operators are performed prior to the logical operators. In this problem all the logical expressions are true except (B). Y.GE.10. is true, hence .NOT.(Y.GE.10.) is false. The last portion of the expression Z.LT.20. is true.

Answer is (B)

PROBLEM 3-27

Consider the following FORTRAN program

```
AM=56.0
PM=45.0
CALL GRADE (AM,PM,X)
PRINT*, X
END
SUBROUTINE GRADE (X,Y,Z)
RAW=X+2.0*Y
Z=0.50*RAW
RETURN
END
```

The printed value of X is closest to

(A) 45.0
(B) 56.0
(C) 73.0
(D) 79.0
(E) 101.0

Solution

The argument list in the main program subroutine call and the argument list in the subroutine must agree in number, order, and type. The argument list names do not have to match. Hence

Main program	Subroutine
AM = 56.0	X = 56.0
PM = 45.0	Y = 45.0
X	Z

The value of Z that is computed in the subroutine is returned to the main program as X and printed. $X = Z = 0.50*(56.0 + 2.0*45.0) = 73.0$.

Answer is (C)

PROBLEM 3-28

Given the FORTRAN statements

```
READ 2, I, X
2 FORMAT(1X,I4,F5.2)
```

and the following data line:

0254000004000

The value of X read is closest to

(A) 0.004
(B) 0.04
(C) 0.40
(D) 4.00
(E) 40.000

Solution

The value of X is read from columns 6 through 10 on the data line, or 00004. Thus the value is 0.04.

Answer is (B)

PROBLEM 3-29

Given the equation

$$X = \frac{-B + B^2 - 4AC}{2A}$$

Which one of the following FORTRAN statements best represents the equation?

(A) X=-B+(B**2.0-4.0*A*C)/2.0*A
(B) X=-(B)+B*B-4.0*A*C/2.0*A
(C) X=(-B+(B**2.0-4.0*A*C))/(2.0*A)
(D) X=-(B+(B**2.0-4.0*A*C))/2.0A
(E) X=-B+B**2.0-4.0*A*C/2.0A

Solution

To divide the entire numerator by $2A$, one must have parentheses around the entire numerator. Only answers (C) and (D) provide the required parentheses. But answer (D) results in a negative value, and its denominator is incorrect.

Answer is (C)

PROBLEM 3-30

Which one of the following separate lines of computer code is *not* correct FORTRAN?

(A) GOSUB 40
(B) IF (A.LT.B) GO TO 12
(C) REAL I
(D) READ(10,*) BB
(E) DO 15 I=5,15,5

Solution

BASIC has GOSUB, while FORTRAN has CALL.

Answer is (A)

PROBLEM 3-31

Consider the following FORTRAN program.

```
A=2.7
B=4.6
J=10
IF(B.LE.A)THEN
J=B/A
ELSEIF(.NOT.(B.EQ.A))THEN
J=A/B
ELSE
J=J/A
ENDIF
PRINT*, J
END
```

The value of J printed by the program is closest to

(A) 0
(B) 1
(C) 2
(D) 3
(E) 4

Solution

Since .NOT.(B.EQ.A) means .NOT.(false)=true, J=A/B= 2.7/4.6 = 0.

Answer is (A)

PROBLEM 3-32

The area of a circle is

$$A = \frac{1}{4} \pi D^2$$

Which FORTRAN statement best represents this situation?

(A) A=(1/4)πD**2
(B) A=(1/4)*π*D**2
(C) A=(1/4)*3.14159*D**2
(D) A=(1/4)*3.14159*D*2
(E) A=3.14159/4*D**2

Solution

In (A) and (B), π is not a FORTRAN character. In the first four answers (1/4) represents integer division which results in the truncation of 1/4 to zero. Answer (E) has mixed mode division, but the result in FORTRAN 77 is a real number. Only (E) correctly computes the area of a circle.

Answer is (E)

PROBLEM 3-33

Consider the FORTRAN statement

WRITE (8,*) A, B

Which of the following best describes the statement?

(A) The values of A and B are written to the output buffer. They may then be output with a PRINT statement.
(B) The values of A and B are sent to the standard output device.
(C) The asterisk (*) must have previously been defined in a DATA statement.
(D) The WRITE statement writes A and B into data file 8.
(E) The WRITE statement writes A and B according to the FORMAT statement labelled 8.

Solution

The form of the given output statement writes information into a data file. The unit number (8 in this case) must previously have been assigned to the proper data file by an OPEN statement.

Answer is (D)

4

STATICS

Statics is the subdivision of mechanics that considers the equilibrium of stationary or uniformly translating particles or rigid bodies. By this definition a unifying characteristic of *all* statics problems is the absence of any acceleration.

A variety of forces that may act on a body are considered when statics is studied. The forces may either act at specific points on the body or be distributed over a region in space; the weight of a body or a distributed pressure are examples of the latter. In connection with distributed forces it is common to discuss the determination of centroids, centers of gravity, and moments of inertia. This is done here. We defer the subject of shear and bending moment diagrams, which are related to the internal static equilibrium of a body, to Chapter 6. Fluid statics problems are covered in Chapter 7.

EQUILIBRIUM

A body is in a state of static equilibrium when no net or unbalanced resultant force $\mathbf{R}$ acts on the body and, in addition, the forces on the body create no net tendency toward rotation or couple $\mathbf{M}$. These requirements for equilibrium, restated, are

$$\mathbf{R} = 0 \qquad \mathbf{M} = 0 \qquad (4\text{-}1)$$

The sense of the vector that represents a moment is determined by use of the right-hand rule.

For the Cartesian (x, y, z) coordinate system shown in Fig. 4-1, Eqs. (4-1) become

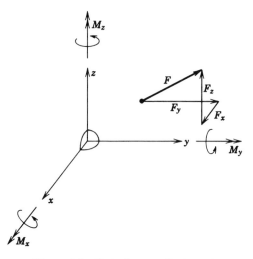

Figure 4-1 Cartesian coordinate system

$$|\mathbf{R}| = \left[\left(\sum F_x\right)^2 + \left(\sum F_y\right)^2 + \left(\sum F_z\right)^2\right]^{1/2} = 0$$

$$|\mathbf{M}| = \left[\left(\sum M_x\right)^2 + \left(\sum M_y\right)^2 + \left(\sum M_z\right)^2\right]^{1/2} = 0$$

(4-2)

which is equivalent to the requirements

$$\sum F_x = 0 \qquad \sum M_x = 0$$
$$\sum F_y = 0 \qquad \sum M_y = 0 \qquad (4\text{-}3)$$
$$\sum F_z = 0 \qquad \sum M_z = 0$$

For a three-dimensional statically determinate problem we thus have six independent equations for equilibrium. For the corresponding two-dimensional situation, the three equations are

$$\sum F_x = 0 \qquad \sum F_y = 0 \qquad \sum M_z = 0 \qquad (4\text{-}4)$$

Commonly encountered examples of bodies in static equilibrium are the two- and three-force bodies.

The two-force body is one that is acted on by concentrated forces that are applied at only two points on the body. Equations (4-4) can be used to show that these two forces have the same line of action, the same magnitude, and act in opposite directions. This information is particularly useful in the solution of statically determinate truss problems.

Concentrated forces acting at three points on a body cause the body to be called a three-force body. For equilibrium these forces must either (*a*) have lines of action that all pass through the same point so that no couple acts on the body or (*b*) act along parallel lines of action.

Example 1

A 4-ft × 10-ft block that weighs 1600 lb is held in a horizontal position by a force *P* and hinge *A*, as shown in Fig. 4-2*a*. Determine the force *P* and the resultant hinge reaction R_A.

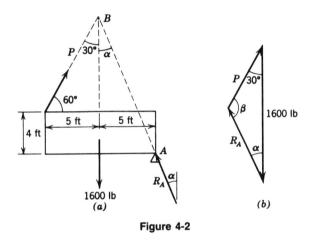

Figure 4-2

Solution

The block, which weighs 1600 lb, is held in equilibrium by forces *P* and R_A, thus forming a three-force body. The point common to all three lines of action is point *B*, which is directly above the location of the equivalent concentrated 1600-lb weight. A force triangle, shown in Fig. 4-2*b*, can easily be drawn.

From Fig. 4-2*a* we have

$$\tan \alpha = \frac{5}{4 + 5 \tan 60°} = \frac{5}{4 + 5\sqrt{3}} = 0.395$$

$$\alpha = 21°33'$$

Hence

$$\beta = 180° - 30° - \alpha = 128°27'$$

Using the law of sines,

$$\frac{P}{\sin \alpha} = \frac{R_A}{\sin 30°} = \frac{1600}{\sin \beta} = 2046$$

$$P = 2046 \sin 21°33' = 752 \text{ lb}$$

$$R_A = 2046 \sin 30° = 1023 \text{ lb} \quad \text{acting at an angle } \alpha$$

FREE-BODY DIAGRAM

In Fig. 4-2a we have drawn a free-body diagram of the block. Relevant dimensions and angles are also shown. It is helpful to prepare a free-body diagram during the course of solving most problems in mechanics. The diagram should clearly show the essential elements of the problem and no other information. Sometimes, in the interest of clarity, dimensions are also excluded from the free-body diagram when they would tend to clutter the drawing. The use of a free-body diagram in finding the reactions for simple bodies is further shown in the solved problems.

FRAMES AND TRUSSES

Certain useful procedures have evolved from the analysis of the forces in frames and trusses. A truss is an assemblage of pieces that may be accurately represented as two-force members; a frame is composed of members that each are usually acted on by applied loads at more than two points. The analysis of a statically determinate frame is usually accomplished by drawing a free-body diagram of each member and then writing the appropriate equilibrium equations for each diagram. These equations are then solved for the unknown forces and reactions. The analysis of trusses, however, usually proceeds by some combination of the methods known as the method of joints or the method of sections.

Example 2

Determine the force in members *BH*, *BC*, and *GD* of the truss shown in Fig. 4-3. Note that the truss is composed of triangles 7.5 ft : 10.0 ft : 12.5 ft, so that they are 3:4:5 right triangles.

Solution

First we solve for the reactions R_L and R_R:

$$\sum M_E = 0 \qquad 40R_L = 30(300) + 20(400)$$

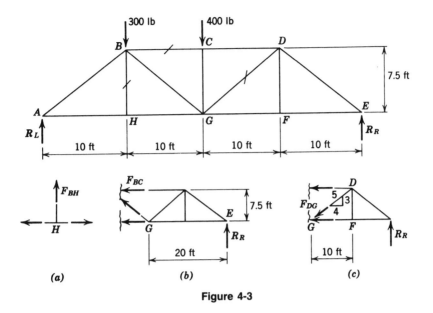

Figure 4-3

$$R_L = 425 \text{ lb}$$

$$\sum F_v = 0 \qquad R_R = 300 + 400 - 425 = 275 \text{ lb}$$

The method of joints considers the equilibrium of each pinned connection between members. Because of the pin, no moment can be transmitted through the joint. For a two-dimensional truss, joint equilibrium requires that $\sum F_h = 0$ and $\sum F_v = 0$ be satisfied for the forces acting at each joint. We have an especially simple case in this problem, as shown in Fig. 4-3a. All bar forces are shown to be in tension. Summing forces vertically,

$$\sum F_v = 0 = F_{BH}$$

The method of joints is most efficiently used (a) when the forces in all members of a truss are desired or (b) when special cases such as the example arise.

The method of sections is normally more efficient than the method of joints when only a few selected bar forces must be found. If the truss as a whole is in equilibrium, then any segment or section of the structure must also be in equilibrium. This principle is put to use by selecting an "appropriate" section of a structure and applying the principles of statics. In most cases the appropriate section is one that (a) severs the member of interest and (b) results in a free body acted on by only three unknown forces. Numerous exceptions to the second requirement exist, however.

Selecting a section as shown in Fig. 4-3b, we write

$$\sum M_G = 0 \qquad 7.5F_{BC} = -20R_R$$

$$F_{BC} = \frac{-20(275)}{7.5} = 733 \text{ lb compression}$$

Referring to Fig. 4-3c,

$$\sum F_v = 0 \qquad \tfrac{3}{5}F_{DG} = R_R$$

$$F_{DG} = \tfrac{5}{3}(275) = 458 \text{ lb tension}$$

FRICTION

In many situations forces due to friction cause a body to remain in static equilibrium. In these problems it is important to recall that frictional forces always act to oppose any actual or impending motion. For most cases of dry friction the frictional force F is simply proportional to the normal force N, or $F = \mu N$ where μ is the coefficient of either static or kinetic friction. (A few more specialized friction problems, involving rolling friction or belt friction, are presented in the statics problem section.)

Example 3

A 500-lb block rests on a 30° plane (Fig. 4-4). If the coefficient of static friction is 0.30 and the coefficient of kinetic friction is 0.20, what is the value of P needed to

(a) prevent the block from sliding down the plane?
(b) start the block moving up the plane?
(c) keep the block moving up the plane?

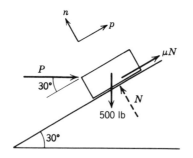

Figure 4-4

Solution

Normal to the plane, $\Sigma F_n = 0$

$$N - P \sin 30° - 500 \cos 30° = 0$$

$$N = \frac{P}{2} + 433$$

(a) Parallel to the plane, $\Sigma F_p = 0$

$$\mu N + P \cos 30° - 500 \sin 30° = 0$$

$$0.3\left(\frac{P}{2} + 433\right) + 0.866P - 250 = 0$$

$$P = 118.2 \text{ lb}$$

(b) Here we have impending motion *up* the plane so the direction of μN in Fig. 4-4 must be directed down the plane to oppose the motion. This can be accomplished mathematically by replacing μN by $-\mu N$ in the earlier equation. Thus

$$-0.3\left(\frac{P}{2} + 433\right) + 0.866P - 250 = 0$$

$$P = 531 \text{ lb}$$

(c) The problem is unchanged from (b) except that we now consider the kinetic ($\mu = 0.2$) rather than the stationary ($\mu = 0.3$) case.

$$-0.2\left(\frac{P}{2} + 433\right) + 0.866P - 250 = 0$$

$$P = 439 \text{ lb}$$

CENTROID, CENTER OF GRAVITY, MOMENT OF INERTIA

The center of gravity (also called center of mass) and mass moment of inertia are physical properties of a body. If the density is uniform throughout the body, these properties exactly coincide with the associated, but purely geometric, properties of a body which are called the centroid and moment of inertia. The equations for the center of gravity of a body are

$$\bar{X} = \frac{1}{M} \int_{V} \rho x \, dV \qquad \bar{Y} = \frac{1}{M} \int_{V} \rho y \, dV \qquad \bar{Z} = \frac{1}{M} \int_{V} \rho z \, dV. \qquad (4\text{-}5)$$

where $M = \int_{V} \rho \, dV$ is the mass of the body composed of volume elements

dV that have density ρ. These equations apply equally well for rods, plane areas, and volumes of any shape; one need only take care to select an appropriate volume element. The integrals may be replaced by finite sums for single or composite bodies whenever the component volume and the centroid or center of gravity of each individual element are already known. Note that the density ρ will cancel in Eqs. (4-5) when it is constant throughout a body, and we then have the equations for the centroid of the body. These principles are illustrated in Example 4. Appendix B gives the centroidal coordinates for some common geometric shapes.

Example 4

Locate the X, Y, Z coordinates of the centroid for the slender rod $ABCD$ of constant density shown in Fig. 4-5. The semicircular portion ABC has a radius of π m and lies in the Y–Z plane. The straight portion is 10 m long and lies in the X–Y plane. The point D is located at $X = 6$, $Y = 8$.

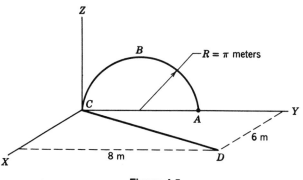

Figure 4-5

Solution

Length $L = \pi R + \sqrt{6^2 + 8^2} = 9.87 + 10 = 19.87$ m

$$\bar{X} = \frac{1}{L} \int x \, dL = \frac{1}{L} \sum_{i=1}^{n} x_i L_i = \frac{0 \times 9.87 + 3 \times 10}{19.87} = 1.51 \text{ m}$$

$$\bar{Y} = \frac{1}{L} \sum_{i=1}^{n} y_i L_i = \frac{\pi \times 9.87 + 4 \times 10}{19.87} = 3.57 \text{ m}$$

The center of gravity of a semicircular rod, measured from the Y-axis, is

$$\frac{2R}{\pi} = \frac{2\pi}{\pi} = 2.00 \text{ m}$$

$$\bar{Z} = \frac{1}{L} \sum_{i=1}^{n} z_i L_i = \frac{2.00 \times 9.87 + 0 \times 10}{19.87} = 0.99 \text{ m}$$

The moment of inertia (or second moment) of an area, with respect to some axis $r = 0$, is

$$I = \int_A r^2 \, dA \tag{4-6}$$

In this expression r is the distance from the axis $r = 0$ to the centroid of the area element dA. Usually r is replaced by either x or y so that the moment of inertia is found with respect to the x or y coordinate axis. The mass moment of inertia is similarly defined as

$$I_m = \int_{\Psi} r^2 \, dm = \int_{\Psi} r^2 \rho \, d\Psi \tag{4-7}$$

If the moment of inertia of a body around its centroidal axis I_0 is known, the moment of inertia around any axis parallel to this centroidal axis may be found from the parallel-axis theorem

$$I = I_0 + Ad^2 \tag{4-8}$$

where A is the area and d is the distance between the two parallel axes. For mass moments of inertia, A is replaced by the mass M in Eq. (4-8). Appendix B gives the moment of inertia I_0 for some common areas.

Example 5

Find the moment of inertia of a rectangle of base b and height h (Fig. 4-6)

    (a) about the centroidal axis,
    (b) about the base of the rectangle.

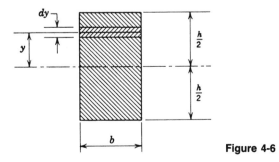

Figure 4-6

Solution

(a) We select a differential area of width b and height dy that has a distance y to its centroid, or $dA = b\,dy$. Applying Eq. (4-6) with $y = 0$ at the centroidal axis,

$$I_0 = \int_{-h/2}^{h/2} y^2 (b\,dy)$$

Since b is constant,

$$I_0 = b \int_{-h/2}^{h/2} y^2\,dy = b[\tfrac{1}{3} y^3]_{-h/2}^{h/2}$$

$$= \frac{b}{3}\left(\frac{h^3}{8} + \frac{h^3}{8}\right) = \frac{b}{3}\left(\frac{h^3}{4}\right) = \frac{bh^3}{12}$$

(b) Applying Eq. (4-8) gives

$$I = I_0 + Ad^2 = \frac{bh^3}{12} + (bh)\left(\frac{h}{2}\right)^2$$

$$= \frac{bh^3}{3}$$

PROBLEM 4-1

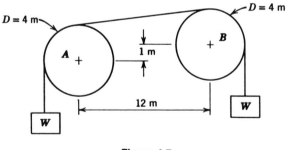

Figure 4-7

For the system shown (Fig. 4-7), choose the one true statement concerning the bearing reactions at A and B if the system is in equilibrium.

(A) Both reactions are vertical.
(B) Neither reaction is vertical.
(C) The reaction at A is vertical.
(D) The reaction at B is vertical.

(E) Since there are two unknown components at A and B, respectively, the system is statically indeterminate and the reactions cannot be described using methods of statics.

Solution

Draw the free-body diagram:

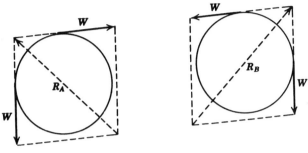

Figure 4-8

Hence neither reaction A nor B is vertical.

Answer is (B)

PROBLEM 4-2

The value of reaction R (Fig. 4-9) in newtons is

(A) 9.6 N
(B) 20.0 N
(C) 26.4 N
(D) 36.0 N
(E) 44.0 N

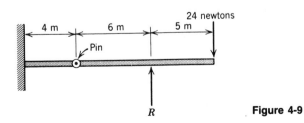

Figure 4-9

Solution

$\Sigma M_{pin} = 0$

$$-24N \times 11 + 6R = 0 \qquad R = \frac{24 \times 11}{6} = 44\,N$$

Answer is (E)

PROBLEM 4-3

The moment at reaction R (Fig. 4-10) is

   (A) Unknown; the structure is statically indeterminate
   (B) 0
   (C) Pa

   (D) $Pa\left(\dfrac{b+c}{a+b+c}\right)$

   (E) $P\left(\dfrac{bc}{a+b}\right)$

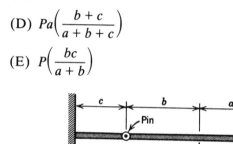

Figure 4-10

Solution

The moment at any point on a beam may be determined by calculating the moments on either side of the point—hence $M_R = Pa$.

Answer is (C)

PROBLEM 4-4

For the beam loaded as shown in Fig. 4-11, reaction R is

   (A) $\dfrac{Pa}{L}$

   (B) $\dfrac{Pb}{L}$

(C) $\dfrac{PL}{a}$

(D) $\dfrac{PL}{b}$

(E) $\dfrac{PL}{ab}$

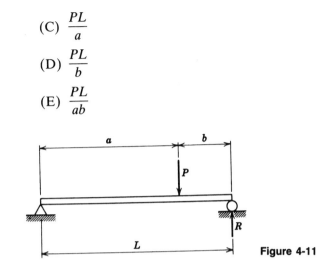

Figure 4-11

Solution

$\Sigma M_{\text{left reaction}} = 0$

$$+RL - Pa = 0 \qquad R = \dfrac{Pa}{L}$$

Answer is (A)

PROBLEM 4-5

Determine the magnitude of the beam reaction marked R in Fig. 4-12. The value is nearest to

(A) 0
(B) $0.5P$
(C) $1.0P$
(D) $1.5P$
(E) $2.0P$

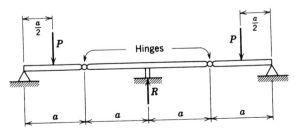

Figure 4-12

Solution

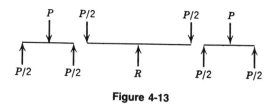

Figure 4-13

We can see that $R = 2(P/2) = P$.

Answer is (C)

PROBLEM 4-6

Determine the magnitude of the reaction marked R in Fig. 4-14. The value is nearest to

    (A) 0
    (B) P
    (C) $2P$
    (D) $3P$
    (E) $4P$

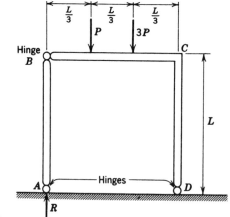

Figure 4-14

Solution

Since joint C in Fig. 4-14 is rigid, we have a stable rigid frame. $\Sigma M_D = 0$.

$$+3P\left(\frac{L}{3}\right) + P\left(\frac{2L}{3}\right) - RL = 0 \qquad PL + \tfrac{2}{3}PL - RL = 0$$

$$R = 1\tfrac{2}{3}P$$

Answer is (C)

PROBLEM 4-7

For the structure loaded as shown in Fig. 4-15, reaction R is

(A) $\dfrac{PL}{H}$

(B) $\dfrac{2PL}{3H}$

(C) $\dfrac{PL}{3H}$

(D) $\dfrac{3PH}{L}$

(E) $\dfrac{3PH}{2L}$

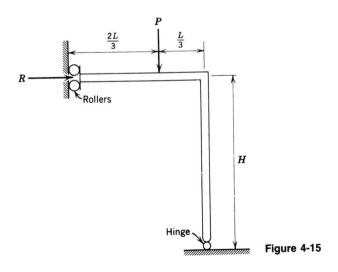

Figure 4-15

Solution

$$\Sigma M_{\text{base hinge}} = 0$$

$$-RH + P\frac{L}{3} = 0 \qquad R = \frac{PL}{3H}$$

Answer is (C)

PROBLEM 4-8

In the vector diagram (Fig. 4-16) $\bar{R}$ represents

    (A) $\bar{A} + \bar{B}$
    (B) $\bar{A} - \bar{B}$
    (C) $\bar{B} - \bar{A}$
    (D) $\bar{A} \cdot \bar{B}$
    (E) $\sqrt{\bar{A}^2 + \bar{B}^2}$

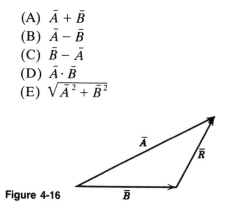

Figure 4-16

Solution

$$\bar{B} + \bar{R} = \bar{A} \qquad \bar{R} = \bar{A} - \bar{B}$$

Answer is (B)

PROBLEM 4-9

Given a 50-N pulley, supported as shown (Fig. 4-17) and carrying a cable supporting an additional 50-N load. The force the beam exerts on the pulley is

    (A)   50 N up
    (B) 100 N up
    (C) 100 N down
    (D) 150 N up
    (E) 150 N down

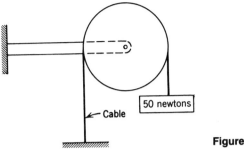

Figure 4-17

Solution

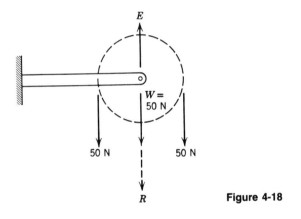

Figure 4-18

The resultant of three forces $= 50 + 50 + 50 = 150$ N down. $\Sigma F_y = 0$.

$$+E - 50 - 50 - 50 = 0$$
$$E = +150 \text{ N equilibrant}$$

Therefore the force the beam exerts on the pulley is 150 N up.

Answer is (D)

PROBLEM 4-10

A 5000-N sphere rests against a smooth plane inclined at 45° to the horizontal and against a smooth wall as shown in Fig. 4-19. The magnitude of the reaction R_B is nearest to

(A) 0 N
(B) 2500 N
(C) 3500 N
(D) 5000 N
(E) 7000 N

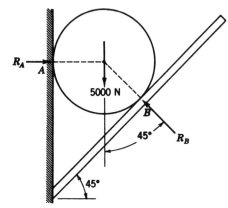

Figure 4-19

Solution

$\Sigma F_V = 0$

$$-5000 + R_B \cos 45° = 0 \qquad R_B = \frac{5000}{0.707} = 7070 \text{ N}$$

Answer is (E)

PROBLEM 4-11

In Fig. 4-19 a 5000-N sphere rests against a smooth plane inclined at 45° to the horizontal and against a smooth vertical wall. What is the magnitude of the reaction R_A?

(A) 0 N
(B) 2500 N
(C) 3535 N
(D) 5000 N
(E) 7070 N

Solution

In Problem 4-10 we found that $R_B = 7070$ N. $\Sigma F_H = 0$.

$$R_A - R_B \sin 45° = 0 \qquad R_A = 7070(0.707) = 5000 \text{ N}$$

Answer is (D)

PROBLEM 4-12

A 1200-lb solid steel triangular prism is supported by three vertical cables as shown in Fig. 4-20. One cable is at the middle of one edge, with the other two at the corners of the opposite edge. All three cables are the same length and are of the same size. Which one of the following statements is correct concerning the tensile forces in the cables?

(A) The tensile force in cable C_1 is twice that in cable C_2.
(B) The tensile force in cable C_1 is half that in cable C_2.
(C) The tensile force in cable C_3 is equal to the sum of the tensile forces in cables C_1 and C_2.
(D) All three cables have the same tensile force.
(E) None of the statements (A) through (D) is correct.

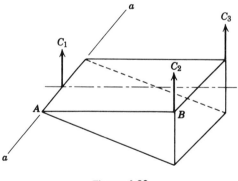

Figure 4-20

Solution

Owing to symmetry, $C_2 = C_3$. $\Sigma M_a = 0$.

$$(C_2 + C_3)L - \tfrac{2}{3}L(1200) = 0 \qquad C_2 + C_3 = 800$$

Therefore

$$C_2 = C_3 = 400 \text{ lb}$$

$$\Sigma F_V = 0$$

$$400 + 400 + C_1 - 1200 = 0$$

Thus

$$C_1 = 400 \text{ lb}$$

Therefore all cable tensions are equal and have a value of 400 lb.

Answer is (D)

PROBLEM 4-13

The tension in the cable supporting the beam loaded as shown in Fig. 4-21 is approximately

(A) 5 kips
(B) 10 kips
(C) 15 kips
(D) 20 kips
(E) 25 kips

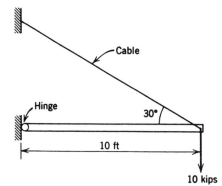

Figure 4-21

Solution

From the free-body diagram, Fig. 4-22, we can see for $\Sigma F_y = 0$ that

$$C_v - 10 \text{ kips} = 0 \qquad C_v = 10 \text{ kips}$$

$$\sin 30° = \frac{C_v}{C} = \frac{1}{2} \qquad C = 2C_v = 2 \times 10 \text{ kips} = 20 \text{ kips}$$

Answer is (D)

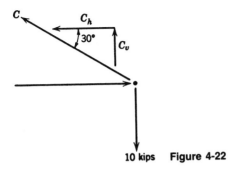

10 kips   **Figure 4-22**

PROBLEM 4-14

Two weights are suspended on a cord as shown in Fig. 4-23. The angle θ at equilibrium is nearest to

(A) 35°
(B) 45°
(C) 55°
(D) 65°
(E) 75°

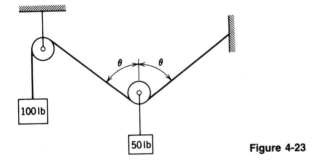

Figure 4-23

Solution

From the diagram we can see that the tension in the cord is 100 lb. We can then draw the free-body diagram and the force triangle.

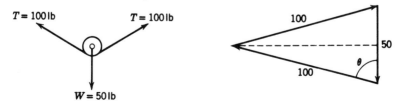

Figure 4-24

$$\theta = \cos^{-1} \tfrac{25}{100} = \cos^{-1} 0.250 = 75.5°$$

Answer is (E)

PROBLEM 4-15

The center of gravity of a log, 10 ft long and weighing 100 lb, is 4 ft from one end of the log. It is to be carried by two men. If one is at the heavy end, how far from the other end does the second man have to hold the log if each is to carry 50 lb?

(A) at the end
(B) 2 ft
(C) 4 ft
(D) 5 ft
(E) 6 ft

Solution

This problem can be treated as a beam with a concentrated load of 100 lb 4 ft from one end. For both men to carry an equal weight of 50 lb, they must be the same distance from the concentrated load.

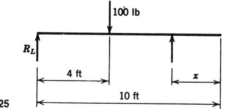

Figure 4-25

The second man is $10 - 4 - 4 = 2$ ft from the other end.

Alternative solution

$$\Sigma M_{R_L} = 0$$

$$+4 \times 100 - 50(10 - x) = 0$$
$$400 - 500 + 50x = 0$$
$$50x = 100$$
$$x = 2 \text{ ft}$$

Answer is (B)

PROBLEM 4-16

The theoretical mechanical advantage of the system shown in the figure is nearest to

(A) 5
(B) 7
(C) 12
(D) 15
(E) 20

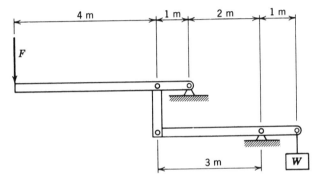

Figure 4-26

Solution

The problem can be solved by considering it as two separate problems:

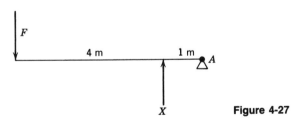

Figure 4-27

$$\Sigma M_A = 0$$

$$X = \frac{5F}{1} = 5F$$

Thus the mechanical advantage for this section is 5.

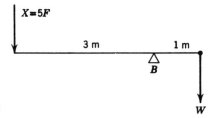

X=5F

3 m   1 m

B

Figure 4-28   W

$$\Sigma\, M_B = 0$$

$$(5F)(3) = W(1)$$

$$\frac{W}{F} = 15 = \text{mechanical advantage}$$

Answer is (D)

PROBLEM 4-17

The beam ABC is loaded as shown. The equilibrium is maintained by a 4000-lb weight suspended from bar DE. Neglect the weight of the members. The required length L of bar DE is nearest to

(A)   2 ft
(B)   4 ft
(C)   6 ft
(D)   8 ft
(E)   10 ft

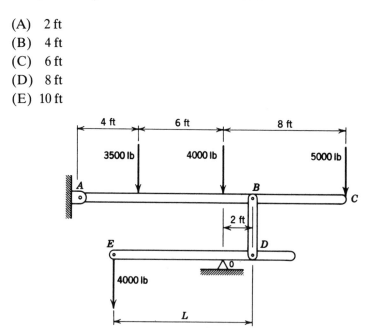

Figure 4-29

Solution

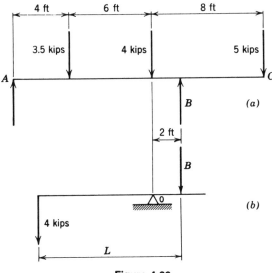

Figure 4-30

Applying $\Sigma M_A = 0$ to Fig. 4-30a,

$$\frac{(4 \times 3500) + (10 \times 4000) + (18 \times 5000)}{12} = \text{force at } B = 12,000 \text{ lb}$$

From Fig. 4-30b, $\Sigma M_0 = 0$.

$$12,000 \times 2 = 4000(L - 2) \qquad L = 8 \text{ ft}$$

Answer is (D)

PROBLEM 4-18

A homogeneous body is composed of a semicylinder and a rectangular parallelepiped as shown in Fig. 4-31. Find the maximum value of h such that the system will be in a *stable* equilibrium on a horizontal plane. Assume no sliding between plane and cylinder. The value of h is nearest to

(A) $0.50R$
(B) $0.70R$
(C) $0.82R$
(D) $1.00R$
(E) $1.25R$

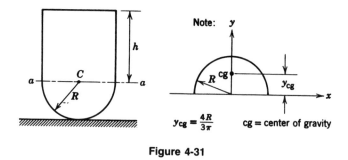

Figure 4-31

Solution

To have stable equilibrium, the composite centroid must be in the semicylindrical portion; thus the maximum value for h occurs when the centroid is on line $a-a$. One of the properties of the centroid is that the first moment of area about it is equal to zero. Hence

$$0 = \text{moment of semicylinder} - \text{moment of rectangle}$$

Having the coordinates of the centroid of a semicircle given, this equation becomes

$$\overset{\text{area} \cdot \text{arm}}{\frac{\pi R^2}{2}\left(\frac{4R}{3\pi}\right)} - \overset{\text{area} \cdot \text{arm}}{2Rh\left(\frac{h}{2}\right)} = 0 \qquad \frac{2R^3}{3} = Rh^2$$

$$h^2 = \tfrac{2}{3}R^2 \qquad h = R\sqrt{\tfrac{2}{3}} = 0.82R$$

Answer is (C)

PROBLEM 4-19

The three-hinged arch ABC is loaded as shown. Neglect the weight of the arch. The hinge force at B is nearest to

(A) 1.0 kips
(B) 1.2 kips
(C) 1.4 kips
(D) 1.6 kips
(E) 1.8 kips

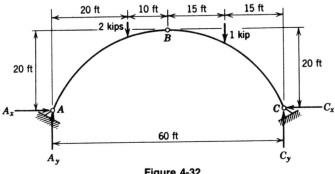

Figure 4-32

Solution

$\Sigma M_A = 0$

$$-2 \times 20 - 1 \times 45 + 60C_y = 0$$

$$C_y = \frac{85}{60} = 1.42 \text{ kips}$$

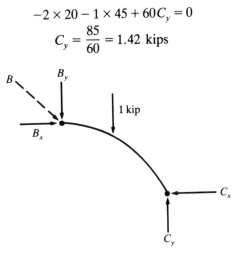

Figure 4-33

Now use Fig. 4-33. $\Sigma M_B = 0$.

$$-1 \times 15 + 30C_y - 20C_x = 0$$

$$-15 + 30(1.42) - 20C_x = 0$$

$$C_x = 1.38 \text{ kips}$$

$\Sigma F_y = 0$

$$C_y - B_y - 1 = 0$$

$$B_y = 0.42 \text{ kip}$$

$$\Sigma F_x = 0$$

$$B_x - C_x = 0$$
$$B_x = 1.38 \text{ kips}$$
$$B = (1.38^2 + 0.42^2)^{1/2} = 1.44 \text{ kips}$$

Answer is (C)

PROBLEM 4-20

A solid steel bar leans against a smooth vertical wall, with its lower end on a smooth, level floor. A stop on the floor prevents the bar from slipping. The bar is of uniform cross section and weighs 800 N. Assume the load acts at the center of gravity of the bar. The value of B_h is nearest to

(A)   0 N
(B) 200 N
(C) 400 N
(D) 600 N
(E) 800 N

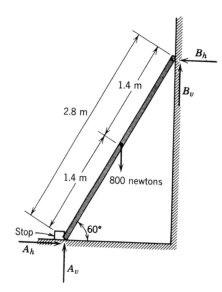

Figure 4-34

Solution

The wall is said to be smooth, hence there is no friction, so $B_v = 0$.
$\Sigma M_A = 0$

$$-800(1.4 \cos 60°) + B_h(2.8 \sin 60°) = 0$$

$$B_h = \frac{560}{2.42} = 231 \text{ N}$$

Answer is (B)

PROBLEM 4-21

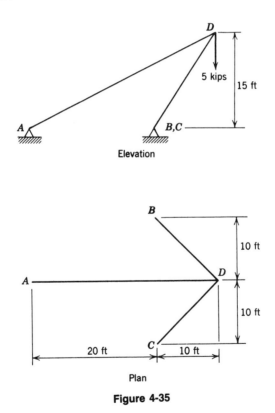

Figure 4-35

The force in member AD of the space frame shown in the plan and elevation views of Fig. 4-35 is nearest to

(A) 4.5 kips
(B) 5.0 kips
(C) 5.6 kips
(D) 6.5 kips
(E) 7.6 kips

Solution

The easiest method of solution is to determine the force in the frame assuming members BD and CD are replaced by a single member ED.

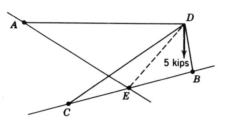

Figure 4-36

Length of members:

$$DE = (15^2 + 10^2)^{1/2} = (325)^{1/2} = 18.0 \text{ ft}$$
$$AD = (15^2 + 30^2)^{1/2} = (1125)^{1/2} = 33.5 \text{ ft}$$

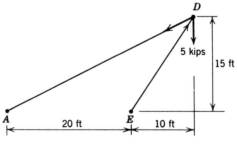

Figure 4-37

$\Sigma F_y = 0$ (assume DE in compression and AD in tension)

$$-\frac{15}{33.5} AD + \frac{15}{18} DE - 5 = 0$$

or

$$-0.537AD + DE - 6.0 = 0 \tag{1}$$

$$\Sigma F_x = 0$$

$$-\frac{30}{33.5} AD + \frac{10}{18} DE = 0$$

or

$$-1.612 AD + DE = 0 \qquad (2)$$

Subtracting (2) from (1) yields

$$1.075 AD - 6.0 = 0$$

$$AD = 5.58 \text{ kips (tension)}$$

Answer is (C)

PROBLEM 4-22

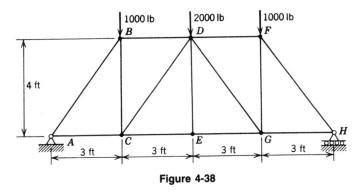

Figure 4-38

Given the pin-jointed structure shown. The force in member CE is nearest to

(A) 0 kips
(B) 0.75 kips
(C) 1.50 kips
(D) 2.25 kips
(E) 3.00 kips

Solution

The structure and the loading pattern are symmetrical.

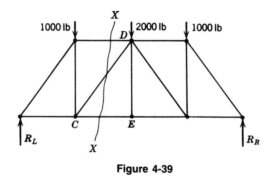

Figure 4-39

Therefore

$$R_L = R_R = \frac{4000}{2} = 2000 \text{ lb}$$

To find the force in member CE, take a section $X - X$ and analyze either side. $\Sigma M_D = 0$.

$$2 \text{ kips} \times 6 \text{ ft} - 1 \text{ kip} \times 3 \text{ ft} - F_{CE} \times 4 \text{ ft} = 0$$

$$F_{CE} = \frac{12 - 3}{4} = 2.25 \text{ kips (tension)}$$

Answer is (D)

PROBLEM 4-23

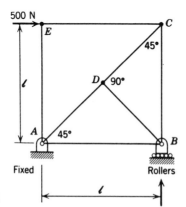

Figure 4-40

The force in member BC of the truss shown is nearest to

(A)   0 N
(B) 250 N
(C) 350 N
(D) 500 N
(E) 700 N

Solution

$$\Sigma\, M_A = 0, \qquad B_v \times \ell = 500 \times \ell, \qquad B_v = 500N\!\uparrow.$$

Now we will apply the method of joints at D and B.

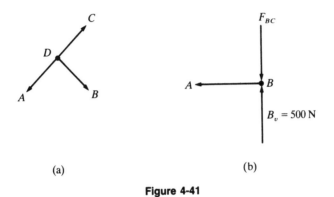

(a)   (b)

Figure 4-41

By inspection of Fig. 4-41a, $F_{DB} = 0$. In Fig. 4-41b, bar BD is not shown since it carries no load. Hence $\Sigma\, F_v = 0$ at joint B shows that $F_{BC} = 500$ N compression.

Answer is (D)

PROBLEM 4-24

A pin-connected truss has a horizontal load of 2000 lb and a vertical load of 1200 lb as shown. The total axial force in member CD is closest to

(A) 1200 lb
(B) 1950 lb
(C) 2600 lb
(D) 3350 lb
(E) 4000 lb

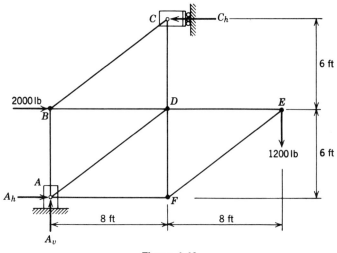

Figure 4-42

Solution

$\Sigma F_y = 0$

$$A_v - 1200 = 0 \qquad A_v = 1200 \text{ lb}$$

$\Sigma M_A = 0$

$$-1200 \times 16 + C_h \times 12 - 2000 \times 6 = 0$$

$$C_h = \frac{19,200 + 12,000}{12} = 2600 \text{ lb}$$

Note that all triangles have a $3:4:5$ relationship.

$\Sigma F_y = 0$

$$CD = \tfrac{3}{5} BC \qquad BC = \tfrac{5}{3} CD$$

$\Sigma F_x = 0$

$$\tfrac{4}{5} BC = C_h \qquad \tfrac{4}{5}(\tfrac{5}{3} CD) = 2600$$

$$CD = 1950 \text{ lb tension}$$

Answer is (B)

PROBLEM 4-25

Two blocks, A and B, are connected by a cord passing over a smooth pulley. The coefficient of friction under block A is 0.25 and that under block B is 0.40. Which one of the following statements is correct?

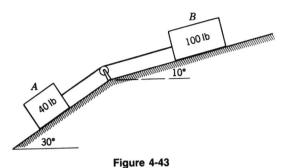

Figure 4-43

(A) Block A will move, but not block B.
(B) Block B will move, but not block A.
(C) Block A and block B will both move.
(D) Neither block A nor block B will move.
(E) None of the above statements is correct.

Solution

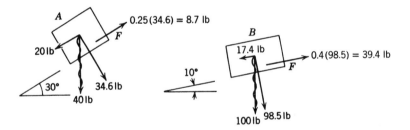

Figure 4-44

	Block A		Block B		Both blocks
Forces down planes	20.0	+	17.4	=	37.4
Friction forces	8.7	+	39.4	=	48.1

Block A will not move when the cord tension $T \geqslant 20.0 - 8.7 = 11.3$ lb. When $T = 11.3$ lb, the 39.4 lb frictional force that could be developed on Block B exceeds the net force $(17.4 + 11.3)$ down the plane; hence Block B will not move.

Answer is (D)

PROBLEM 4-26

A hangar door weighs 1200 lb and is supported on two rollers. The rollers have rusted, causing the door to slide along the track when moved. Assume the coefficient of friction is 0.40. If the door must be opened in an emergency by a materials-handling machine pushing horizontally at point P, the maximum distance d that will not cause one roller to leave the track is nearest to

(A)   0 ft
(B)   5.0 ft
(C)   7.5 ft
(D) 10.0 ft
(E) 12.5 ft

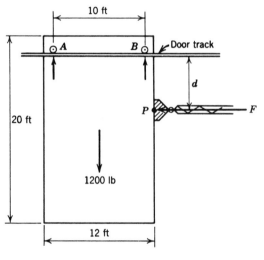

Figure 4-45

Solution

To slide the door along the track, a force $F = \mu N$ must be applied.

$$F = 0.40 \times 1200 = 480 \text{ lb}$$

Taking moments about roller B, we see that roller A will lift off the track only when the moment Fd is greater than 1200 lb × 5 ft.

$$d_{\max} = \tfrac{6000}{480} = 12.5 \text{ ft}$$

Answer is (E)

PROBLEM 4-27

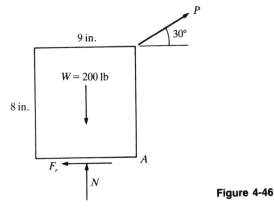

Figure 4-46

A solid block 9 in. × 9 in. × 8 in. high weighs 200 lb. If μ equals 0.25, the force P required to cause the block to slide is closest to

(A) 50 lb
(B) 60 lb
(C) 70 lb
(D) 80 lb
(E) block will overturn before it slides

Solution

$$\Sigma F_H = 0 \qquad F_r = P \cos 30° = 0.866P$$
$$\Sigma F_V = 0 \qquad N = 200 - P \sin 30° = 200 - 0.5P$$

Now use the friction relation $F_r = \mu N$:

$$0.866P = 0.25(200 - 0.5P)$$
$$= 50 - 0.125P$$
$$P = 50.5 \text{ lb}$$

Check to ensure that the block does not overturn around point A when $P = 50.5$ lb.

Overturning moment = 50.5 cos 30° × 8 in. = 350 in.-lb

Righting moment   = 200 × 4½ in. = 900 in.-lb

The block will not overturn; it will slide.

Answer is (A)

PROBLEM 4-28

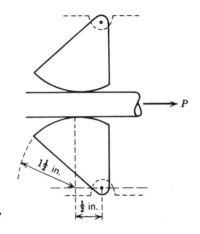

Figure 4-47

The cam arrangement shown was designed to develop large friction forces on a cable that is subjected to a tension force. The coefficient of friction is 0.3. The maximum value of P at which the cable will not slip is nearest to

(A)   0 lb
(B) 200 lb
(C) 400 lb
(D) 600 lb
(E) 800 lb

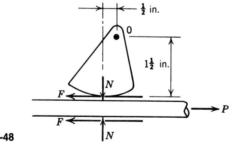

Figure 4-48

Solution

The tension in the cable equals P. Therefore F equals $0.5P$. From the cam, $\Sigma M_0 = 0$:

$$\tfrac{1}{2} \cdot N = 1\tfrac{1}{2} \cdot F$$

$$N = 3F = 3(0.5P) = 1.5P$$

The maximum friction possible $F = \mu N = 0.3(1.5P) = 0.45P$

$$2F = 0.90P$$

The tension in the cable is P; the resisting force is $0.9P$. This indicates that the cable will slip for all values of P.

Answer is (A)

PROBLEM 4-29

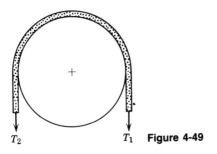

T_2 T_1 **Figure 4-49**

For a belt passing over a pulley, the ratio of the forces is given as

$$\frac{T_2}{T_1} = e^{\mu\beta}$$

where μ is the coefficient of friction and β is the angle of contact. For the pulley shown and $\mu = 0.30$, the ratio T_2/T_1 is nearest to

(A) 1.0
(B) 2.5
(C) 3.1
(D) 4.5
(E) 6.2

Solution

The angle of contact β is expressed in radians in this equation. Since there are 2π rad in 360° and the angle of contact in this case is 180°, $\beta = \pi$ rad.

$$\frac{T_2}{T_1} = e^{0.3\pi} = 2.57$$

Answer is (B)

PROBLEM 4-30

In a problem involving rolling friction, the value of it would be given as

(A) an angle ϕ
(B) μ_s, the tangent F_s/N of an angle ϕ (F_s = static friction)
(C) μ_k, the tangent F_k/N of an angle ϕ (F_k = kinetic friction)
(D) r_f, the radius of the friction circle ($r_f = r \sin \phi$)
(E) b, a deformation, as a linear dimension

Solution

In rolling friction we encounter a situation where the "ideal" situation fails to give us any appreciation of the actual situation. If we were to place a rigid wheel on a rigid smooth surface and set the wheel in motion, the forces would appear as in Fig. 4-50.

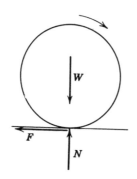

Figure 4-50

In the absence of retarding forces, the wheel would theoretically roll forever. In the actual case there is deformation of the surface and the wheel.

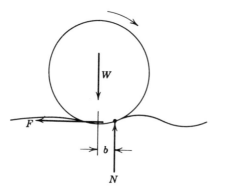

Figure 4-51

The result is that the normal force N acts ahead of the line of action of the weight W. This small distance (b) is called the coefficient of rolling friction and is a linear dimension.

Answer is (E)

PROBLEM 4-31

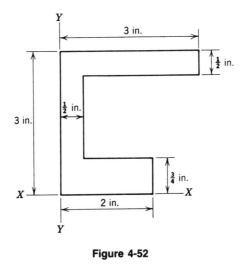

Figure 4-52

The centroid of the figure shown has a value $\bar{Y}$ nearest to

(A) 1.00 in.
(B) 1.25 in.
(C) 1.50 in.
(D) 1.75 in.
(E) 2.00 in.

Solution

Although only $\bar{Y}$ is required in the problem, both $\bar{X}$ and $\bar{Y}$ will be calculated to show the method of computation.

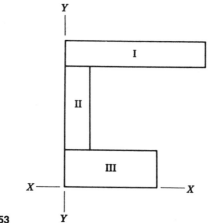

Figure 4-53

Section	b (in.)	h (in.)	A (in.²)	X (in.)	Y (in.)	AX (in.³)	AY (in.³)
I	3.0	0.5	1.5	1.5	2.75	2.25	4.13
II	0.5	1.75	0.875	0.25	1.625	0.22	1.42
III	2.0	0.75	1.5	1.0	0.375	1.5	0.56
			3.875			3.97	6.11

$$\bar{X} = \frac{\Sigma \, AX}{\Sigma \, A} = \frac{3.97}{3.875} = 1.02 \text{ in.}$$

$$\bar{Y} = \frac{\Sigma \, AY}{\Sigma \, A} = \frac{6.11}{3.875} = 1.58 \text{ in.}$$

Answer is (C)

PROBLEM 4-32

A circular disk of uniform density has a hole cut out of it, as shown. $\bar{X}$ of the center of mass is nearest to

(A) 8 cm
(B) 10 cm
(C) 12 cm
(D) 14 cm
(E) 16 cm

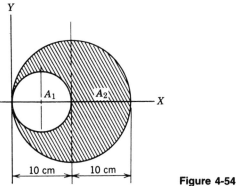

Figure 4-54

Solution

The center of mass, or center of gravity as it probably is more frequently called, is the point through which the resultant of the total weight of the object will pass regardless of the orientation of the object.

Since the disk has an axis of symmetry, the center of mass is on that axis and $\bar{Y} = 0$.

$$\bar{X} = \frac{\int X \, dA}{\int dA} = \frac{\sum AX}{\sum A}$$

$$A_1 = \frac{\pi}{4} \, 10^2 = 25\pi$$

$$A_2(\text{total area of disk incl. } A_1) = \frac{\pi}{4} \, 20^2 = 100\pi$$

$$\bar{X} = \frac{A_2 X_2 - A_1 X_1}{A_2 - A_1} = \frac{100\pi \times 10 - 25\pi \times 5}{100\pi - 25\pi}$$

$$= \frac{1000\pi - 125\pi}{75\pi} = \frac{875\pi}{75\pi} = 11\frac{2}{3} \text{ cm}$$

$$\bar{X} = 11\frac{2}{3} \text{ cm}$$

Answer is (C)

PROBLEM 4-33

The formula $I = \int y^2 \, dA$ represents the

(A) product of inertia

(B) section modulus
(C) area of cross section
(D) moment of inertia
(E) modulus of elasticity

Solution

The formula represents the moment of inertia.

Answer is (D)

PROBLEM 4-34

The term $I = bh^3/12$ refers to the

(A) radius of gyration
(B) section modulus
(C) instantaneous center
(D) moment of inertia
(E) product of inertia

Solution

$I = bh^3/12$ is the equation for the moment of inertia for a rectangular cross section.

Answer is (D)

PROBLEM 4-35

The moment of inertia of the area shown in the figure about the $x-x$ axis is

(A) 28 in.^{4}
(B) 40 in.^{4}
(C) 64 in.^{4}
(D) 110 in.^{4}
(E) 256 in.^{4}

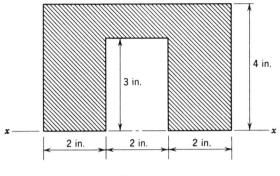

Figure 4-55

Solution

The moment of inertia $I_x = \int y^2 \, dA$. Thus the moment of inertia of a cross-sectional area is equal to the sum of the differential areas dA multiplied by the square of their moment arms about the reference axis.

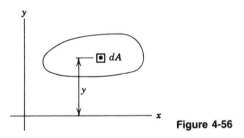

Figure 4-56

This problem can be solved in either of two ways:

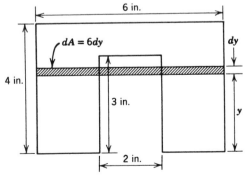

Figure 4-57

1. Integration of the differential areas.

$$I_x = \int y^2 \, dA$$

$$= \int_0^4 y^2 6 \, dy - \int_0^3 y^2 2 \, dy$$

$$= 6\left[\frac{y^3}{3}\right]_0^4 - 2\left[\frac{y^3}{3}\right]_0^3$$

$$= 6(\tfrac{64}{3}) - 2(\tfrac{27}{3}) = 128 - 18$$

$$= 110 \text{ in.}^4$$

2. Transfer of the moment of inertia, with respect to the centroid, to a parallel axis. In this method three facts are utilized:

(*a*) The moment of inertia of an object is the sum of the moments of inertia of its individual parts (all referred to the same axis).

(*b*) The moment of inertia of a rectangle with respect to its centroid is $I_{x_0} = bh^3/12$, where b is the width of the rectangle parallel to the centroidal axis and h is its depth.

(*c*) The transfer formula for obtaining the moment of inertia with respect to an axis parallel to the centroidal axis is $I_x = I_{x_0} + Ad^2$, where A is the cross-sectional area and d is the distance between the axes.

For a rectangle

$$I_x = \frac{bh^3}{12} + bh\left(\frac{h}{2}\right)^2 = \frac{bh^3}{3}$$

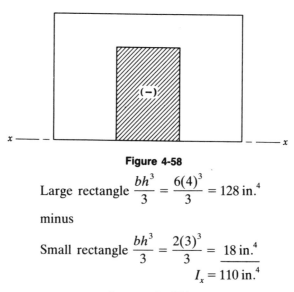

Figure 4-58

Large rectangle $\dfrac{bh^3}{3} = \dfrac{6(4)^3}{3} = 128 \text{ in.}^4$

minus

Small rectangle $\dfrac{bh^3}{3} = \dfrac{2(3)^3}{3} = 18 \text{ in.}^4$

$$I_x = 110 \text{ in.}^4$$

Answer is (D)

PROBLEM 4-36

The moment of inertia of a rectangle with respect to an axis passing through its base is

(A) $\dfrac{bh^3}{3}$

(B) $\dfrac{bh^3}{12}$

(C) $\dfrac{bh^2}{3}$

(D) $\dfrac{bh}{2}$

(E) none of the above

Solution

Using the transfer formula for the moment of inertia $I_x = I_0 + Ad^2$, where $I_0 = $ moment of inertia about the centroid, $A = $ area $= bh$, and $d = h/2$, we have

$$I_x = \frac{bh^3}{12} + bh\left(\frac{h}{2}\right)^2 = \frac{bh^3}{12} + \frac{bh^3}{4} = \frac{bh^3}{3}$$

Solution using calculus: The general equation for the moment of inertia is $I = \int y^2 \, dA$.

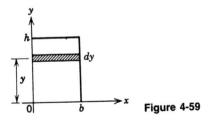

Figure 4-59

$$dA = b \, dy$$

$$dI_x = y^2(b \, dy) = by^2 \, dy$$

$$I_x = b \int_0^h y^2 \, dy = b\left[\frac{y^3}{3}\right]_0^h = \frac{bh^3}{3}$$

Answer is (A)

PROBLEM 4-37

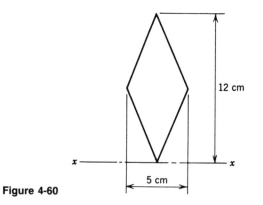

12 cm

5 cm

Figure 4-60

The moment of inertia about the $x-x$ axis is

(A) 420 cm^4
(B) 1230 cm^4
(C) 1260 cm^4
(D) 1380 cm^4
(E) 2460 cm^4

Solution

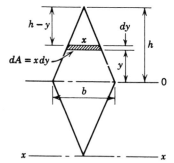

Figure 4-61

From similar triangles in Fig. 4-61,

$$\frac{b}{h} = \frac{x}{h-y} \qquad x = \frac{b}{h}(h-y)$$

$$I_0 = \int y^2 \, dA = 2 \int_0^h y^2 x \, dy = 2 \int_0^h y^2 \frac{b}{h}(h-y) \, dy$$

$$= \frac{2b}{h} \left(\int_0^h hy^2 \, dy - \int_0^h y^3 \, dy \right)$$

$$= \frac{2b}{h} \left[\frac{hy^3}{3} - \frac{y^4}{4} \right]_0^h = \frac{2b}{h} \left(\frac{h^4}{3} - \frac{h^4}{4} \right) = \frac{2bh^4}{12h}$$

$$I_0 = \frac{bh^3}{6}$$

But we want I_x. Using the transfer formula $I_x = I_0 + Ad^2$, where $A = 2(b/2)h = bh$ and $d = h$, we have

$$I_x = \frac{bh^3}{6} + bh(h^2) = \frac{7bh^3}{6}$$

In our case $h = 6$ and $b = 5$. Therefore

$$I_x = \frac{7 \times 5 \times 6^3}{6} = 1260 \text{ cm}^4$$

Answer is (C)

PROBLEM 4-38

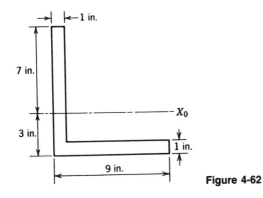

Figure 4-62

The moment of inertia of the angle section shown about the X_0 axis is nearest to

(A)   60 in.^{4}
(B)   90 in.^{4}
(C) 120 in.^{4}
(D) 150 in.^{4}
(E) 180 in.^{4}

Solution

$$I_x = \sum (I_0 + Ad^2) = \frac{1(10)^3}{12} + 1(10)(2)^2 + \frac{8(1)^3}{12} + 8(1)(2.5)^2$$

$$= 83.3 + 40 + 0.7 + 50 = 174 \text{ in.}^4$$

Answer is (E)

PROBLEM 4-39

This is a problem set containing six questions. Given the pin-connected frame shown in Fig. 4-63 with a 10-kip load at point D and a 5-kip load at point F.

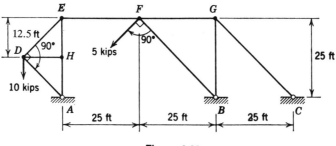

Figure 4-63

Question 1. The magnitude of reaction A is closest to

(A) 10 kips
(B) 11 kips
(C) 12 kips
(D) 13 kips
(E) 14 kips

Solution

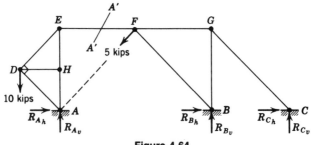

Figure 4-64

Taking the left side of Section $A'-A'$:

$$\Sigma M_E = 0 \qquad 12.5(10) + R_{A_h}(25) = 0$$

$$R_{A_h} = -5 \text{ kips or } 5 \text{ kips} \leftarrow$$

$$\Sigma F_v = 0 \qquad R_{A_v} - 10 \text{ kips} = 0$$

$$R_{A_v} = 10 \text{ kips} \uparrow$$

$$\text{Reaction } A = (5^2 + 10^2)^{1/2} = 11.2 \text{ kips}$$

Answer is (B)

Question 2. The angular direction of reaction A, measured clockwise from the horizontal, is closest to

(A) 50°
(B) 60°
(C) 70°
(D) 80°
(E) 90°

Solution

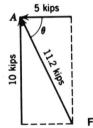

Figure 4-65 Reaction A

$$\theta_A = \tan^{-1} \frac{10}{5} = 63.5°$$

Answer is (B)

Question 3. The force in bar EF is nearest to

(A) 0 kips
(B) 5 kips
(C) 10 kips
(D) 15 kips
(E) 20 kips

Solution

Taking the left side of Section $A'-A'$:

$$\Sigma M_A = 0 \qquad 10(12.5) - 25 F_{EF} = 0$$

$$F_{EF} = 5 \text{ kips (tension)}$$

Answer is (B)

Question 4. The magnitude of reaction C is closest to

(A) 9 kips
(B) 11 kips
(C) 13 kips
(D) 15 kips
(E) 17 kips

Solution

Right side of Section $A'-A'$:

$$\Sigma M_B = 0 \qquad F_{EF} \times 25 + F_{AF} \times (25^2 + 25^2)^{1/2} + R_{C_v} \times 25 = 0$$

$$R_{C_v} = \frac{-5 \times 35.36 - 5 \times 25}{25} = -12.07 \text{ kips} = 12.07 \text{ kips } \downarrow$$

$$\Sigma F_v = 0 \qquad -12.07 \text{ kips} + R_{B_v} - \frac{5}{\sqrt{2}} = 0$$

$$R_{B_v} = 15.61 \text{ kips } \uparrow$$

At joint C:

$$\Sigma F_h = 0 \qquad -\frac{F_{GC}}{\sqrt{2}} + R_{C_h} = 0$$

$$\Sigma F_v = 0 \qquad \frac{F_{GC}}{\sqrt{2}} = 12.07 \text{ kips}$$

$$R_{C_h} = 12.07 \text{ kips} \rightarrow$$

Reaction $C = (12.07^2 + 12.07^2)^{1/2} = 17.07 \text{ kips}$

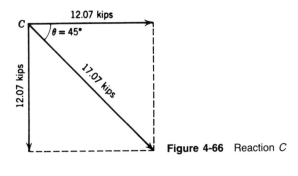

12.07 kips

$\theta = 45°$

12.07 kips

17.07 kips

Figure 4-66 Reaction C

Answer is (E)

Question 5. The magnitude of reaction B is closest to

(A) 10 kips
(B) 11 kips
(C) 12 kips
(D) 13 kips
(E) 14 kips

Solution

Total structure:

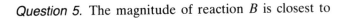

$$\sum F_h = 0 \qquad -5 + R_{B_h} + 12.07 \text{ kips} - \frac{5}{\sqrt{2}} = 0$$

$$R_{B_h} = -3.53 \text{ kips} = 3.53 \text{ kips} \leftarrow$$

$$\text{Reaction } B = (15.61^2 + 3.53^2)^{1/2} = 16 \text{ kips}$$

Answer is (E)

Question 6. The angular direction of reaction B, measured clockwise from the horizontal, is closest to

(A) 30°
(B) 45°
(C) 60°
(D) 75°
(E) 90°

Solution

Figure 4-67   Reaction *B*

$$\theta_B = \tan^{-1} \frac{15.61}{3.53} = 77.3°$$

Answer is (D)

5

DYNAMICS

Dynamics is the study of the motion of nondeformable bodies. This study is normally divided into *kinematics*, the study of acceleration-velocity-displacement relations, and *kinetics*, which relates these motions and the forces causing, and caused by, them. Statics may be regarded as merely a special (but important) subdivision of dynamics where forces are in equilibrium and no accelerations are present.

KINEMATICS

Kinematics ignores the forces causing motion and considers only the motion itself. The acceleration a, velocity v, and displacement x of a body are related by the very basic expressions

$$a = \frac{dv}{dt} \qquad v = \frac{dx}{dt} \tag{5-1}$$

where t is time. When the acceleration is a known function of time, velocity, or displacement, Eqs. (5-1) may be integrated to give direct velocity-time or displacement-time relations. For the important case of rectilinear motion beginning at a point x_0 with a constant acceleration a_0 and initial velocity v_0, the displacement at any later time, by integration, is

$$x(t) = x_0 + v_0 t + \tfrac{1}{2} a_0 t^2 \tag{5-2}$$

Since we also have

$$a_0 = \frac{dv}{dt} = \frac{dv}{dx}\frac{dx}{dt} = v\frac{dv}{dx} = \frac{d}{dx}\left(\frac{v^2}{2}\right)$$

157

we also find by integration that the velocity and displacement are related by

$$\tfrac{1}{2}(v^2 - v_0^2) = a_0(x - x_0) \tag{5-3}$$

Example 1

A driver sees a stoplight when his car is traveling at 55 miles/hr. If it takes him 0.6 sec to apply the brakes, and the brakes produce a deceleration of 15 ft/sec^2 in the car, how many feet does the car travel before coming to a stop?

Solution

Using the conversion factor 88 ft/sec = 60 miles/hr, we find the initial speed $v_0 = 55(88/60) = 80.7$ ft/sec. During the 0.6-sec reaction period the car moves a distance $s_1 = v_0 t = (80.7)(0.6) = 48.4$ ft.

During the deceleration period $a_0 = -15$ ft/sec^2 while the velocity decreases from v_0 to zero. From Eq. (5-3), the distance traveled $s_2 = x - x_0$ is

$$s_2 = \frac{1}{2a_0}(v^2 - v_0^2) = \frac{1}{2(-15)}[0 - (80.7)^2] = 217 \text{ ft}$$

Thus the total distance traveled is

$$s = s_1 + s_2$$
$$s = 48.4 + 217 = 265 \text{ ft}$$

In addition to rectilinear or straight-line motions there are curvilinear motions. Equations (5-1) apply in a vectorial sense. For two- or three-dimensional motion, the trajectory and velocity of the body are often expressed by a set of parametric equations, usually with time t as the parameter. The trajectory of a body in a gravity field is a good example. Choosing x and y to be horizontal and vertical displacements, respectively,

$$x = x_0 + (v_0 \cos \theta_0)t$$
$$y = y_0 + (v_0 \sin \theta_0)t - \tfrac{1}{2}gt^2 \tag{5-4}$$

when the body is released at $t = 0$ from the point (x_0, y_0) with the initial velocity v_0 and orientation θ_0 from the horizon. Here the location of the body is defined in terms of the time parameter.

The accelerations of a body experiencing curvilinear motion are often split into tangential and normal components. For the case of circular motion of radius r at velocity v, the tangential acceleration a_t and normal acceleration a_n are

$$a_t = \frac{dv}{dt} \qquad a_n = \frac{v^2}{r} \tag{5-5}$$

If the speed of the body is unchanging, $a_t = 0$ but a normal acceleration still exists. In terms of the angular velocity ω, $v = \omega r$ and $a_n = \omega^2 r$.

KINETICS

Here we consider the kinetics of bodies whose dynamic behavior is adequately described by reference to only one body property, the mass, which is assumed to be concentrated at a point. Mass is the proportionality constant that relates the acceleration of a body to the net force that acts on it. The Newtonian law expresses this as

$$\sum \mathbf{F} = m\mathbf{a} \tag{5-6}$$

for a body of constant mass m. This vector equation is usually written in component scalar form for use in obtaining numerical answers, however.

Example 2

The blocks A and B, of weights $W_A = 500$ lb and $W_B = 100$ lb, are connected by a rope that passes over a small pulley (Fig. 5-1). No friction is present. When the system is released from rest, what is the acceleration a of the bodies and the tension T in the rope?

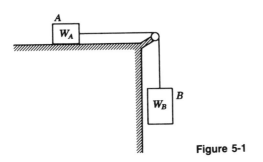

Figure 5-1

Solution

Drawing a diagram of each block (Fig. 5-2) shows that the blocks are clearly not in equilibrium but will accelerate to the right (block A) and downward (block B) at the same rate a. Applying the Newtonian equation, we have for block A

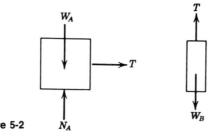

Figure 5-2 N_A

$$\Sigma F_x = ma$$

$$T = \frac{W_A}{g} a$$

and for block B

$$\Sigma F_y = ma$$

$$W_B - T = \frac{W_B}{g} a$$

Hence

$$W_B - \frac{W_A}{g} a = \frac{W_B}{g} a$$

$$a = g \frac{W_B}{W_A + W_B}$$

$$= 32.2 \frac{100}{500 + 100} = 5.37 \text{ ft/sec}^2$$

$$T = \frac{W_A}{g} a = \frac{500}{32.2} (5.37) = 83.3 \text{ lb}$$

The normal and tangential components of Eq. (5-6) are useful in situations involving curvilinear motion. The components are

$$\Sigma F_t = m \frac{dv}{dt} \qquad \Sigma F_n = m \frac{v^2}{r} \qquad (5\text{-}7)$$

The normal force component is also called the centrifugal force.

Example 3

A segment of a flywheel weighs 1300 lb, and its center of gravity is 6 ft from the center of the shaft. The wheel rotates at 125 rpm. What is the pull on the arm supporting the segment?

Solution

The angular velocity $\omega = 125(2\pi/60) = 13.1$ rad/sec. The speed of the segment is $v = \omega r$, so the normal force, Eq. (5-7), is

$$F_n = m \, \frac{v^2}{r} = mr\omega^2 = \left(\frac{1300}{32.2}\right)(6)(13.1)^2 = 41,500 \text{ lb}$$

ENERGY CONSERVATION

If Newton's law of motion is rearranged and integrated with respect to distance between two points, we find that the work done in moving a body from one point to the other is equal to the change in kinetic energy KE of that body. If the amount of work done depends only on the end points of the path, that is, if a conservative force produces the work, then the work can be expressed as a change in potential energy PE. Common conservative forces include the weight of a body and the force due to an elastic spring. For conservative forces we may then say that the total mechanical energy of a system is conserved, or

$$PE + KE = \text{constant} \tag{5-8}$$

for a process. This may be restated in terms of changes of kinetic and potential energy as

$$\Delta(KE) = -\Delta(PE) \tag{5-9}$$

For *non*conservative forces, such as those due to friction, total mechanical energy is not conserved, and Eqs. (5-8) and (5-9) should not be used then.

Example 4

A block of mass m is released from rest on an inclined frictionless plane, as in Fig. 5-3. What is the block's speed when it has dropped a vertical distance of 4 ft?

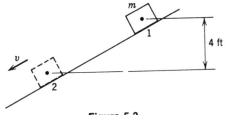

Figure 5-3

Solution

Here we apply Eq. (5-9):

$$\Delta(KE) = -\Delta(PE)$$
$$KE_2 - KE_1 = -(PE_2 - PE_1)$$
$$\tfrac{1}{2}mv^2 - 0 = -(0 - mgh)$$
$$v^2 = 2gh$$
$$v^2 = 2(32.2)(4)$$
$$v = 16.05 \text{ ft/sec}$$

Example 5

The spring of a spring gun has an uncompressed length of 8 in. The modulus of the spring $k = 1$ lb/in. The spring is compressed to a length of 4 in., and a ball weighing 1 oz is put in the barrel against the compressed spring, as in Fig. 5-4. If the spring is then released, find the velocity with which the ball leaves the gun. Neglect friction.

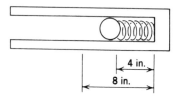

4 in.

8 in.

Figure 5-4

Solution

The force involved in the compression of an elastic spring is a conservative force whose magnitude is proportional to the distance x that the spring is compressed, that is, $F = kx$, where k is the spring constant. The work done on the spring is

$$W = \int_0^x F \, dx = \int_0^x kx \, dx = \tfrac{1}{2}kx^2$$

In our case

$$W = \tfrac{1}{2}(1 \text{ lb/in.})(4 \text{ in.})^2 = 8 \text{ in.-lb} = \tfrac{2}{3} \text{ ft-lb}$$

This work represents stored or potential energy. When the spring is released, this is converted into the kinetic energy of the ball, which was initially at rest.

$$\Delta(KE) = -\Delta(PE)$$

$$\tfrac{1}{2}mv^2 = W$$

$$v = \left(\frac{2W}{m}\right)^{1/2}$$

$$v = \left[\frac{2(\tfrac{2}{3})}{(\tfrac{1}{16})/32.2}\right]^{1/2} = 26.2 \text{ ft/sec}$$

MOMENTUM CONSERVATION

Newton's law of motion, when viewed another way, yields a useful principle that relates impulse and momentum. Expressing Eq. (5-6) as

$$\sum \mathbf{F} = \frac{d}{dt}(m\mathbf{v}) \tag{5-10}$$

we may integrate with respect to time to obtain

$$\sum \int \mathbf{F}\,dt = (m\mathbf{v})_2 - (m\mathbf{v})_1 \tag{5-11}$$

The left term represents an external impulse that changes the momentum $m\mathbf{v}$ of a body from that of state 1 to that of state 2. In the absence of impulsive external forces, we note that momentum is conserved for the body. Extended to a system of several bodies, we may state again, if the net external impulsive force on the system is zero, that system momentum is conserved, or

$$\left(\sum m\mathbf{v}\right)_1 = \left(\sum m\mathbf{v}\right)_2 \tag{5-12}$$

This is true even though individual bodies within a system may impact with one another; the reason is that the impulsive forces are internal, not external, to the system.

The nature of the impact that occurs between two bodies is important even when no external impulsive forces act. An index of the kind of impact that occurs between two bodies is the coefficient of restitution e, which is the ratio of the relative velocity between the two bodies after impact to the relative velocity before impact. Two cases are of particular interest.

1. The case $e = 1$ represents elastic impact where the relative velocities before and after impact are equal. This is the only impact situation where energy is conserved; all other impact cases involve a change in total mechanical energy.

2. The other extreme $e = 0$ represents inelastic or plastic impact. After impact the two bodies move together with a common velocity. Energy is not conserved.

Different parts of a single problem commonly use the principles of momentum and energy conservation and the equation of motion together to achieve a solution.

Example 6

A bullet of mass m, traveling at velocity v_1, impacts inelastically with a simple pendulum composed of a mass M on the end of a flexible cord. If the impact occurs a distance L below the pendulum's suspension point, as illustrated in Fig. 5-5, through what angle θ will the pendulum move? (Assume the impact occurs at a right angle to the vertical.)

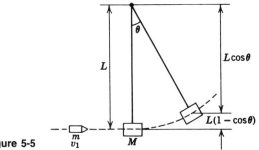

Figure 5-5

Solution

By the principle of conservation of momentum for an inelastic impact, the sum of the momenta of the bullet and the pendulum before impact equals the momentum of the combination after impact, or

$$mv_1 + 0 = (m + M)V_c$$

where V_c is the velocity of the combination immediately after impact. Thus

$$V_c = \frac{mv_1}{m + M}$$

Using the energy conservation principle for the motion after the impact, we write

$$Wh = \tfrac{1}{2}mV^2$$

which in this case gives

$$g(m + M)L(1 - \cos \theta) = \tfrac{1}{2}(m + M)V_c^2 = \tfrac{1}{2}(m + M)\left(\frac{mv_1}{m + M}\right)^2$$

Solving for θ,

$$\theta = \cos^{-1}\left[1 - \frac{1}{2gL}\left(\frac{mv_1}{m + M}\right)^2\right]$$

RIGID BODY DYNAMICS

The motion of some bodies cannot be properly analyzed by assuming the mass of the body is concentrated at a point and analyzing it as a particle. In particular this is true when bodies execute rotational motion. We consider briefly the analysis of such motions now.

A general plane motion can always be regarded as the sum of a translational motion and a rotational motion. We have already examined translational motion. For rotational motion the velocity v of a point which is a distance r from the center of rotation on a body rotating with an angular velocity ω is $v = \omega r$, and the normal and tangential acceleration components are $a_n = \omega^2 r$, $a_t = \alpha r$, where $\alpha = d\omega/dt$ is the angular acceleration of the body. These quantities are often easier to calculate when one first finds the instantaneous center of rotation, the point around which the body appears to rotate at a given instant.

Example 7

A 5-in. radius cylinder rolls to the right at a constant rate of 10 in./sec. For the instant shown in Fig. 5-6, what is the velocity at point A?

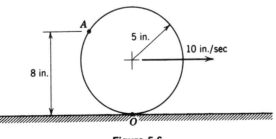

Figure 5-6

Solution

We first note that the point on the cylinder that is in contact with the plane, point O, is the instantaneous center of rotation. At a given instant it is stationary, and all other points are executing a purely rotational motion about point O.

From Fig. 5-7 the angular velocity ω of the body is

$$\omega = \frac{v}{r} = \frac{10}{5} = 2 \, \text{rad/sec}$$

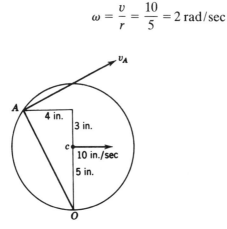

Figure 5-7

Also the distance $OA = d = (8^2 + 4^2)^{1/2} = 8.94 \, \text{in}$. The velocity $v_A = \omega d = (2)(8.94) = 17.88 \, \text{in./sec}$. The instantaneous center of rotation for velocity is generally *not* also the instantaneous center for acceleration.

The kinetics of the plane motion of rigid bodies is quite similar to the kinetics of particles. The differences arise because we are now analyzing the motion of distributed masses rather than point masses.

To study the plane motion of a rigid body directly, we must supplement the equation of motion, Eq. (5-6), with an equation that says the net external moment on a body is equal to the product of the moment of inertia I and the angular acceleration α of the body, or

$$\sum M_0 = I_0 \alpha \qquad (5\text{-}13)$$

where the subscript denotes some common reference axis.

For plane motion Eq. (5-9), which says that energy is conserved, remains valid so long as we now express the kinetic energy as the sum of the translational kinetic energy $\frac{1}{2}mV^2$ and the rotational kinetic energy $\frac{1}{2}I_0\omega^2$.

The momentum of a body may be treated as the sum of linear momentum and angular momentum just as a general motion is the sum of a translational and a rotational motion. The treatment of linear momentum conservation is unchanged from the presentation given in Eqs. (5-10)–(5-12). Conservation of angular momentum may be expressed as

$$I_0\omega_1 + \int M_0 \, dt = I_0\omega_2 \qquad (5\text{-}14)$$

In this equation the $I_0\omega$ terms are the angular momentum of the body before and after an angular impulse, given by the middle term, is applied to the body. M_0 is the net moment, computed with respect to axis 0, of the forces acting on the body at an instant. One then integrates over the time period of interest to find the total angular impulse. In the absence of an applied angular impulse, the angular momentum $I_0\omega$ is constant.

PROBLEM 5-1

A pebble is dropped in a well, and it is found that 4.25 sec elapse after release of the pebble before the splash is heard. If the velocity of sound in the well is 1030 ft/sec, the depth to the water surface is most nearly

(A) 230 ft

(B) 260 ft

(C) 290 ft

(D) 450 ft

(E) 550 ft

$$x = x_1 + v_1 t + \tfrac{1}{2} g t^2$$

Solution

The pebble is initially at rest ($V_0 = 0$) and thereafter falls with a constant acceleration due to gravity g so that

$$V(t) = \frac{dy}{dt} = gt$$

The fall distance is then $y = \tfrac{1}{2}gt_1^2$ by integration, where t_1 is the time for the pebble to reach the water. The sound wave will return the same distance y at a constant velocity so that $y = 1030t_2$, where t_2 is the time for the sound to travel from the base to the top of the well. By the problem statement,

$$t_1 + t_2 = 4.25 \text{ sec}$$

and we also have

$$\tfrac{1}{2}gt_1^2 = 1030t_2$$

Solving these equations simultaneously,

$$\tfrac{1}{2}(32.2)t_1^2 - 1030t_2 = 0$$

$$1030t_1 + 1030t_2 = 4.25(1030)$$

$$\overline{16.1t_1^2 + 1030t_1 - 4380} = 0$$

or
$$t_1^2 + 64.0t_1 - 272 = 0$$

$$t_1 = \tfrac{1}{2}[-64.0 \pm \sqrt{64.0^2 + 4(272)}] = \tfrac{1}{2}(-64.0 \pm 72.0)$$

Since t_1 must be positive, $t_1 = 4.0$ sec, and

$$y = \tfrac{1}{2}gt_1^2 = \tfrac{1}{2}(32.2)(4.0)^2 = 258 \text{ ft}$$

Answer is (B)

PROBLEM 5-2

A body moves so that the x component of acceleration is given by the equation $6 - t$ and the y component of acceleration is given by $6 + t$. If the initial x and y components of velocity are both 2, the speed of the body at the end of 2 sec is closest to

(A) 12
(B) 14
(C) 16
(D) 18
(E) 20

Solution

$$a_x = \frac{d^2x}{dt^2} = 6 - t \qquad a_y = \frac{d^2y}{dt^2} = 6 + t$$

Integrating these expressions once,

$$V_x = \frac{dx}{dt} = 6t - \frac{1}{2}t^2 + C_1 \qquad V_y = \frac{dy}{dt} = 6t + \frac{1}{2}t^2 + C_2$$

At $t = 0$, $V_x = 2$ and $V_y = 2$ so $C_1 = C_2 = 2$ by substitution into the velocity expressions. At the end of 2 sec we have

$$V_x = 6(2) - \tfrac{1}{2}(2)^2 + 2 \qquad V_y = 6(2) + \tfrac{1}{2}(2)^2 + 2$$

$$V_x = 12 \qquad\qquad V_y = 16$$

The speed is

$$V = (V_x^2 + V_y^2)^{1/2} = (12^2 + 16^2)^{1/2}$$

$$V = 20$$

Answer is (E)

PROBLEM 5-3

The driver of a car traveling 30 miles/hr suddenly applies the brakes and skids 60 ft before coming to a stop.

Question 1. If the car weighs 1600 lb and a constant rate of deceleration is assumed, the rate of deceleration is nearest to

(A) -8.0 ft/sec^2
(B) -16.1 ft/sec^2
(C) -24.1 ft/sec^2
(D) -32.2 ft/sec^2
(E) -40.2 ft/sec^2

$x = x_1 + V_1 t + \frac{1}{2} g t^2$

$= 30 t + \frac{1}{2} g t^2$

$60 = 30 t + \frac{1}{2} g t^2$

Solution

The general displacement equation for rectilinear motion can be obtained by integrating the expression $d^2x/dt^2 = a$ twice. The result is

$$x = x_0 + V_0 t + \tfrac{1}{2} a t^2$$

where x_0 and V_0 are the initial displacement and velocity, respectively, and a is the (constant) acceleration. If t is measured from the moment the brakes are applied and the initial location of the car is $x_0 = 0$, then $V_0 = 30$ miles/hr $= 44$ ft/sec. When the car stops at time t later, $x = 60$ ft, so

$$60 = 44t + \tfrac{1}{2} a t^2$$

or, solving for a,

$$a = \frac{120}{t^2} - \frac{88}{t}$$

Since the deceleration is to be constant,

$$\frac{da}{dt} = 0 = \frac{120(-2)}{t^3} - \frac{88(-1)}{t^2}$$

yielding $t = 240/88 = 2.73$ sec as the time to stop. Hence

$$a = \frac{120}{(2.73)^2} - \frac{88}{2.73} = -16.1 \text{ ft/sec}^2$$

(The minus sign denotes deceleration.)

Answer is (B)

Question 2. If the car weighs 1600 lb and a constant rate of deceleration is assumed, the average coefficient of sliding friction is closest to

(A) 0.3
(B) 0.4
(C) 0.5
(D) 0.6
(E) 0.7

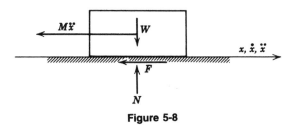

Figure 5-8

Solution

For sliding,

$$F = -\mu N = -1600\mu$$

Also

$$F = M\ddot{x} = Ma$$

Equating these expressions,

$$-1600\mu = \left(\frac{1600}{32.2}\right)(-16.1) \quad \text{and} \quad \mu = 0.5$$

Answer is (C)

PROBLEM 5-4

An aircraft begins its take-off run with an acceleration of $4\,\text{m/sec}^2$, which then decreases uniformly to zero in 15 sec, at which time the craft becomes airborne.

Question 1. The take-off speed, in meters per second, is closest to

(A) 60
(B) 50
(C) 40
(D) 30
(E) 20

Solution

The aircraft's acceleration, as a function of time, is

$$a = \frac{d^2x}{dt^2} = 4 - \frac{4}{15}t = 4\left(1 - \frac{t}{15}\right)$$

At the initial instant when $t = 0$, $V = 0$, and $x = 0$. By integration, the velocity is

$$V = \frac{dx}{dt} = 4\left(t - \frac{t^2}{30}\right) + C_1$$

At $t = 0$, $V = 0$, so $C_1 = 0$. At $t = 15$ sec the velocity is then

$$V = 4\left[15 - \frac{(15)^2}{30}\right] = 4[15 - 7.5]$$

$$V = 30 \text{ m/sec}$$

Answer is (D)

Question 2. The length (in meters) of the take-off run is nearest to

(A) 150
(B) 225
(C) 300
(D) 375
(E) 450

Solution

Integrating once more, the position $x(t)$ is

$$x = 4\left(\frac{t^2}{2} - \frac{t^3}{90}\right) + C_2$$

At $t = 0$, $x = 0$, hence $C_2 = 0$. At $t = 15$ sec the length of the take-off run is

$$x = 4\left[\frac{(15)^2}{2} - \frac{(15)^3}{90}\right] = 4[112.5 - 37.5]$$

$$x = 300 \text{ m}$$

Answer is (C)

PROBLEM 5-5

A body weighing 322 lb is subjected to an acceleration (in the positive x direction) which is a linearly decreasing function of the velocity. The body is stationary at $x = 0$ when $t = 0$, and the force acting on the body at this instant is 100 lb. The acceleration is zero when the velocity reaches 100 ft/sec.

Question 1. The differential equation that expresses this situation mathematically is

    (A) $\ddot{x} + 0.1\dot{x} - 10 = 0$
    (B) $\ddot{x} + 0.01\dot{x} - 100 = 0$
    (C) $\ddot{x} - \dot{x} = 0$
    (D) $\ddot{x} + 0.1\dot{x} + 10 = 0$
    (E) $\ddot{x} + \dot{x} - 100 = 0$

Solution

Using $F = Ma$, the initial acceleration is

$$a(0) = \ddot{x} = \frac{F}{M} = \frac{F}{W/g} = \frac{100(32.2)}{322} = 10 \text{ ft/sec}^2$$

The other initial conditions are $\dot{x}(0) = 0$, $x(0) = 0$. The acceleration as a function of time is of the form

$$\ddot{x} = -K\dot{x} + C \qquad K > 0$$

At $t = 0$, $\ddot{x} = 10$ when $\dot{x} = 0$, so $C = 10$. Also when $\dot{x} = 100$, $\ddot{x} = 0$, or

$$0 = -100K + 10 \qquad K = 0.1$$

Hence

$$\ddot{x} + 0.1\dot{x} - 10 = 0$$

Answer is (A)

Question 2. The displacement $x(t)$ is

    (A) $100t$
    (B) $100(t + 10e^{-t})$
    (C) $1000(e^{-t/10} - 1)$
    (D) $-100(1 - e^{t/10})$
    (E) $1000(e^{-0.1t} - 1) + 100t$

Solution

We seek the general solution to the differential equation for x, which is the sum of a particular solution and the complementary solution to

$$\ddot{x} + 0.1\dot{x} = 10$$

By inspection, a particular solution is $x = 100t$. To find the complementary solution, that is, the solution to $\ddot{x} + 0.1\dot{x} = 0$, assume a solution of the form

$$x = Ae^{\alpha t} + B$$

Then

$$\dot{x} = A\alpha e^{\alpha t} \quad \text{and} \quad \ddot{x} = A\alpha^2 e^{\alpha t}$$

Substituting into the homogeneous differential equation, we obtain

$$A\alpha^2 e^{\alpha t} + 0.1 A\alpha e^{\alpha t} = 0$$

or

$$\alpha(\alpha + 0.1) = 0$$

The nontrivial solution is $\alpha = -0.1$. The general solution is then

$$x = Ae^{-0.1t} + B + 100t$$

subject to the initial conditions $\dot{x} = x = 0$ at $t = 0$.

$$x = 0 = A + B$$
$$\dot{x} = 0 = -0.1A + 100$$

The solution is $A = 1000$, $B = -1000$. Hence

$$x(t) = 1000(e^{-0.1t} - 1) + 100t$$

Answer is (E)

PROBLEM 5-6

A weight is attached to one end of a 53-ft rope passing over a small pulley 29 ft above the ground. A man whose hand is 5 ft above the ground grasps the other end of the rope and walks away at the rate of 5 ft/sec. When the man is 7 ft from a point directly under the pulley, the rate at which the weight is rising is closest to

(A) 1.0 ft/sec
(B) 1.4 ft/sec
(C) 2.5 ft/sec
(D) 3.6 ft/sec
(E) 5.0 ft/sec

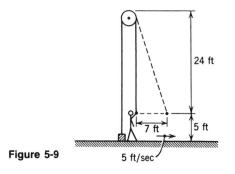

24 ft

7 ft   5 ft

Figure 5-9   5 ft/sec

Solution

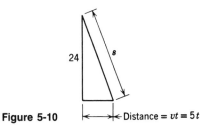

24   8

Figure 5-10   Distance = $vt = 5t$

At $t = 0$ we note that the rope is taut since 29 ft + 24 ft = 53 ft. Let s be the distance between the pulley and the man's hand:

$$s = [24^2 + (5t)^2]^{1/2} = (576 + 25t^2)^{1/2}$$

$$v = \frac{ds}{dt} = \tfrac{1}{2}(576 + 25t^2)^{-1/2} \times 50t$$

When the horizontal distance $5t = 7$, $t = 1.4$ sec, and

$$v = \tfrac{1}{2}[576 + 25(1.4)^2]^{-1/2} \times 50(1.4) = 1.4 \text{ ft/sec}$$

The rope is therefore moving at 1.4 ft/sec, and consequently the weight is rising at this same velocity.

Answer is (B)

PROBLEM 5-7

A river flows north with a speed of 3 miles/hr. A man rows a boat across the river. His speed relative to the water is 4 miles/hr. What is his velocity relative to the earth?

(A) 4 miles/hr
(B) 3 miles/hr
(C) 7 miles/hr
(D) 1 miles/hr
(E) 5 miles/hr

Solution

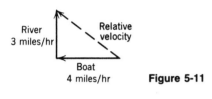

River
3 miles/hr

Relative
velocity

Boat
4 miles/hr

Figure 5-11

$$V = (3^2 + 4^2)^{1/2} = 5 \text{ miles/hr}$$

Answer is (E)

PROBLEM 5-8

A car jumps across a 10-ft-wide ditch with a constant velocity V. The ditch is 6 in. lower on the far side. The minimum velocity (in miles per hour) that will keep the car from falling into the ditch is nearest to

(A) 35
(B) 40
(C) 45
(D) 50
(E) 55

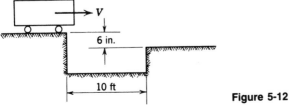

V

6 in.

10 ft

Figure 5-12

Solution

The vertical drop $y = \frac{1}{2}gt^2 = 0.5$ ft. Hence

$$t = \left(\frac{2y}{g}\right)^{1/2} = \left[\frac{2(0.5)}{32.2}\right]^{1/2} = 0.176 \text{ sec}$$

The horizontal motion is not accelerated and is

$$x = Vt = 10 \text{ ft}$$

$$V = \frac{10}{t} = \frac{10}{0.176} = 56.8 \text{ ft/sec}$$

Since 60 miles/hr = 88 ft/sec,

$$V = \left(\frac{60}{88}\right)(56.8) = 38.8 \text{ miles/hr}$$

Answer is (B)

PROBLEM 5-9

A batted baseball leaves the bat at an angle of 30° above the horizontal and is caught by an outfielder 400 ft from home plate. Assume the ball's height when hit is the same as when caught and that air resistance is negligible. The initial velocity of the ball is closest to

(A) 200 ft/sec
(B) 175 ft/sec
(C) 150 ft/sec
(D) 125 ft/sec
(E) 100 ft/sec

Solution

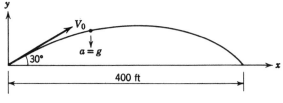

Figure 5-13

Since we neglect air resistance, the only force acting on the ball is due to its weight. Hence the acceleration of the ball at all times is g directed downward.

The horizontal distance transversed is $x = V_x t$, where $V_x = V_0 \cos 30°$ and $x = 400$ ft.

$$400 = V_0 \frac{\sqrt{3}}{2} t \quad \text{or} \quad t = \frac{800}{V_0\sqrt{3}} \text{ sec}$$

The displacement vertically is

$$y = V_y t - \tfrac{1}{2} g t^2$$

or

$$0 = V_0 \sin 30°t - \tfrac{1}{2} g t^2$$

Substituting for t,

$$0 = \frac{800}{\sqrt{3}} \sin 30° - \frac{1}{2} g \left(\frac{800}{V_0\sqrt{3}}\right)^2$$

$$V_0^2 = 32.2 \left(\frac{800}{\sqrt{3}}\right) \qquad V_0 = 122 \text{ ft/sec}$$

Answer is (D)

PROBLEM 5-10

A satellite travels in a perfectly circular orbit around the earth at an altitude of 1000 miles. Assume the earth is a perfect sphere with a radius of 4000 miles and that the force of the earth's gravity g at the height of the satellite is 20.6 ft/sec^2. The speed of the satellite is nearest to

(A) 10,000 miles/hr
(B) 12,000 miles/hr
(C) 14,000 miles/hr
(D) 16,000 miles/hr
(E) 18,000 miles/hr

Solution

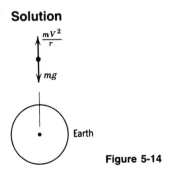

Figure 5-14

The two forces acting on the satellite are the attractive force of the earth mg, and the centrifugal force mV^2/r. If the satellite is to remain a constant distance from earth, these two forces must be in equilibrium. Hence

$$\frac{mV^2}{r} = mg$$

$$V = (rg)^{1/2}$$

$$= [(4000 + 1000) \text{ miles} \times (5280 \text{ ft/mile})(20.6 \text{ ft/sec}^2)]^{1/2}$$

$$= 2.33 \times 10^4 \text{ ft/sec} = 15{,}900 \text{ miles/hr}$$

Answer is (D)

PROBLEM 5-11

The distance of the planet Neptune from the sun is 30 times that of the earth from the sun. The force of attraction between two bodies is directly proportional to the product of their masses and inversely proportional to the square of the distance between them. The approximate period of Neptune's revolution about the sun is nearest to

(A) 900 yr
(B) 160 yr
(C) 90 yr
(D) 30 yr
(E)   5 yr

Solution

Mathematically, the attractive force F is

$$F = k \frac{MM'}{R^2}$$

Between the sun and earth it is

$$F = k \frac{M_s M_E}{R^2}$$

and between the sun and Neptune

$$F = k \frac{M_s M_N}{(30R)^2}$$

For uniform circular motion the force F causes a centripetal acceleration $a = V^2/R$, where $V = 2\pi R/T$ and T is the period of revolution. Hence

$$a_E = \frac{4\pi^2 R}{T_E^2} \quad \text{and} \quad a_N = \frac{4\pi^2 (30R)}{T_N^2}$$

Using $F = ma$,

$$k\frac{M_S M_E}{R^2} = \frac{M_E 4\pi^2 R}{T_E^2} \quad \text{and} \quad k\frac{M_S M_N}{(30R)^2} = \frac{M_N 4\pi^2 (30R)}{T_N^2}$$

or

$$\frac{R^3}{T_E^2} = \frac{kM_S}{4\pi^2} = \frac{(30R)^3}{T_N^2}$$

The period of the earth's revolution about the sun T_E is 1 yr, so the period for Neptune's revolution is

$$T_N = (30)^{3/2} \text{ yr} = 164.3 \text{ yr}$$

(Using Kepler's third law of planetary motion would yield the same result.)

Answer is (B)

PROBLEM 5-12

A traveling crane lifts a 1000-lb load on a 20-ft hoisting cable. The maximum horizontal acceleration that the crane may have without producing a devia-tion of the cable of more than 30° from the vertical is closest to

(A) 19 ft/sec²
(B) 25 ft/sec²
(C) 32 ft/sec²
(D) 43 ft/sec²
(E) 55 ft/sec²

Solution

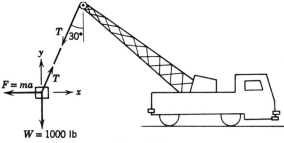

Figure 5-15

Writing Newton's equation of motion in the x and y directions for the weight

$$\Sigma F_y = ma_y \qquad T \cos 30° - W = 0$$

$$\Sigma F_x = ma_x \qquad T \sin 30° = \frac{W}{g} a$$

Eliminating W from these equations and simplifying gives

$$a = g \tan 30° = (32.2)/\sqrt{3} = 18.6 \text{ ft/sec}^2$$

Answer is (A)

PROBLEM 5-13

An elevator, which with its load weighs 70,000 N (newtons), is descending at a speed of 4.5 m/sec. If the load on the cable must not exceed 125,000 N, the shortest distance in which the elevator should be stopped is most nearly

(A) 0.6 m
(B) 1.0 m
(C) 1.3 m
(D) 2.0 m
(E) 2.6 m

Solution

The maximum allowable inertia force is the difference between the maximum cable load and the elevator weight; in this case the maximum inertia force to stop the elevator is $125,000 - 70,000 = 55,000$ N. Writing the inertia force as

$$F = -ma = -\frac{W}{g} a$$

$$55,000 \text{ N} = -\frac{70,000 \text{ N}}{9.81 \text{ m/sec}^2} a$$

and the maximum allowable deceleration is $a = -7.71 \text{ m/sec}^2$. Since

$$a = \frac{dv}{dt} = \frac{dv}{ds}\frac{ds}{dt} = \frac{d}{ds}\left(\frac{v^2}{2}\right)$$

$$a\,ds = \tfrac{1}{2}(V_2^2 - V_1^2)$$

The initial elevator velocity $V_1 = 4.5$ m/sec, and the final velocity $V_2 = 0$. Hence the minimum stopping distance s is

$$s = \frac{1}{2(-7.71)} (-4.5)^2 = -1.31 \text{ m}$$

Answer is (C)

PROBLEM 5-14

The system of pulleys carries two loads as shown in Fig. 5-16. The acceleration of the 8-lb weight is closest to

(A) 38.6 ft/sec^2
(B) 32.2 ft/sec^2
(C) 17.6 ft/sec^2
(D)   9.2 ft/sec^2
(E)   7.4 ft/sec^2

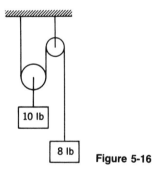

Figure 5-16

Solution

The pulleys are assumed to be weightless and frictionless. We will select the downward direction to be positive in each free-body diagram. For the movable pulley,

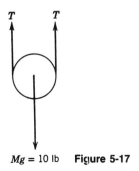

$Mg = 10$ lb   **Figure 5-17**

$$\sum F = -2T + Mg = Ma_1 \tag{1}$$

and for the 8-lb weight

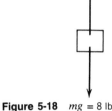

Figure 5-18 $mg = 8$ lb

$$\sum F = -T + mg = ma_2 \tag{2}$$

The accelerations a_1 and a_2 are related by the pulley geometry since the 8-lb weight will rise twice the distance the 10-lb weight will fall; hence

$$a_1 = \frac{-a_2}{2} \tag{3}$$

Equations (1), (2), and (3) are three equations for the unknowns T, a_1, and a_2. First eliminate a_1:

$$2T - Mg = \frac{Ma_2}{2}$$

$$-2T + 2mg = 2ma_2$$

Now add:

$$a_2\left(2m + \frac{M}{2}\right) = 2mg - Mg$$

$$a_2 = \left(\frac{2mg - Mg}{2mg + Mg/2}\right)g$$

$$a_2 = \left[\frac{2(8) - 10}{2(8) + 10/2}\right](32.2)$$

Downward acceleration $= a_2 = 9.20 \text{ ft/sec}^2$

Answer is (D)

PROBLEM 5-15

A 16-lb weight and an 8-lb weight resting on a horizontal frictionless surface are connected by a cord A and are pulled along the surface with a uniform acceleration of 4 ft/sec^2 by a second cord attached to the 16-lb weight. The tension in cord A is closest to

(A) 4 lb
(B) 1 lb
(C) 2 lb
(D) 3 lb
(E) 32 lb

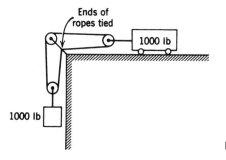

8 lb A 16 lb   **Figure 5-19**

Solution

The presence of the 16-lb weight does not affect the tension in A. The tensile force is that needed to accelerate the 8-lb weight.

$$F = ma = \frac{W}{g}\, a = \left(\frac{8}{32.2}\right)(4) \approx 1 \text{ lb}$$

Answer is (B)

PROBLEM 5-16

Ends of ropes tied

1000 lb

1000 lb

Figure 5-20

The tension in the rope for the arrangement shown in Fig. 5-20 (assume no friction and weightless pulleys) is nearest to

(A) 1000 lb
(B)   500 lb
(C)   333 lb
(D)   250 lb
(E)   200 lb

Solution

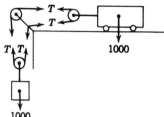

Figure 5-21   1000

For the suspended weight, $\Sigma F_y = ma$

$$1000 - 2T = \frac{1000}{32.2} a \tag{1}$$

For the cart, $\Sigma F_x = ma$

$$2T = \frac{1000}{32.2} a \tag{2}$$

Solving Eqs. (1) and (2) for T,

$$1000 - 2T = 2T \qquad T = 250 \text{ lb}$$

Answer is (D)

PROBLEM 5-17

Two 3-N weights are connected by a massless string hanging over a smooth frictionless peg. If a third weight of 3 N is added to one of the weights and the system is released, the amount of the increased force on the peg is most nearly

(A) 0 N
(B) 1.5 N
(C) 2 N
(D) 3 N
(E) 6 N

Solution

The initial system is in static equilibrium with $T = 3 \text{ N}$, $2T = 6 \text{ N}$. The additional weight is then added. Applying $\Sigma F = ma$ successively to the 6- and 3-N weights, we obtain

$$6 - T_1 = \frac{6}{g} a \tag{1}$$

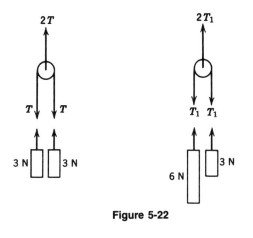

Figure 5-22

$$T_1 - 3 = \frac{3}{g} a \qquad (2)$$

The increase in force on the peg is represented by the quantity $2T_1 - 2T$, so $2T_1$ must be found. Doubling Eq. (2) and comparing it with Eq. (1),

$$6 - T_1 = 2(T_1 - 3)$$
$$3T_1 = 12 \quad \text{and} \quad 2T_1 = 8$$

Hence the increase in force on the peg $= 8 - 6 = 2\text{N}$.

Answer is (C)

PROBLEM 5-18

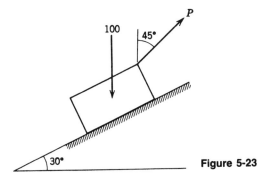

Figure 5-23

This system is initially at rest. The force P that will be required to give the 100-lb block a velocity V of 10 ft/sec up the plane in a time interval of 5 sec is most nearly

(A) 360 lb
(B) 310 lb
(C) 116 lb
(D) 58 lb
(E) 45 lb

Solution

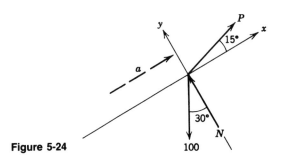

Figure 5-24

Since no coefficient of friction is mentioned, let us assume the plane to be smooth. Considering the block to be a particle, we choose the coordinate system to be as shown and write the equation of motion

$$\Sigma F_x = \frac{W}{g} a_x$$

$$P \cos 15° - 100 \sin 30° = \frac{100}{32.2} a_x$$

To determine a_x, $V = V_0 + at$ with $V_0 = 0$. At $t = 5$ sec,

$$10 \text{ ft/sec} = a(5 \text{ sec})$$

$$a = 2 \text{ ft/sec}^2$$

Hence

$$P = \left[\frac{100}{32.2} (2) + 100(0.500) \right] \Big/ 0.966$$

from the earlier equation, or $P = 58.2$ lb.

Answer is (D)

PROBLEM 5-19

A locomotive weighing 120 tons is coupled to, and pulls, a car weighing 40 tons. The resistances to motion on a level track are $\frac{1}{100}$ th of its weight for the locomotive and $\frac{1}{160}$ th of its weight for the car. The tractive force exerted by the locomotive is 8000 lb. The tension T in the coupling is nearest to

(A) 1775 lb
(B) 2225 lb
(C) 4500 lb
(D) 5100 lb
(E) 5775 lb

Solution

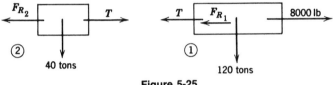

Figure 5-25

The diagram on the right represents the locomotive; the car is to the left. The forces resisting motion are

$$F_{R_1} = \frac{W_1}{100} = \frac{120(2000)}{100} = 2400 \text{ lb}$$

$$F_{R_2} = \frac{W_2}{160} = \frac{40(2000)}{160} = 500 \text{ lb}$$

Successively applying $\Sigma F_x = ma_x$ to the two bodies,

$$8000 - 2400 - T = \frac{120(2000)}{g} a \tag{1}$$

$$T - 500 = \frac{40(2000)}{g} a \tag{2}$$

We wish to solve for the tension T. Since the common factor $2000a/g$ appears on the right side of Eqs. (1) and (2), we can immediately write

$$\frac{5600 - T}{120} = \frac{T - 500}{40}$$

$$T = 1775 \text{ lb}$$

Answer is (A)

PROBLEM 5-20

The coefficient of friction between the 50-lb weight and the 300-lb weight is 0.5. The 300-lb weight is free to roll, and the weight and friction for the

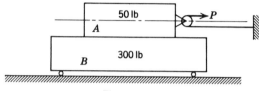

Figure 5-26

pulley are negligible. When $P = 16$ lb, the acceleration of block B is closest to

(A) 1.5 ft/sec^2
(B) 2.7 ft/sec^2
(C) 2.9 ft/sec^2
(D) 3.4 ft/sec^2
(E) 5.4 ft/sec^2

Solution

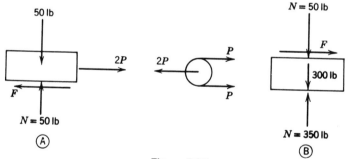

Figure 5-27

Let us first assume that slipping occurs between the blocks. Then the frictional force developed is

$$F = \mu N = 0.5(50) = 25 \text{ lb}$$

Now apply $\Sigma F_x = ma$ to block B,

$$25 = \frac{300}{g} a_B$$

$$a_B = \frac{25}{300} g = 2.68 \text{ ft/sec}^2 \text{ to the right}$$

To check our original assumption, assume the blocks move together. For the two blocks, $\Sigma F_x = ma$

$$32 = 2P = \frac{350}{g} a \qquad a = 2.94 \text{ ft/sec}^2$$

For block B alone, this acceleration requires a frictional force

$$F = \frac{300}{g} (2.94) = 27.4 \text{ lb}$$

However, since this force is greater than the maximum frictional force that can be developed (25 lb), the first assumption was correct. Hence

$$a_B = 2.68 \text{ ft/sec}^2 \text{ to the right}$$

Answer is (B)

PROBLEM 5-21

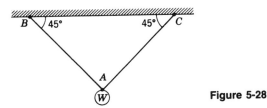

Figure 5-28

A weight W is suspended by two strings, AB and AC, as shown. The ratio of the forces in AC (1) just before, to (2) just after the instant that AB is cut is most nearly

(A) 2.0
(B) 1.6
(C) 1.4
(D) 1.0
(E) 0.5

Solution

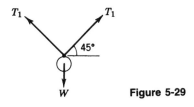

Figure 5-29

Before string AB is cut, the tensile forces in strings AB and AC are equal because of symmetry. Summing forces vertically,

$$2T_1 \sin 45° = W \qquad T_1 = \frac{W}{\sqrt{2}}$$

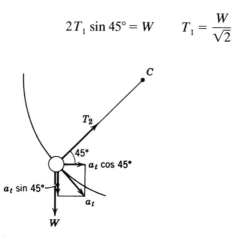

Figure 5-30

At the initial instant after string AB is cut, the body possesses only a tangential acceleration; the radial acceleration becomes nonzero only when the body is moving. Writing the equations of motion,

$$\sum F_y = ma_y \qquad W - T_2 \sin 45° = ma_t \sin 45°$$

$$\sum F_x = ma_x \qquad T_2 \cos 45° = ma_t \cos 45°$$

$$T_2 = ma_t$$

Inserting this relation to eliminate the acceleration term in the first equation, we obtain

$$W - T_2 \sin 45° = T_2 \sin 45°$$

$$W = 2T_2 \sin 45°$$

$$T_2 = \frac{W}{\sqrt{2}}$$

Hence $T_1 = T_2$ and the ratio $T_1/T_2 = 1$.

Answer is (D)

PROBLEM 5-22

A 0.25-kg ball is thrown horizontally with a kinetic energy of 50 joules from a vertical cliff 20 m high.

Question 1. The ball's kinetic energy when it strikes the ground, level with the base of the cliff, is most nearly

(A)   50 J
(B)   55 J
(C)   100 J
(D)   250 J
(E) 5000 J

Solution

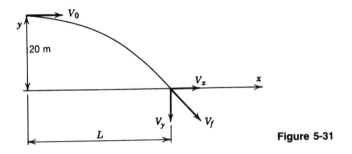

Figure 5-31

The mechanical energy, which is the sum of kinetic and potential energy, of the ball remains constant while the ball falls. If the base of the cliff is chosen as a datum, then relative to this datum the ball will possess only kinetic energy when it hits the ground. Hence

Ground kinetic energy = initial kinetic energy + change in potential energy

$$= 50 + Wh$$
$$= 50 + mgh$$
$$= 50 + (0.25)(9.81)(20) = 99 \text{ J}$$

Answer is (C)

Question 2. The distance from the foot of the cliff to the point where the ball strikes the ground is most nearly

(A) 20 m
(B) 25 m
(C) 30 m
(D) 35 m
(E) 40 m

Solution

Initially $KE = 50 = \frac{1}{2}mV_0^2$

$$V_0^2 = \frac{100}{m} = \frac{100}{0.25} = 400$$

$$V_0 = V_x = 20 \text{ m/sec}$$

The ball falls a vertical distance of 20 m, so

$$20 = \frac{1}{2}gt^2 \quad \text{or} \quad t = \left(\frac{40}{g}\right)^{1/2} = \left(\frac{40}{9.81}\right)^{1/2} = 2.02 \text{ sec}$$

Finally, $L = V_0 t = 20(2.02) = 40.4$ m.

Answer is (E)

PROBLEM 5-23

A static weight W produces a static deflection of 2 in. in a spring having a spring constant k. The mass of the spring is neglected.

Question 1. The maximum deflection x when the weight W is dropped on the spring from a point 6 in. above the free position of the spring is most nearly

(A) 2.0 in.
(B) 4.6 in.
(C) 5.3 in.
(D) 6.6 in.
(E) 7.3 in.

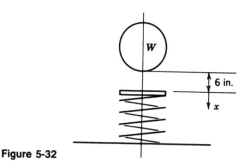

Figure 5-32

Solution

Conservation of the sum of kinetic and potential energy can be expressed here as

$$W(6 + x) = \tfrac{1}{2}mV^2 + \tfrac{1}{2}kx^2 \tag{1}$$

The spring constant can be found from the given data for the static case; then $F = kx = W$, and $k = W/x = W/2$.

The maximum deflection occurs when the weight W has compressed the spring and its velocity has become zero. Then

$$W(6 + x) = \tfrac{1}{2}kx^2 = \frac{1}{2}\left(\frac{W}{2}\right)x^2$$

or

$$x^2 - 4x - 24 = 0$$
$$x = \tfrac{1}{2}[+4 \pm \sqrt{16 + 4(24)}]$$
$$x = 2 \pm \sqrt{28} = 2 \pm 5.3$$

Only the positive root is relevant here: $x = 7.3$ in.

Answer is (E)

Question 2. The maximum velocity of the falling weight is most nearly

(A) 5.5 ft/sec
(B) 5.7 ft/sec
(C) 5.9 ft/sec
(D) 6.1 ft/sec
(E) 6.3 ft/sec

Solution

The maximum velocity occurs when $dV/dx = 0$ in the motion governed by equation (1). By direct differentiation

$$W = mV\,\frac{dV}{dx} + kx$$

Thus the maximum occurs at $x = W/k = 2$ in. $= 0.167$ ft, which also is the point at which the weight and spring are in static equilibrium. From Eq. (1),

$$W(0.5 + 0.167) = \frac{1}{2} \frac{W}{g} V^2 + \frac{1}{2} \left[\frac{W}{0.167} \right] (0.167)^2$$

$$0.667 = \frac{V^2}{2(32.2)} + \frac{0.167}{2} \qquad V = 6.13 \text{ ft/sec}$$

Answer is (D)

PROBLEM 5-24

A projectile weighing 100 lb strikes the concrete wall of a fort with an impact velocity of 1200 ft/sec. The projectile comes to rest in 0.01 sec, having penetrated the 8-ft thick wall to a distance of 6 ft. The average force exerted on the wall by the projectile is closest to

(A) 2×10^5 lb
(B) 3×10^5 lb
(C) 1×10^7 lb
(D) 2×10^8 lb
(E) 1×10^{10} lb

Solution

Knowing that the impulse exerted on the wall is equal to the change of momentum of the projectile, one may write

$$\int F \, dt = m \, \Delta V$$

If F is assumed to be the average force, then it is a constant, and the equation becomes

$$F \, \Delta t = \frac{W}{g} \, \Delta V$$

$$F(0.01) = \frac{100}{32.2} \, (1200)$$

$$F = 3.73 \times 10^5 \text{ lb}$$

Answer is (B)

PROBLEM 5-25

A 10-lb block which is suspended by a long cord is at rest when a 0.05-lb bullet traveling horizontally to the left strikes and is embedded in it. The impact causes the block to swing upward 0.5 ft measured vertically from its lowest position.

Question 1. The velocity of the bullet just before it strikes the block is most nearly

(A)   80 ft/sec
(B)   200 ft/sec
(C) 1135 ft/sec
(D) 1140 ft/sec
(E) 1600 ft/sec

Solution

Let $m =$ mass of bullet, $M =$ mass of block, $V_b =$ velocity of bullet before impact, and $V =$ system velocity immediately after impact. Energy is conserved for the motion of the system after impact so that

$$\Delta(KE) = -\Delta(PE)$$
$$\tfrac{1}{2}(M + m)V^2 = (M + m)gh$$
$$V^2 = 2gh = 2g(0.5)$$
$$V = g^{1/2} \text{ ft/sec}$$

Also momentum is conserved during impact.

$$mV_b = (M + m)V$$
$$V_b = \left(\frac{M}{m} + 1\right)V$$
$$V_b = \left(\frac{10}{0.05} + 1\right)(32.2)^{1/2} = 1140 \text{ ft/sec}$$

Answer is (D)

Question 2. The loss of kinetic energy of the system during impact is most nearly

(A)   5 ft-lb
(B)   30 ft-lb
(C) 1005 ft-lb
(D) 1010 ft-lb
(E) 1980 ft-lb

Solution

The initial kinetic energy of the bullet is

$$\frac{1}{2} mV_b^2 = \frac{1}{2} \left(\frac{0.05}{32.2} \right)(1140)^2 = 1010 \text{ ft-lb}$$

The kinetic energy immediately after impact is

$$\frac{1}{2}(M + m)V^2 = \frac{1}{2}(M + m)g$$
$$= \frac{1}{2}(10 + 0.05) = 5.025 \text{ ft-lb}$$

The system loss of kinetic energy is the difference between these two values, or

$$1010 - 5.025 = 1005 \text{ ft-lb}$$

Answer is (C)

PROBLEM 5-26

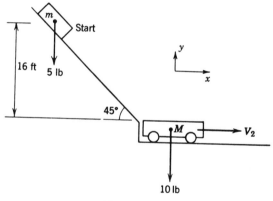

Figure 5-33

A 5-lb block of wood slides down a frictionless inclined plane at an angle of 45° with the horizontal and lands on a 10-lb cart with frictionless wheels. The slant length of the plane is $16\sqrt{2}$ ft. If the block sticks to the cart, the cart and the block will move away from the bottom of the inclined plane at a speed most nearly equal to

(A) 7.6 ft/sec
(B) 10.7 ft/sec
(C) 18.5 ft/sec
(D) 22.7 ft/sec
(E) 32.1 ft/sec

Solution

Since the plane is frictionless, the fall of the block is unimpeded. Equating the changes in kinetic and potential energy between the top and bottom of the plane,

$$\tfrac{1}{2}mV^2 = mgh$$
$$V^2 = 2gh = 2(32.2)(16)$$

Because of the 45° slope, V_x and V_y are equal, or

$$V_x^2 = V_y^2 = \tfrac{1}{2}V^2 = (32.2)(16)$$
$$V_x = 22.7 \text{ ft/sec}$$

Momentum in the x-direction is conserved when the block and cart collide:

$$mV_x = (m + M)V_2$$
$$V_2 = \frac{m}{m + M} V_x = \left(\frac{1}{1 + M/m}\right)V_x$$
$$= \left(\frac{1}{1+2}\right)(22.7)$$
$$V_2 = 7.57 \text{ ft/sec}$$

Answer is (A)

PROBLEM 5-27

A 150-lb man stands at the rear of a 250-lb boat. The distance from the man to the pier is 30 ft, and the length of the boat is 16 ft. Assume no friction between the boat and the water. The distance of the man from the pier after he walks to the front of the boat at a velocity of 3 miles/hr is most nearly

(A) 14 ft
(B) 16 ft
(C) 20 ft
(D) 24 ft
(E) 30 ft

Solution

So that momentum is conserved, $m_m v_m = m_b v_b$. Let $v_{m/b}$ be the velocity of the man relative to the boat so that $v_{m/b} = 3$ miles/hr $= 4.4$ ft/sec. Since

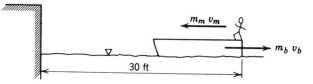

Figure 5-34

$$v_m = v_{m/b} - v_b$$

$$m_m(v_{m/b} - v_b) = m_b v_b$$

or, by rearrangement,

$$v_b = \frac{m_m}{m_m + m_b} v_{m/b} = \frac{150}{400}(4.4) = 1.65 \text{ ft/sec}$$

The man walks to the front of the boat in

$$t = \frac{s_m}{v_m} = \frac{16.0}{4.4} = 3.64 \text{ sec}$$

In this time the boat will move away from the pier a distance

$$s_b = v_b t = (1.65)(3.64) = 6.0 \text{ ft}$$

The rear of the boat is now $30 + 6 = 36$ ft from the pier, and the man is

$$36 - 16 = 20 \text{ ft from the pier}$$

Answer is (C)

PROBLEM 5-28

A pile weighing 2000 lb is driven vertically into the ground by a "monkey" (pile hammer) weighing 6000 lb which falls freely from rest through a height of 16 ft to the head of the pile. The impact of the monkey on the pile is assumed to be inelastic. The resistance of the ground can be assumed uniform and equivalent to a force of 150,000 lb. With each blow the pile moves into the ground a distance that is most nearly

(A) 0.10 ft
(B) 0.25 ft
(C) 0.50 ft
(D) 0.75 ft
(E) 1.00 ft

Solution

For conservation of momentum,

$$M_h V_1 + M_p V_{1_p} = M_h V_{2_h} + M_p V_{2_p}$$

Here $V_{1_p} = 0$ and $V_{2_h} = V_{2_p} = V_2$ for inelastic impact. The equation is now

$$M_h V_1 = (M_h + M_p)V_2$$

From the work-energy relation for a freely falling body (the hammer),

$$V_1 = (2gh)^{1/2} = [2(32.2)(16)]^{1/2} = 32.1 \text{ ft/sec}$$

Hence

$$V_2 = \frac{M_h}{M_h + M_p} V_1 = \frac{6000}{8000} (32.1) = 24.1 \text{ ft/sec}$$

Since the ground resistance F is uniform, the deceleration of the pile-hammer combination will be constant.

$$+8000 - 150,000 = F = (M_h + M_p)a = \frac{8000}{32.2} a$$

or the deceleration $a = -572 \text{ ft/sec.}^2$
    The pile displacement $x = x_0 + V_0 t + \frac{1}{2}at^2$, where $x_0 = 0$, $V_0 = V_2$, and $a = -572 \text{ ft/sec}^2$ in this case. The pile stops moving when $dx/dt = 0$.

$$0 = \frac{dx}{dt} = V_2 + at$$

$$t = -\frac{V_2}{a} = -\frac{24.1}{-572} = 0.0421 \text{ sec}$$

With each blow the pile movement or "set" is

$$x = 24.1(0.0421) - \frac{1}{2}(572)(0.0421)^2 = 0.51 \text{ ft}$$

Answer is (C)

PROBLEM 5-29

The crankshaft of an engine is turning at the rate of 20 rev/sec. The connecting rod is 6 in. long, and the radius is 2 in. When the angle θ is $30°$, the piston is moving at a rate that is most nearly

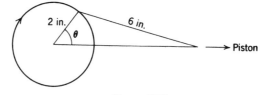

Figure 5-35

(A) 21.0 ft/sec
(B) 10.5 ft/sec
(C) 7.0 ft/sec
(D) 18.2 ft/sec
(E) 13.5 ft/sec

Solution

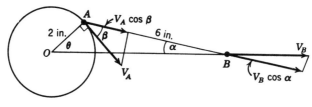

Figure 5-36

The velocity at point A is $V_A = 20(2\pi)2 = 80\pi$ in./sec. The velocity of the piston is V_B. From the diagram $V_B \cos \alpha = V_A \cos \beta$, and β is a function of the given angles α and θ.

Using the law of sines for triangle OAB,

$$\frac{6}{\sin \theta} = \frac{2}{\sin \alpha} \qquad \sin \alpha = \tfrac{1}{3} \sin \theta$$

Also from triangle OAB,

$$\alpha + \theta + 90° + \beta = 180°$$

For $\theta = 30°$,

$$\sin \alpha = \tfrac{1}{3} \sin 30° \qquad \alpha = 9°36'$$

$$\beta = 180° - 90° - 30° - 9°36' = 50°24'$$

and

$$V_B = V_A \frac{\cos \beta}{\cos \alpha} = 80\pi \frac{\cos 50°24'}{\cos 9°36'} = 80\pi \frac{0.637}{0.986}$$

$$V_B = 162 \text{ in.}/\sec = 13.5 \text{ ft}/\sec$$

Answer is (E)

PROBLEM 5-30

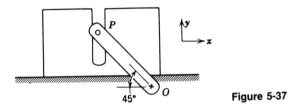

45°   O   **Figure 5-37**

A block weighs 30 N. A pin on the rotating arm OP, which moves in a smooth vertical slot, causes the block to slide on a horizontal plane, as shown in Fig. 5-37. In this 45° position, the 15-cm-long arm OP is rotating clockwise at a constant angular velocity of 5 rad/sec. The coefficient of friction between the block and the plane is 0.2. The torque exerted by the arm OP is most nearly

(A) 0.9 N-m
(B) 1.5 N-m
(C) 1.9 N-m
(D) 2.1 N-m
(E) 2.6 N-m

Solution

Consider first the rotating arm. The normal acceleration of point P is

$$a_n = r\omega^2 = (0.15)(5)^2 = 3.75 \text{ m}/\sec^2$$

For the position shown, the component in the x direction is $(a_n)_x = 3.75 \sin 45° = 2.65 \text{ m}/\sec^2$. The block slides along the plane, creating a frictional force

$$F_r = \mu N = \mu W = (0.2)(30) = 6 \text{ N}$$

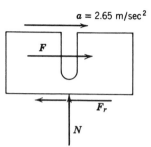

Figure 5-38

Writing the equation of motion in the x direction, $\Sigma F_x = ma_x$ or

$$F - F_r = ma$$

$$F - 6 = \frac{30}{9.81}(2.65) = 8.10$$

$$F = 14.10 \text{ N}$$

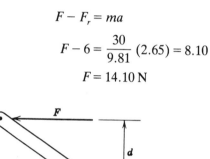

Figure 5-39

The torque on the arm T is

$$T = Fd = (14.10)(0.15 \sin 45°) = 1.50 \text{ N-m}$$

The torque exerted by the arm is equal and opposite to this torque.

$$\text{Torque by arm} = 1.50 \text{ N-m} \circlearrowleft$$

$$\text{Answer is (B)}$$

PROBLEM 5-31

A homogeneous cylinder begins to roll without slipping down a plane that is inclined at an angle θ. The moment of inertia of the cylinder about the axis through O is $I_O = \frac{1}{2}mr^2$.

Question 1. The acceleration of the center O as a function of θ is

(A) $2g \sin \theta$

(B) $\frac{1}{2}g \tan \theta$

(C) $\frac{2}{3} g \tan \theta$

(D) $\frac{2}{3} g \sin \theta$

(E) $\frac{1}{2} g \sin \theta$

Solution

The cylinder rotates about the contact point A between the plane and the cylinder in a case of noncentroidal rotation. The governing equation is

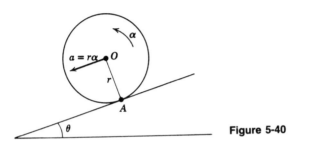

Figure 5-40

$\Sigma M_A = I_A \alpha$. The only force that can produce a nonzero moment about A is the weight W of the cylinder. This moment is equal to rH, where H is the

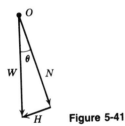

Figure 5-41

component of W parallel to the inclined plane. Note that $H = W \sin \theta = mg \sin \theta$. The moment of inertia I_A, by the parallel-axis theorem, is

$$I_A = I_O + md^2$$
$$= \tfrac{1}{2} mr^2 + mr^2$$
$$= \tfrac{3}{2} mr^2$$

We now have $rmg \sin \theta = \frac{3}{2} mr^2 \alpha$. Using $a = r\alpha$ and simplifying,

$$a = \tfrac{2}{3} g \sin \theta$$

Answer is (D)

Question 2. The minimum coefficient of friction μ to prevent the cylinder from slipping is

(A) $\frac{1}{3}$
(B) $\frac{1}{2}$
(C) $\frac{1}{2} \cos \theta$
(D) $\frac{1}{2} \tan \theta$
(E) $\frac{1}{3} \tan \theta$

Solution

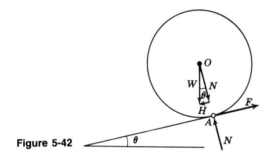

Figure 5-42

Parallel to the inclined plane,

$$\sum F = ma$$

$$H - F = ma$$

$$W \sin \theta - F = \frac{W}{g} \left(\frac{2}{3} g \sin \theta \right)$$

$$F = \frac{W}{3} \sin \theta$$

For impending slipping $F = \mu N = \mu W \cos \theta$. Hence

$$\mu = \frac{F}{W \cos \theta} = \frac{W \sin \theta}{3W \cos \theta} \qquad \mu = \frac{1}{3} \tan \theta$$

Answer is (E)

PROBLEM 5-32

A homogeneous solid cylinder of mass m and radius R has a string wound around it. One end of the string is fastened to a fixed point, and the cylinder is allowed to fall as shown.

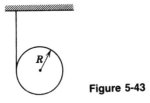

Figure 5-43

The moment of inertia of a cylinder about its axis is $\frac{1}{2}mR^2$. The tension in the string is

(A) $\frac{1}{3}mg$
(B) $\frac{1}{2}mg$
(C) $\frac{3}{5}mg$
(D) $\frac{2}{3}mg$
(E) mg

Solution

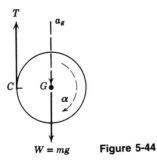

$W = mg$ **Figure 5-44**

For dynamic equilibrium vertically, $\Sigma F_y = ma_y$ or

$$mg - T = ma_g$$

Since the contact point C has no vertical acceleration, $a_g = R\alpha$ and

$$mg - T = mR\alpha \tag{1}$$

Also, about point G, $\Sigma M_G = I_G \alpha$ so that

$$TR = \frac{1}{2}mR^2\alpha \quad \text{or} \quad 2T = mR\alpha \tag{2}$$

Substituting Eq. (2) for $mR\alpha$ into Eq. (1),

$$T = \frac{1}{3}mg$$

Answer is (A)

PROBLEM 5-33

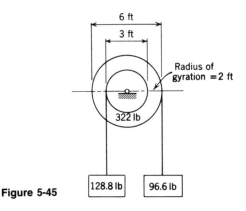

Figure 5-45

If the system shown in Fig. 5-45 is released from rest, the angular acceleration of the circular drum is most nearly

(A) $4.4 \, \text{rad}/\text{sec}^2$
(B) $1.3 \, \text{rad}/\text{sec}^2$
(C) $8.3 \, \text{rad}/\text{sec}^2$
(D) $1.7 \, \text{rad}/\text{sec}^2$
(E) $24.2 \, \text{rad}/\text{sec}^2$

Solution

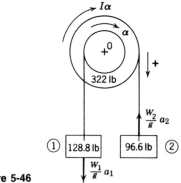

Figure 5-46

For dynamic equilibrium $\Sigma M = I\alpha$. The moment of inertia of the drum is
$I = (W/g)k^2$, where k is the radius of gyration of the drum.

$$I = \frac{322}{32.2} \, (2)^2 = 40 \, \text{lb-ft-sec}^2$$

The acceleration of each weight is $a = r\alpha$. Therefore $a_1 = 1.5\alpha$ and $a_2 = 3\alpha$. Now summing moments about 0 gives

$$3\left(W_2 - \frac{W_2}{g}a_2\right) - 1.5\left(W_1 + \frac{W_1}{g}a_1\right) = I\alpha$$

$$3\left[96.6 - \frac{96.6}{32.2}(3\alpha)\right] - 1.5\left[128.8 + \frac{128.8}{32.2}(1.5\alpha)\right] = 40\alpha$$

$$289.8 - 27\alpha - 193.2 - 9\alpha = 40\alpha$$

$$\alpha = \frac{96.6}{76} = 1.27 \text{ rad/sec}^2$$

Answer is (B)

PROBLEM 5-34

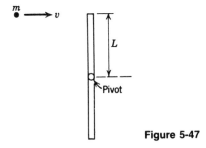

Figure 5-47

A putty ball of mass m, moving with a velocity v, as shown, makes inelastic impact with the end of a long thin bar of length $2L$ that has a moment of inertia I_0 about its center. The bar is frictionlessly pivoted at its center. The expression for the angular velocity of the bar just after the impact occurs is

(A) v/L

(B) mvL/I_0

(C) $mvL/(mL^2 + I_0)$

(D) $[mv^2/(I_0 + mL^2)]^{1/2}$

(E) $v(m/I_0)^{1/2}$

Solution

In this problem we employ the principle of conservation of angular momentum (moment of momentum). This may be expressed as $H_{p1} = H_{p2}$, where H_p is the moment of momentum at any given instant about the pivot.

Before impact, the moment of momentum of the putty ball is $H_{p1} = mvL$. After impact, the moment of momentum of the ball-bar combination is $H_{p2} = mv_2L + I_0\omega$, where $v_2 = L\omega$. Hence

$$mvL = mL^2\omega + I_0\omega$$

$$\omega = \frac{mvL}{mL^2 + I_0}$$

Answer is (C)

PROBLEM 5-35

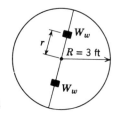

Figure 5-48

A homogeneous cylindrical wheel, of 3-ft radius, weighing 400 lb carries two symmetrically placed weights, each weighing 64.4 lb, attached 2 ft from its center. If each weight moves radially outward 1 ft when the wheel is rotating at 2000 rpm, the change, in radians per second, which will occur in the angular velocity of the wheel is most nearly

(A)   13 rad/sec
(B)   28 rad/sec
(C)   46 rad/sec
(D)   82 rad/sec
(E)   290 rad/sec

Solution

Angular momentum is conserved for the wheel-weight combination. Expressed mathematically, $I_1\omega_1 = I_2\omega_2$. The initial angular velocity ω_1 is

$$\omega_1 = (2000\text{ rpm})(\tfrac{1}{60}\text{ min/sec})(2\pi\text{ rad/rev}) = 210\text{ rad/sec}$$

The polar moment of inertia of the wheel is

$$I_0 = \tfrac{1}{2}mR^2 = \frac{1}{2}\frac{W}{g}R^2 = \frac{1}{2}\left(\frac{400}{32.2}\right)(3)^2 = 56.0\text{ lb-ft-sec}^2$$

The moment of inertia of the combination is then

$$I = I_0 + 2\frac{W_w}{g}r^2$$

For $r = 2$ ft,

$$I_1 = 56.0 + 2\left(\frac{64.4}{32.2}\right)(2)^2 = 72.0 \text{ lb-ft-sec}^2$$

For $r = 3$ ft,

$$I_2 = 56.0 + 2\left(\frac{64.4}{32.2}\right)(3)^2 = 92.0 \text{ lb-ft-sec}^2$$

For conservation of angular momentum,

$$I_1\omega_1 = (72.0)(210) = 92.0\omega_2$$
$$\omega_2 = 164 \text{ rad/sec}$$

Therefore the *decrease* in angular velocity is $210 - 164 = 46 \text{ rad/sec}$.

$$\text{Answer is (C)}$$

PROBLEM 5-36

An unbalanced flywheel has its center of mass 4.00 in. from the axis of rotation. The radius of gyration of the flywheel with respect to an axis through the center of mass parallel to the axis of rotation is 16.00 in. The flywheel, which weighs 145.0 lb, is rotating clockwise about its axis at an angular speed of 3600 rpm when a counterclockwise torque $T = 18.0t^2$ is applied, where T is in pound-feet and t is in seconds. Neglecting friction, the angular speed in revolutions per minute of the flywheel when t is 10.00 sec is closest to

(A) 1,500 rpm clockwise
(B) 2,900 rpm clockwise
(C) 3,600 rpm counterclockwise
(D) 3,100 rpm counterclockwise
(E) 16,600 rpm counterclockwise

Solution

By the parallel-axis theorem, the moment of inertia about axis O is

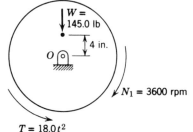

Figure 5-49 $T = 18.0t^2$

$$I_0 = \bar{I} + Md^2 = Mk^2 + Md^2 = M(k^2 + d^2)$$

where M = flywheel mass = $145/32.2 = 4.50$ slugs, k = radius of gyration = 16 in., and d = distance from O to center of mass = 4 in. Thus

$$I_0 = 4.5[(\tfrac{16}{12})^2 + (\tfrac{4}{12})^2] = 8.50 \text{ lb-ft-sec}^2$$

In this problem angular momentum is not constant, but instead the change in angular momentum is equal to the angular impulse caused by the torque T.

$$\text{Angular impulse} = \int T\, dt = \int_0^{10} 18.0t^2\, dt = 6.0t^3 \Big|_0^{10} = 6000 \text{ lb-ft-sec}$$

This is now equated to the change in angular momentum, so

$$6000 = I_0(\omega_2 - \omega_1) = 8.50(\omega_2 - \omega_1)$$

$$\omega_2 = \omega_1 + \frac{6000}{8.50} = \omega_1 + 706$$

Hence

$$N_2 = N_1 + \frac{60}{2\pi}(706)$$

$$= -3600 + 6740 = 3140 \text{ rpm counterclockwise}$$

Answer is (D)

PROBLEM 5-37

A simple spring-mass system possesses a certain natural frequency. If the mass is quadrupled in value, the ratio of the new period of oscillation to the original value is closest to which of the following?

(A) 4
(B) $\frac{1}{4}$
(C) 2
(D) $\frac{1}{2}$
(E) 16

Solution

The period T of a spring may be written as

$$T = 2\pi \left(\frac{m}{k}\right)^{1/2}$$

where m is the mass and k is the spring constant. Hence the ratio

$$\frac{\text{New } T}{\text{Old } T} = \left(\frac{4m}{m}\right)^{1/2} = 2$$

Answer is (C)

6

MECHANICS OF
MATERIALS

This field of mechanics considers the equilibrium behavior of _deformable_
solid bodies. It differs from statics in that statics is primarily concerned with
the action of external forces on *rigid* bodies, whereas mechanics of materials
is primarily concerned with the stress-deformation behavior of nonrigid solid
bodies subjected to these external force systems.

In this chapter we consider some fundamental relations that describe the
behavior of a solid body subjected to axial tension or compression, or to
twisting or torsion, or to transverse bending or flexure. These forces or force
combinations create internal stresses in the body and cause related material
displacements and deformations. A knowledge of these relations is valuable
in the analysis and design of structures and machinery.

AXIAL STRESS AND STRAIN

The axial stress σ is the axial force per unit area acting on a member. If P is
the magnitude of the force and A is the cross-sectional area, then the axial
stress acting on that area is

$$\sigma = \frac{P}{A}$$

(6-1)

When the axial force P acts on a deformable solid bar, the bar changes
length. The elongation (or shortening) per unit length is the strain ε, or

$$\varepsilon = \frac{\delta}{L}$$

(6-2)

where L is the initial length of the bar and δ is the change in that length. For elastic materials, that is, materials that completely return to their initially undeformed state when an applied stress is removed, it is found that stress is proportional to strain

$$\sigma = E\varepsilon \qquad (6\text{-}3)$$

This is Hooke's law; within limits (below the yield point of the material) many common engineering materials follow this law. E is called the modulus of elasticity.

Application of the foregoing equations to an axially loaded bar segment of differential length dx that elongates a distance $d\delta$ shows

$$\delta = \int_0^\delta d\delta = \int_0^L \frac{P}{EA}\, dx \qquad (6\text{-}4)$$

when integrated over the entire bar length L. If P, E, and A are constant for the bar, Eq. (6-4) is easily evaluated to give $\delta = PL/EA$.

When a bar is stretched axially in tension, it is found that the cross-sectional area of the bar is decreased by a small amount. A measure of the behavior is Poisson's ratio μ, sometimes called ν, which is formed by dividing the lateral strain by the axial strain. The range of Poisson's ratio is $0 < \mu < 0.5$, although 0.25–0.30 is typical of most metals.

These very basic stress-deformation relations can be used to solve many statically indeterminate problems which could not be solved by statics alone. Here we use the observation that all body members will deform when stressed and require that the deformations be *geometrically consistent*.

Example 1

A steel rod containing a turnbuckle has its ends attached to rigid walls and is tightened by the turnbuckle in summer when the temperature is 90°F to give a stress of 2000 psi. What is the stress in the rod in winter when its temperature is −20°F? (For steel, $E = 30 \times 10^6$ psi and $\alpha = 6.5 \times 10^{-6}$.)

Solution

The strain ε_t induced in an equivalent unrestrained rod by the temperature change is

$$\varepsilon_t = \alpha\, \Delta T = (6.5 \times 10^{-6})[90 - (-20)] = 7.15 \times 10^{-4}$$

This shortening, however, cannot occur, and instead the stress in the rod increases. The thermally induced additional tensile stress is

$$\sigma_t = \varepsilon_t E = (7.15 \times 10^{-4})(30 \times 10^6) = 21{,}450 \text{ psi}$$

The total stress σ in the rod is

$$\sigma = 2000 + 21{,}450 = 23{,}450 \text{ psi tension}$$

Example 2

A short column is made of a $\frac{1}{2}$-in.-thick pipe (8-in. ID) filled with concrete and capped with a rigid plate in contact with both the steel and the concrete. What load P can be carried by the column if the maximum allowable compressive stress is 15,000 psi in the steel and 900 psi in the concrete? Assume $E_s = 30 \times 10^6$ psi for steel and $E_c = 2 \times 10^6$ psi for concrete.

Solution

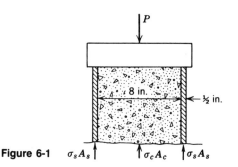

Figure 6-1 $\sigma_s A_s$ ↑   ↑$\sigma_c A_c$ ↑$\sigma_s A_s$

Figure 6-1 shows a free-body diagram of the column. Summing forces vertically,

$$P = \sigma_c A_c + \sigma_s A_s$$

In addition, each material must compress or shorten by the same amount since the rigid plate is in contact with both materials. Thus our consistent deformation requirement requires equal strains $\varepsilon_c = \varepsilon_s$. In terms of stress

$$\frac{\sigma_s}{E_s} = \frac{\sigma_c}{E_c}$$

or

$$\sigma_s = \sigma_c \left(\frac{E_s}{E_c}\right) = \sigma_c \left(\frac{30 \times 10^6}{2 \times 10^6}\right) = 15\sigma_c$$

If the maximum allowable steel stress of 15,000 psi is attained, the resulting concrete stress would be

$$\frac{15{,}000}{15} = 1000 \text{ psi}$$

which is greater than the allowable stress. Hence the concrete stress governs, and

$$P = \sigma_c A_c + 15\sigma_c A_s$$

$$P = (900) \; \frac{\pi}{4} \; (8)^2 + 15(900) \; \frac{\pi}{4} \; (9^2 - 8^2)$$

$$P = 45{,}200 + 180{,}300 = 225{,}500 \text{ lb}$$

TORSION

When a circular shaft or other body is twisted, as in cases where shafts transmit power from one point to another by rotational motion, the body is in a state of torsion; shear stresses τ and angular rotations ϕ are the result. A moment or torque T causes these stresses and deformations. When a circular shaft is in a state of pure torsion (when no axial or bending stresses are present), the shear stress at any point in the shaft is

$$\tau = \frac{Tr}{J} \tag{6-5}$$

Thus the stress increases in direct proportion to the distance r from the shaft axis. For a hollow circular shaft of outer diameter d_o and inner diameter d_i, the polar moment of inertia J of the cross section is

$$J = \frac{\pi}{32} \; (d_o^4 - d_i^4) \tag{6-6}$$

For solid shafts we set $d_i = 0$.

The effect of this torque and resulting shear stress is to cause a relative rotation between different shaft sections. For a shaft section of length L the total angle of twist is

$$\phi = \frac{TL}{GJ} \tag{6-7}$$

where G is the shearing modulus of the material. In these relations the shear equivalent of Hooke's law $\tau = G\gamma$ has been used, where γ is the shear strain.

Example 3

A solid steel shaft 8 ft long is to transmit a torque $T = 20{,}000$ ft-lb. The shear modulus of the material is $G = 12 \times 10^6$ psi, and the allowable shearing stress is 10,000 psi. Compute (*a*) the required shaft diameter

(*b*) the angle of twist between the two ends of the shaft

Solution

(*a*) Since the shear stress is largest when $r = R$, the radius of the shaft, we have

$$\tau_{max} = \frac{TR}{J}$$

For a solid shaft,

$$J = \frac{\pi}{32} d_o^4 = \frac{\pi}{2} R^4$$

and

$$R^3 = \frac{2T}{\pi\tau} = \frac{2(20,000)}{\pi(10,000)(144)} = 0.00884 \text{ ft}^3$$

$$R = 0.207 \text{ ft} = 2.48 \text{ in.}$$

The required shaft diameter is $2R = 4.96$ in. or, practically, 5 in.

(*b*) The twist is

$$\phi = \frac{TL}{GJ}$$

$$\phi = \frac{TL}{G\left(\frac{\pi}{2} R^4\right)}$$

$$\phi = \frac{(20,000)(8)}{(12 \times 10^6)(144)\frac{\pi}{2}(0.207)^4}$$

$$\phi = 0.0321 \text{ rad} = 0.0321\left(\frac{180}{\pi}\right) = 1.84°$$

BEAM EQUILIBRIUM

The principles of statics are adequate to analyze the external equilibrium of any statically determinate beam for any combination of concentrated or distributed loads. Here, however, we wish to determine the shear force V and bending moment M which exist at a point *in* the beam so that the beam is internally in equilibrium. With the aid of a systematic sign convention, the same statics principles may still be directly used. This information is commonly displayed in diagrams which give V and M for every cross section of the beam. This information can then be used to determine stresses and deformations.

The shear force V, bending moment M, and distributed beam loading w (directed downward) are related by the expressions

$$\frac{dV}{dx} = -w \qquad \frac{dM}{dx} = V \qquad (6\text{-}8)$$

where w, V, and M are all functions of the distance along the beam x. Here we have adopted the sign convention shown in Fig. 6-2 for positive shear and moment.

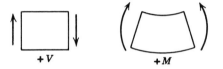

Figure 6-2 Sign convention for positive shear and moment.

The shear diagram can be constructed a bit more easily when one proceeds from left to right along the beam, for then the loads on the beam act in the same direction as the change in the shear ordinate. From Eqs. (6-8), the slope in the shear diagram is equal to minus the loading intensity at that point. Wherever concentrated loads are applied, the shear ordinate abruptly changes by that amount. Also from Eqs. (6-8), a positive shear gives a positive slope to the moment diagram, and a zero shear occurs where the moment M is a maximum or minimum, since the ordinate of the shear diagram equals the slope of the moment diagram. Finally, if Eqs. (6-8) are integrated, we find that (a) minus the area under the load curve between two points equals the change in shear between these points, and (b) the area under the shear curve between two points equals the change in the bending moment between these same two points.

Example 4

The loading diagram for a beam is shown in Fig. 6-3. The beam is 6 in. × 12 in. and is placed on edge with respect to the loads. Determine the reactions R_1 and R_2 and construct the shear and bending moment diagrams.

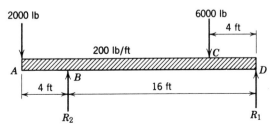

Figure 6-3

Solution

To find R_2 we sum moments around point D.

$$\sum M_D = 0 = 2000(20) + 200(20)(10) + 6000(4) - 16R_2$$
$$R_2 = \tfrac{1}{16}(40{,}000 + 40{,}000 + 24{,}000)$$
$$R_2 = 6500 \text{ lb}$$

Summing forces vertically,

$$\sum F_v = 0$$
$$R_1 = 2000 + 6000 + 200(20) - 6500$$
$$R_1 = 5500 \text{ lb}$$

Diagrams for the shear force V and bending moment M are given in Fig. 6-4.

The shear curve drops continuously down to the right at a slope of 200 lb/ft, the magnitude of the distributed loading. At points A, B, and C the shear ordinate jumps by the amount, and in the direction, of the

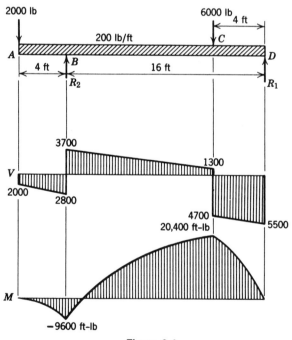

Figure 6-4

concentrated loads at those points. At D the 5500-lb reaction returns the shear ordinate to zero.

The moment diagram M can be constructed directly from the shear diagram since no concentrated couples act on the beam. The change in the moment ordinate equals the area under the shear curve. From A to B we find

$$M_B = -\tfrac{1}{2}(4)(2000 + 2800) = -9600 \text{ ft-lb}$$

since $M_A = 0$. From B to C the shear is positive, giving

$$M_C = M_B + \tfrac{1}{2}(12)(3700 + 1300)$$
$$= -9600 + 30,000 = 20,400 \text{ ft-lb}$$

From C to D the shear is negative, which causes the moment to decrease to zero at D. The slope of the moment diagram is steepest where the ordinate of the shear diagram is largest. Finally, we note that the maximum positive moment is at C where the shear curve passes through zero. [As an alternative to this procedure, we could mathematically integrate Eqs. (6-8) and plot them.]

BEAM STRESSES AND DEFLECTIONS

The existence of a shearing force or bending moment at an internal beam cross section creates axial and shearing stresses at the section.

The bending stress σ at a point is given by the flexure formula

$$\sigma = \frac{My}{I} \tag{6-9}$$

where M is the bending moment acting on a beam cross section, I is the moment of inertia around the neutral axis of the section, and y is the distance from the neutral axis to the point. The largest bending stress thus occurs in the beam fibers most distant from the neutral axis.

A shear force V acting on a section causes an internal longitudinal (horizontal) shear stress distribution to be set up across the section. The magnitude of this shear stress τ at some distance y from the neutral axis is

$$\tau = \frac{VQ}{Ib} \tag{6-10}$$

Here I is again the section moment of inertia and b is the section width at height y. The quantity Q is the first moment of the area of a portion of the beam cross section. The area is that portion of the section which lies outside the distance y from the neutral axis. If c denotes the extreme fiber in the section, then Q is

$$Q = \int_y^c y \, dA \tag{6-11}$$

This equation shows immediately that the extreme fibers in a beam carry zero shear stress since $Q = 0$ when $y = c$.

Example 5

For the beam in Example 4 determine

(a) the location and magnitude of the maximum bending stress
(b) the location and magnitude of the maximum shear stress

Solution

(a) The maximum bending moment occurs at point C and is $M_C = 20,400$ ft-lb. The moment of inertia of the 6 in. × 12 in. section is $I = bh^3/12 = (6)(12)^3/12 = 864$ in.^{4}. The maximum bending stress occurs farthest from the neutral axis where $y = h/2 = 6$ in. Equation (6-9) now gives the maximum stress as

$$\sigma = \frac{My}{I} = \frac{(20,400)(12)(6)}{864} = 1700 \text{ psi}$$

(b) Since cross-sectional properties are constant throughout the length of the beam, the maximum shear stress will occur at some point in the section where the shear force V is a maximum. From the shear diagram in Example 4 the shear is a maximum at the right support where $V = 5500$ lb. (Notice here that the maximum shear ordinate does not occur at the point of application of the largest concentrated force.) The maximum stress occurs at the point in this section where Q is maximized. First we evaluate Eq. (6-11) for any point y on the rectangular beam section of height h and width b:

$$Q = \int_y^{h/2} y(b \, dy) = \frac{b}{2}\left(\frac{h^2}{4} - y^2\right)$$

Thus Q, and consequently τ, is a maximum at the neutral axis $y = 0$.

$$\tau = \frac{VQ}{Ib} = \frac{V(bh^2/8)}{(bh^3/12)b} = \frac{3V}{2bh} = \frac{3V}{2A}$$

is the maximum shear stress in a rectangular beam. Hence

$$\tau = \frac{3(5500)}{2(6)(12)} = 114.6 \text{ psi}$$

For small deflections of beams the basic relation between the beam deflection y, curvature $1/\rho$, and moment M at a point is

$$\frac{d^2y}{dx^2} = \frac{1}{\rho} = \frac{M}{EI} \tag{6-12}$$

Here we assume that y is positive upward when M is positive according to the earlier sign convention for moment diagrams. If M is written as a function of x, Eq. (6-12) may be integrated twice. The constants of integration are evaluated by noting the slope and deflection constraints for the particular beam.

The moment-area method is a rapid, efficient alternative method of interpreting and using the basic deflection equation. Integrated once between points A and B on the beam, Eq. (6-12) gives

$$\theta_B - \theta_A = \int_A^B \frac{M}{EI}\, dx \tag{6-13}$$

which shows that the change in slope between A and B equals the area of the M/EI diagram between A and B. It can also be shown that the deflection δ of point A from the tangent to point B is

$$\delta = \int_A^B \frac{M}{EI} x\, dx \tag{6-14}$$

The integral is the first moment of the M/EI diagram about point A. In applying the moment-area principles, one should treat positive and negative sections of the moment diagram separately.

Since Eq. (6-12) is a linear equation, the principle of linear superposition is valid for finding slopes and deflections for a beam which is simultaneously acted upon by several loads. This principle is therefore a useful tool in the solution of statically indeterminate beam problems.

Example 6

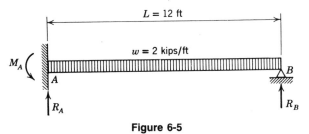

Figure 6-5

A propped cantilever beam 12 ft long carries a uniformly distributed load of 2000 lb/ft (2 kips/ft), as shown in Fig. 6-5. Calculate the reactions.

Solution

The beam is statically indeterminate to the first degree. We consider the problem as the sum of two statically determinate problems and require the net deflection of the propped end (point B) to be zero. Shown beneath each statically determinate cantilever in Fig. 6-6 is its associated bending moment

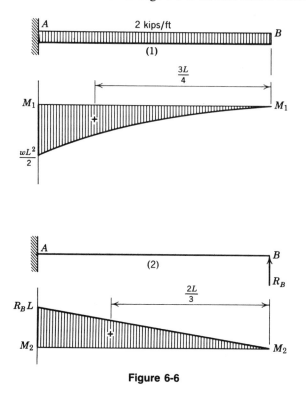

Figure 6-6

diagram; also shown is the distance to the centroid of the moment diagram from the free end of the beam. For this problem we note that EI is constant throughout the beam. According to Eq. (6-14), the deflection at B for case 1 is

$$\delta_{B_1} = \frac{1}{EI} \text{ (area of } M_1 \text{ diagram) } \bar{x}_1$$

$$\delta_{B_1} = \frac{1}{EI} \left(\frac{-wL^3}{6} \right) \frac{3L}{4} = \frac{-wL^4}{8EI}$$

For case 2,

$$\delta_{B_2} = \frac{1}{EI} \left(\frac{+R_B L^2}{2} \right) \frac{2L}{3} = \frac{+R_B L^3}{3EI}$$

Since point B cannot deflect,

$$\delta_B = \delta_{B_1} + \delta_{B_2} = 0$$

$$\frac{R_B L^3}{3EI} - \frac{wL^4}{8EI} = 0$$

$$R_B = \frac{+3}{8} wL = \frac{+3}{8} (2)(12) = +9 \text{ kips}$$

Now, by using statics, $\Sigma F_v = 0$.

$$R_A + R_B = wL$$

$$R_A = \tfrac{5}{8} wL = \tfrac{5}{8}(2)(12) = +15 \text{ kips}$$

At point A, $\Sigma M_A = 0$.

$$M_A + R_B L - \frac{wL^2}{2} = 0$$

$$M_A = \frac{wL^2}{2} - \frac{3}{8} wL^2$$

$$M_A = \frac{wL^2}{8} = \frac{2(12)^2}{8} = +36 \text{ kip-ft}$$

The plus sign indicates that the moment acts in the indicated direction. The shear and moment diagrams for the propped cantilever beam can now be constructed directly, if desired.

EULER COLUMN BUCKLING

Columns are slender members which carry primarily axial loads. The critical load P_{cr} which will cause lateral buckling, an instability phenomenon, in long slender columns is given by the Euler column formula, which may be expressed as

$$P_{cr} = \frac{\pi^2 EI}{L_1^2} \tag{6-15}$$

where L_1 is related to the length of the column. The moment of inertia may be written $I = Ar^2$, r being the radius of gyration. Euler's formula is applicable only when the modified slenderness ratio L_1/r is greater than 100, approximately. Shorter columns do not fail by buckling, which is described by the Euler equation. The relation between L_1 and the column length L depends on the end conditions of the column. For both ends fixed $L_1/L = \tfrac{1}{2}$;

for one end fixed and one end pinned $L_1/L = 0.7$; for both ends pinned $L_1/L = 1$; and for one end fixed and one end free $L_1/L = 2$.

Example 7

A $\frac{3}{4}$-in.-diameter solid steel rod ($E = 30 \times 10^6$ psi) is 5 ft long and is pin-connected at its ends. How large a compressive load can be applied before buckling occurs?

Solution

The moment of inertia for a circular rod of diameter d is

$$I = \frac{\pi}{64} d^4 = Ar^2$$

The radius of gyration is

$$r = \left[\frac{(\pi/64)d^4}{(\pi/4)d^2} \right]^{1/2} = \frac{d}{4} = \frac{3}{16} \text{ in.}$$

For pinned ends $L_1 = L$ so

$$\frac{L_1}{r} = \frac{5(12)}{\frac{3}{16}} = 320 > 100$$

and the Euler formula applies. The critical load is

$$P_{cr} = \frac{\pi^2 EI}{L_1^2}$$

$$P_{cr} = \frac{\pi^2 E}{L^2} \left(\frac{\pi}{64} d^4 \right)$$

$$P_{cr} = \frac{\pi^3 (30 \times 10^6)(0.75)^4}{64[5(12)]^2}$$

$$P_{cr} = 1277 \text{ lb}$$

PROBLEM 6-1

The rod in Fig. 6-7a of a material having the stress-strain curve shown in Fig. 6-7b has a spring attached at one end. The spring constant of this spring is 20,000 lb/in. The rod has a cross-sectional area of 1.00 in.^{2} and is 20.00 in. long. The load F is increased until the spring has elongated 0.75 in. and then decreased to zero. The length of the rod, in inches, after the load is removed is closest to

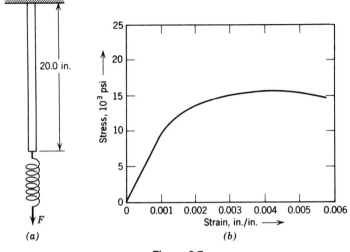

Figure 6-7

(A) 20.000
(B) 20.003
(C) 20.030
(D) 20.060
(E) rod will break

Solution

The spring tells us the magnitude of the applied load. The load F is increased to $0.75 \times 20,000 = 15,000$ lb, then removed. For a rod of 1.00 in.^{2} the applied stress is

$$\sigma = \frac{P}{A} = \frac{15,000}{1.00} = 15,000 \text{ psi}$$

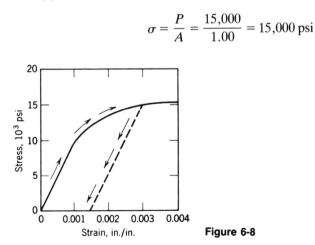

Figure 6-8

Up to 10,000 psi on the stress-strain diagram there is a linear relationship between stress and strain (Hooke's law applies). For stresses above the elastic limit (10,000 psi) there is a permanent deformation. As the load is increased, the stress-strain relation is as represented by the curve. Beyond the elastic limit the stress-strain curve is not retraced as the load is removed but instead decreases with a slope equal to that below the elastic limit. The result is a permanent elongation of the rod.

$$\Delta L = \varepsilon L = 0.0015 \times 20.00 = 0.03 \text{ in.}$$

Thus the new rod length will be

$$20.00 + 0.03 = 20.03 \text{ in.}$$

$$\text{Answer is (C)}$$

PROBLEM 6-2

The rails of a tramway are welded together at $+50°F$. Assume the coefficient of linear expansion of rails is 70×10^{-7} in./in.-°F and the modulus of elasticity is 30×10^6 lb/in². The stress, in psi, produced in the rails when heated by the sun to $+100°F$ is nearest to

(A)   0
(B) 10,500
(C) 32,000
(D) 35,000
(E) 52,000

Solution

If the rails were free to expand, then their increase per unit length caused by temperature would be

$$\varepsilon = \alpha \, \Delta T$$
$$= 70 \times 10^{-7} \times (100 - 50) = 350 \times 10^{-6} \text{ in./in.}$$

Then the stress developed in the rail is given by

$$\sigma = \varepsilon E = 350 \times 10^{-6} \times 30 \times 10^6 = 10{,}500 \text{ psi}$$

$$\text{Answer is (B)}$$

PROBLEM 6-3

A $\frac{1}{2}$-in. OD brass rod is 100 ft 5 in. long and a 4-in. nominal cast-iron pipe (4.80 in. OD, 0.38 in. wall thickness) is 100 ft $5\frac{1}{2}$ in. long when both are at the same temperature (60°F). Given

α brass $= 10 \times 10^{-6}$ units per unit length per degree Fahrenheit

α cast iron $= 6 \times 10^{-6}$ units per unit length per degree Fahrenheit

The rod and the pipe are the same length at a temperature, in °F, closest to

(A) -40
(B)   0
(C)   80
(D) 120
(E) 160

Solution

Since brass has a greater coefficient of expansion than cast iron and the brass piece is presently shorter than the cast-iron one, they will be the same length at some elevated temperature.

$$\Delta T(\alpha_{\text{brass}})(\text{length}_{\text{brass}}) = 0.5 \text{ in.} + \Delta T(\alpha_{c-i})(\text{length}_{c-i})$$

$$\Delta T = \frac{0.5 \text{ in.}}{(\alpha_{\text{brass}})(\text{length}_{\text{brass}}) - (\alpha_{c-i})(\text{length}_{c-i})}$$

$$\Delta T = \frac{0.5 \text{ in.}}{(10 \times 10^{-6})(1205 \text{ in.}) - (6 \times 10^{-6})(1205.5 \text{ in.})}$$

$$\Delta T = \frac{0.5}{0.01205 - 0.00723} = \frac{0.5}{0.00482} = 103.7°\text{F}$$

Therefore the temperature at which both pieces are the same length is $60 + 103.7 = 163.7°\text{F}$.

Answer is (E)

PROBLEM 6-4

A $\frac{1}{2}$-in.-diameter steel tie rod, 18 ft in length, is joined to two rigid walls in such a way that an axial tensile stress of 20,000 psi is induced in the rod. Assume a Poisson's ratio of 0.25 and $E = 30 \times 10^6$ psi.

The change in the diameter of the rod, in inches, caused by the application of this tensile load is closest to

(A) 6.66×10^{-4}
(B) 1.66×10^{-4}
(C) 8.33×10^{-5}
(D) 6.50×10^{-6}
(E) No change in the diameter

Solution

$$\text{Unit strain } \varepsilon = \frac{\sigma}{E} = \frac{20,000}{30 \times 10^6} = \frac{2}{3} \times 10^{-3}$$

Unit change in diameter $= \mu\varepsilon = \frac{1}{4}\varepsilon = \frac{1}{6} \times 10^{-3}$

Total change in diameter $= d\mu\varepsilon = \frac{1}{2} \times \frac{1}{6} \times 10^{-3} = 8.33 \times 10^{-5}$ in.

Answer is (C)

PROBLEM 6-5

The stress in an elastic material is

(A) inversely proportional to the material's yield strength
(B) inversely proportional to the force acting
(C) proportional to the displacement of the material acted on by the force
(D) inversely proportional to the strain
(E) proportional to the length of the material subject to the force

Solution

According to Hooke's law, stress is directly proportional to strain. Strain is the deformation (or displacement) of the material per unit length. Thus we can say that stress is proportional to the displacement of the material acted on by the force.

Answer is (C)

PROBLEM 6-6

What load, in lb, must be applied to a 1-in. round steel bar 8 ft long $(E = 30 \times 10^6 \text{ psi})$ to stretch the bar 0.05 in.?

(A) 7,200
(B) 9,850
(C) 8,600
(D) 12,250
(E) 15,000

Solution

$$\delta = \frac{PL}{AE} \qquad P = \frac{AE\delta}{L} = \frac{\pi}{4}(1)^2 \frac{30 \times 10^6(0.05)}{8 \times 12} = 12{,}250 \text{ lb}$$

Answer is (D)

PROBLEM 6-7

A steel bar having a 1-in.^{2} cross section is 150 in. long when lying on a horizontal surface. Assume $E = 30 \times 10^6$ psi and $W = 0.283$ lb/in^3. The increase in the length of the bar, in inches, when it is suspended vertically from one end is nearest to

(A) 1×10^{-5}
(B) 1×10^{-4}
(C) 1×10^{-3}
(D) 1×10^{-2}
(E) 1×10^{-1}

Solution

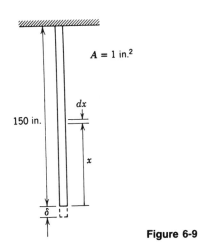

Figure 6-9

$$d\delta = \frac{P(x)\,dx}{AE} \quad \text{where } P(x) = WAx$$

$$\int_0^\delta d\delta = \int_0^L \frac{WAx\,dx}{AE}$$

$$\int_0^\delta d\delta = \frac{W}{E}\int_0^L x\,dx$$

$$\delta = \frac{W}{E}\frac{L^2}{2}$$

$$\delta = \frac{0.283\ \text{lb/in.}^3}{30\times 10^6\ \text{lb/in.}^2}\times\frac{150^2\ \text{in.}^2}{2}$$

$$\delta = 1.06\times 10^{-4}\ \text{in.}$$

Answer is (B)

PROBLEM 6-8

A steel test specimen is $\frac{5}{8}$ in. in diameter at the root of the thread. It is to be stressed to 50,000 psi tension. What load, in lb, must be applied?

(A) 12,790

(B) 15,340

(C) 16,320

(D) 25,600

(E) 31,250

Solution

Stress $= P/A$. Here $A = (\pi/4)(\frac{5}{8})^2$ and stress $= 50,000$ psi.

$$P = 50,000\ \frac{\pi}{4}\left(\frac{5}{8}\right)^2 = 15,340\ \text{lb}$$

Answer is (B)

PROBLEM 6-9

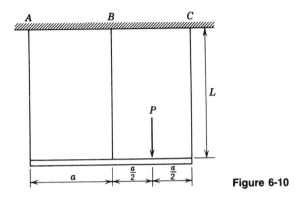

Figure 6-10

Given a rigid bar hanging from three wires of length L, modulus of elasticity E, cross-sectional area A, spaced a distance a apart as shown in the figure. Neglect the weight of the bar. The force exerted by wire A is closest to

(A) $P/3$
(B) $P/4$
(C) $P/5$
(D) $P/8$
(E) $P/12$

Solution

When load P is applied, we know that $P = P_A + P_B + P_C$, and since A, B, and C are wires, none are in compression. Further, we know that each wire will elongate; hence

$$\text{length } A = L + \Delta_A$$
$$\text{length } B = L + \Delta_B$$
$$\text{length } C = L + \Delta_C$$

As the bar is rigid, B will have a deflection intermediate to that of A and C, or

$$\Delta_B = \frac{\Delta_A + \Delta_C}{2}$$

But since all wires are identical, load is proportional to deflection, so

$$P_B = \frac{P_A + P_C}{2}$$

Taking moments about wire A, $\Sigma M_A = 0$.

$$aP_B - 1.5aP + 2aP_C = 0$$

Dividing by a,

$$P_B - 1.5P + 2P_C = 0$$

Thus we have three equations in three unknowns:

$$P = P_A + P_B + P_C \qquad (1)$$

$$0 = P_A - 2P_B + P_C \qquad (2)$$

$$1.5P = +P_B + 2P_C \qquad (3)$$

Solving simultaneously,

(1) $-P = -P_A - P_B - P_C$
(2) $0 = P_A - 2P_B + P_C$
$\overline{\ -P = \ -3P_B}$ $\qquad P_B = \dfrac{P}{3}$

(3) $1.5P = + \dfrac{P}{3} + 2P_C$ $\qquad P_C = \dfrac{1.5P - P/3}{2} = \dfrac{7P}{12}$

(1) $P = P_A + \dfrac{P}{3} + \dfrac{7P}{12}$ $\qquad P_A = P - \dfrac{4P + 7P}{12} = \dfrac{P}{12}$

Answer is (E)

PROBLEM 6-10

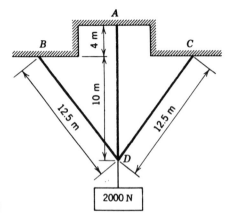

Figure 6-11

Three rods, each 2 cm in diameter, labeled AD, BD, and CD, support the 2000-N load as shown. The load in rod A, in N, is nearest to

(A)   580
(B)   670
(C)   730
(D)   820
(E)   2000

Solution

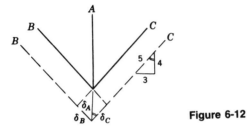

Figure 6-12

By geometry, $\delta_B = \delta_C = \frac{4}{5}\delta_A$. By symmetry, $P_B = P_C$. Using $\delta = PL/AE$,

$$\frac{4}{5}\frac{P_A(14)}{AE} = \frac{P_B(12.5)}{AE} \qquad P_A = 1.12P_B = 1.12P_C$$

$$\sum F_v = 0 \qquad \tfrac{4}{5}P_B + P_A + \tfrac{4}{5}P_C = 2000$$

$$\tfrac{4}{5}P_B + 1.12P_B + \tfrac{4}{5}P_B = 2000$$

$$P_B = P_C = 735 \text{ N (tension)}$$

$$P_A = 823 \text{ N (tension)}$$

Answer is (D)

PROBLEM 6-11

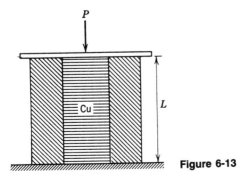

Figure 6-13

Two concentric cylinders of length L are loaded between rigid smooth plates as shown. The inner cylinder is copper and the outer cylinder is steel.
Let

$$E_s = \text{modulus of elasticity of steel}$$
$$E_c = \text{modulus of elasticity of copper}$$
$$A_s = \text{cross-sectional area of steel cylinder}$$
$$A_c = \text{cross-sectional area of copper cylinder}$$
$$P = \text{total applied load } P = P_c + P_s$$

Which one of the following represents the formula for stress in the copper cylinder?

(A) $\sigma_c = \dfrac{P}{A_c}$

(B) $\sigma_c = \dfrac{E_c P}{A_c E_s + A_s E_c}$

(C) $\sigma_c = \dfrac{P_s}{A_c}$

(D) $\sigma_c = \dfrac{E_c P}{A_c E_c + A_s E_s}$

(E) None of the above is correct

Solution

$$\text{Stress} = \frac{\text{force}}{\text{area}} = \frac{P}{A} \qquad \text{Deflection} = \text{length} \times \text{strain} = L\varepsilon = \delta$$

$$\text{Modulus of elasticity} = \frac{\text{stress}}{\text{strain}} = \frac{\sigma}{\varepsilon} \qquad \delta = L\varepsilon = \frac{L\sigma}{E} = \frac{LP}{AE}$$

But $\delta_s = \delta_c$. Therefore

$$\frac{LP_s}{A_s E_s} = \frac{LP_c}{A_c E_c} \qquad P_s = \frac{P_c A_s E_s}{A_c E_c} \qquad P = P_s + P_c$$

$$P = \frac{P_c A_s E_s}{A_c E_c} + P_c \qquad P = P_c\left(1 + \frac{A_s E_s}{A_c E_c}\right)$$

Therefore

$$P_c = \frac{A_c E_c P}{A_c E_c + A_s E_s}$$

But

$$\sigma_c = \frac{P_c}{A_c} \qquad \sigma_c = \frac{E_c P}{A_c E_c + A_s E_s}$$

Answer is (D)

PROBLEM 6-12

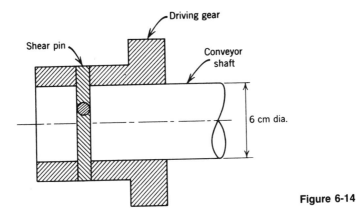

Figure 6-14

A shear pin 1 cm in diameter, as shown in the figure, is used on a screw conveyor to protect the mechanism when it jams. If the screw conveyor revolves with a torque of 520 N-m, the unit shearing stress, in Pa, on the shear pin is closest to

(A) 2×10^3
(B) 2×10^4
(C) 5×10^5
(D) 1×10^7
(E) 1×10^8

Solution

The *pascal* (Pa) is 1 newton per square meter.

Torque (newton-meters) = force (newtons) × distance (meters)

Here

$$T = \tau A \times r$$
$$520 = \tau \times 2\left(\frac{\pi}{4} \times 0.01^2\right) \times 0.03$$

$$520 = \tau \times 4.71 \times 10^{-6}$$

$$\tau = \frac{520}{4.71 \times 10^{-6}} = 110.3 \times 10^{6} \text{ Pa}$$

Answer is (E)

PROBLEM 6-13

The maximum shear stress in a solid round shaft subjected only to torsion occurs

(A) on principal planes
(B) on planes containing the axis of the shaft
(C) on the surface of the shaft
(D) only on planes perpendicular to the axis of the shaft
(E) at the neutral axis

Solution

$$\tau = \frac{Tc}{J}$$

where T = torque, c = distance from the center of the shaft, and J = polar moment of inertia. Thus the torsional shearing stress at any point is proportional to its distance from the center of the shaft. The maximum shearing stress, therefore, is at the surface of the shaft.

Answer is (C)

PROBLEM 6-14

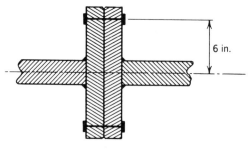

6 in.

Figure 6-15

A shaft coupling is to be designed, using 1-in. diameter bolts at a distance of 6 in. from the center of the shaft. Allowable shearing stress on the bolts is 15,000 psi. If the shaft is to transmit 5800 hp at a speed of 1200 rpm, how many bolts are needed in the connection?

(A) 2
(B) 3
(C) 4
(D) 5
(E) 6 or more

Solution

$$\text{Work/revolution} = \text{force} \times \text{distance} = F(2\pi R) = 2\pi T$$

where T = torque in pound-feet. At 1200 rpm

$$\text{Work/minute} = 1200 \times 2\pi T \text{ lb-ft/min}$$

Since 1 hp = 33,000 ft-lb/min,

$$\text{hp} = \frac{1200 \times 2\pi T}{33,000} = 5800$$

$$\text{Torque} = \frac{33,000 \times 5800}{1200 \times 2\pi} = 25,400 \text{ lb-ft}$$

Assuming that the shearing stress is uniform over the bolt cross section, we can determine the torsional resistance per bolt.

$$\text{Torque} = \text{force} \times \text{lever arm} = \tau A \times \text{lever arm}$$

$$= 15,000 \text{ psi} \times \frac{\pi}{4} 1^2 \text{ in. } 2 \times 0.5 \text{ ft}$$

$$= 5890 \text{ lb-ft}$$

$$\text{No. of bolts required} = \frac{25,400}{5890} = 4.31$$

Five bolts are required.

Answer is (D)

PROBLEM 6-15

In an I-beam subjected to simple bending, the maximum bending stress occurs

(A) at the neutral axis
(B) in the web above the neutral axis
(C) in the web below the neutral axis
(D) at the top and bottom surfaces of the beam
(E) where the web joins the lower flange

Solution

In simple bending, the stress at each point in the cross section of a beam is directly proportional to that point's distance from the neutral axis. For a symmetrical section, like an I-beam, the maximum bending stress occurs at both the top and bottom surfaces of the beam.

Answer is (D)

PROBLEM 6-16

The bending moment of a beam

(A) depends on the modulus of elasticity of the beam
(B) is minimum where the shear is zero
(C) is maximum at the free end of a cantilever
(D) is plotted as a straight line for a simple beam with a uniformly distributed load
(E) may be determined from the area of the shear diagram

Solution

The shear at any point along the beam is

$$V = \frac{dM}{dx}$$

so

$$M = \int V \, dx$$

or the bending moment is the area of the shear diagram.

Answer is (E)

PROBLEM 6-17

The moment diagram for a simple beam with a concentrated load at midspan takes the shape of a

(A) triangle
(B) semicircle
(C) semiellipse
(D) parabola
(E) rectangle

Solution

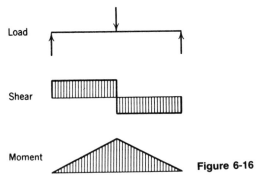

Load

Shear

Moment   **Figure 6-16**

Answer is (A)

PROBLEM 6-18

The moment diagram for a beam uniformly loaded with a concentrated load in the center is the sum of

(A) two triangles
(B) a rectangle and a triangle
(C) a parabola and a rectangle
(D) a parabola and a triangle
(E) a rectangle and a trapezoid

Solution

The moment diagram for a beam with a uniform load is a parabola, and the moment diagram for a concentrated load is a triangle. By superposition, the combined diagram is the sum of the two individual diagrams.

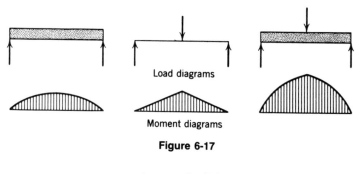

Load diagrams

Moment diagrams

Figure 6-17

Answer is (D)

PROBLEM 6-19

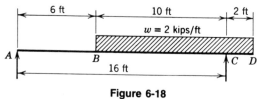

Figure 6-18

The maximum moment, in kip-ft, in the beam AD is nearest to

- (A) 4
- (B) 18
- (C) 36
- (D) 45
- (E) 66

Solution

$\Sigma M_A = 0$

$$+16R_C - (12 \times 2)(12) = 0 \qquad R_C = \frac{24 \times 12}{16} = 18 \text{ kips}$$

$\Sigma F_y = 0$

$$R_A + 18 - (12 \times 2) = 0 \qquad R_A = 6 \text{ kips}$$

The shear is zero at a point 3 ft to the right of B. The moment at that point is

$$M_0 = 6 \text{ kips} \times 6 \text{ ft} + (6 \text{ kips} \times 3 \text{ ft})/2 = 45 \text{ kip-ft}$$

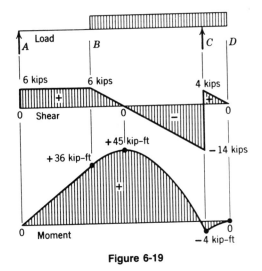

Figure 6-19

Answer is (D)

PROBLEM 6-20

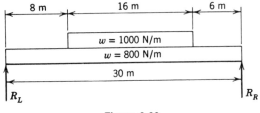

Figure 6-20

A simple beam is loaded as shown. The value of the maximum moment, in N-m, is closest to

(A) 18,000
(B) 36,000
(C) 72,000
(D) 144,000
(E) 177,000

Solution

Solve first for the left reaction R_L and the right reaction R_R. $\Sigma M_{R_L} = 0$.

$$+30R_R - 800 \times 30 \times 15 - 1000 \times 16 \times 16 = 0$$

$$+30R_R - 360,000 - 256,000 = 0 \qquad R_R = \frac{616,000}{30} = 20,500 \text{ N}$$

$\Sigma F_y = 0$

$$R_L + R_R - 800 \times 30 - 1000 \times 16 = 0$$

$$R_L + 20,500 - 24,000 - 16,000 = 0 \qquad R_L = 19,500 \text{ N}$$

Then draw the shear diagram:

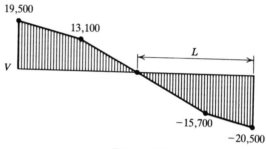

Figure 6-21

$$L = 6 + \frac{15,700}{1800} = 14.72 \text{ m}$$

The maximum moment occurs at the point where the shear is zero, or 14.72 m from the right end of the beam.

$$M_{max} = +14.72 \times 20,500 - 14.72 \times 800 \times \frac{14.72}{2} - 8.72 \times 1000 \times \frac{8.72}{2}$$

$$= 177,100 \text{ N-m}$$

Answer is (E)

PROBLEM 6-21

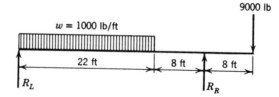

Figure 6-22

For the beam loaded as shown, the value of the maximum moment, in kip-ft, is nearest to

(A) 11
(B) 19
(C) 66
(D) 72
(E) 104

Solution

$\Sigma M_{R_L} = 0$

$$22 \times 1 \text{ kip} \times 11 + 9 \text{ kips} \times 38 - 30R_R = 0$$

$$+242 + 342 - 30R_R = 0 \qquad R_R = \frac{242 + 342}{30} = 19.47 \text{ kips}$$

$\Sigma F_y = 0$

$$R_L + 19.47 - 22 \times 1 \text{ kip} - 9 \text{ kips} = 0 \qquad R_L = 11.53 \text{ kips}$$

The moment is a maximum at the point where the shear is zero. Expressing the shear as a function of x from the left end, $V(x) = 11.53 \text{ kips} - x \text{ kips} = 0$ when $x = x_0$. $x_0 = 11.53$ ft.

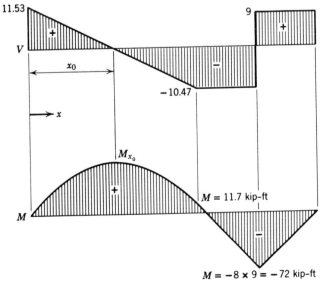

Figure 6-23

$$M(x_0) = M_{x_0} = x_0 R_L - \frac{wx_0^2}{2} = 11.53(11.53) - \frac{1(11.53)^2}{2}$$

$$= \frac{11.53^2}{2} = 66.5 \text{ kip-ft}$$

Thus the maximum moment is a negative moment of 72 kip-ft, located at the right support, 8 ft to the left of the right end of the beam.

Answer is (D)

PROBLEM 6-22

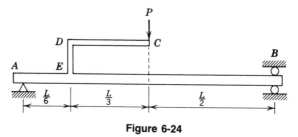

Figure 6-24

Question 1. For beam AB loaded as shown the value of the right reaction R_B is

(A) $\frac{1}{6}P$
(B) $\frac{1}{3}P$
(C) $\frac{1}{2}P$
(D) $\frac{2}{3}P$
(E) $\frac{5}{6}P$

Solution

$\Sigma M_{R_A} = 0$

$$-P\left(\frac{L}{6} + \frac{L}{3}\right) + LR_B = 0 \qquad R_B = \frac{0.5LP}{L} = 0.5P$$

Answer is (C)

Question 2. The maximum bending moment in the beam is

(A) $\frac{1}{12}PL$
(B) $\frac{1}{4}PL$
(C) $\frac{1}{3}PL$
(D) $\frac{5}{12}PL$
(E) $\frac{1}{2}PL$

Solution

Although the solution could be obtained more directly, it is instructive to draw both the shear and moment diagrams:

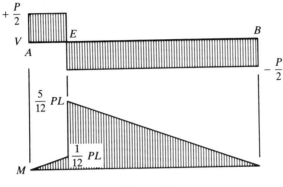

Figure 6-25

From Fig. 6-25 we see that the shear curve passes through zero at point E, a distance $L/6$ from the left support. The area under the shear curve between E and B is

$$\left(\frac{P}{2}\right)\left(\frac{5}{6}L\right) = \frac{5}{12}PL$$

Answer is (D)

PROBLEM 6-23

Which is the proper shape of the moment diagram for the cantilever beam AB loaded as shown?

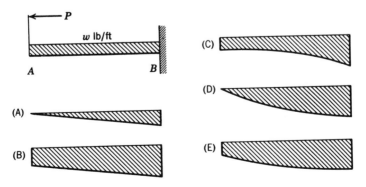

Figure 6-26

Solution

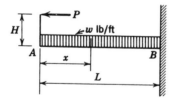

Figure 6-27

The force P contributes a constant moment $-PH$ from A to B. The uniform load w, at any point x, produces a moment $-wx^2/2$. The total moment diagram is the sum of the moment diagrams of the component loads.

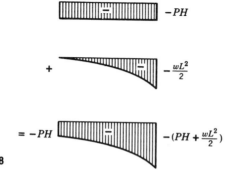

$-PH$

$+$

$-\frac{wL^2}{2}$

$= -PH$

$-(PH + \frac{wL^2}{2})$

Figure 6-28

Answer is (C)

PROBLEM 6-24

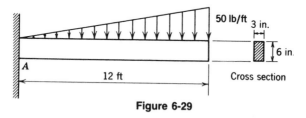

Figure 6-29

A cantilever beam 3 in. $\times$ 6 in. supports a uniformly varying load as shown. Neglect the weight of the beam. The maximum bending stress, in psi, in the beam is nearest to

(A)   135
(B)   1350
(C)   1600
(D)   2250
(E)   4500

Solution

The moment at any section is the algebraic sum of the moments of all external forces on one side of the section.

$$M_{max} = M_A = (\tfrac{1}{2} \times 12 \times 50)(\tfrac{2}{3} \times 12)$$
$$= 300 \times 8 = 2400 \text{ lb-ft}$$

Since the lower fibers of the beam are in compression, the moment is negative.

The maximum bending stress is

$$\sigma = \frac{M_{max}c}{I}$$

where σ = bending stress in pounds per square inch, c = distance from neutral axis to extreme fiber = 3 in., M_{max} = maximum moment in inch-pounds = 2400×12, and I = moment of inertia ($= bh^3/12$ for rectangular beams).

$$\sigma = \frac{2400 \times 12 \times 3}{(3 \times 6^3)/12}$$
$$= \frac{2400 \times 12^2}{6^3} = 1600 \text{ psi}$$

Answer is (C)

PROBLEM 6-25

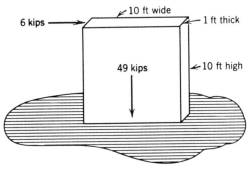

6 kips

10 ft wide

1 ft thick

49 kips

10 ft high

Figure 6-30

A solid steel block, which weighs 49 kips, rests on a level concrete slab. If a horizontal force of 6 kips is applied to the top, the minimum pressure, in lb/ft², under the base of the block is equal to

(A)   0
(B) 1300
(C) 3600
(D) 4900
(E) 8500

Solution

$$\sigma_{max} = \frac{P}{A} + \frac{My}{I} = \frac{49,000}{10 \times 1} + \frac{6000 \times 10 \times 5}{\frac{1}{12} \times 1 \times 10^3} = 4900 + 3600 = 8500 \text{ lb/ft}^2$$

$$\sigma_{min} = \frac{P}{A} - \frac{My}{I} = 4900 - 3600 = 1300 \text{ lb/ft}^2$$

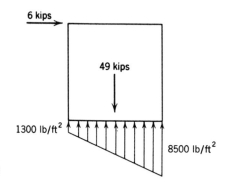

6 kips

49 kips

1300 lb/ft^2

8500 lb/ft^2

Figure 6-31

Answer is (B)

PROBLEM 6-26

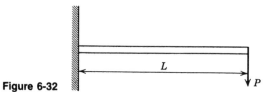

Figure 6-32

L

P

A cantilever beam of length L is loaded at the free end by a force P. Assume the modulus of elasticity $E = 20 \times 10^6$ psi, the cross-sectional area of the beam $A = 6$ in.^{2}, the moment of inertia $I = 2$ in.^{4}, and $L = 30$ in. The force P, in lb, that produces a 0.3-in. deflection of the beam at the free end is nearest to

(A) 333
(B) 667
(C) 1000
(D) 1333
(E) 1667

Solution

To solve this problem, we must first derive the equation relating P and beam deflection. The second moment-area proposition says that the vertical displacement Δ of point A from the tangent to the elastic curve at B equals the moment (with respect to A) of the area of the bending moment diagram between A and B divided by EI.

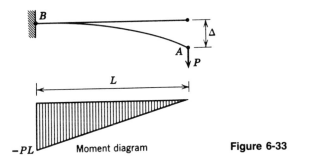

Figure 6-33

$$\Delta = \frac{\text{area of } M \text{ diagram} \times \text{lever arm}}{EI}$$

$$= \frac{-PL(L/2) \times \frac{2}{3}L}{EI} = \frac{-PL^3}{3EI}$$

For $\Delta = 0.3$,

$$\frac{P \times 30^3}{3(20 \times 10^6) \times 2} = 0.3$$

$$P = \frac{0.3 \times 2 \times 3 \times 20 \times 10^6}{30^3} = 1333 \text{ lb}$$

Answer is (D)

PROBLEM 6-27

A helical spring has a natural length of 6 in. It requires a force of 20 lb to hold it extended to a length of 12 in. Assume the spring does not exceed its elastic limit. The work done in inch-pounds to stretch the spring from a total length of 9 in. to 11 in. is nearest to

(A) 5
(B) 10
(C) 15
(D) 20
(E) 25

Solution

$$\text{Spring constant } k = \frac{\text{force}}{\text{deflection}} = \frac{20 \text{ lb}}{(12-6) \text{ in.}} = \frac{20 \text{ lb}}{6 \text{ in.}}$$

The energy required to extend a spring a distance dx is $F\,dx$.

$$\text{Force } F = \text{spring constant } k \times \text{distance extended } x$$

For a displacement $S_2 - S_1$,

$$\text{Total work} = \int_{S_1}^{S_2} kx\,dx = \tfrac{1}{2}kx^2 \Big|_{S_1}^{S_2} = \tfrac{1}{2}k(S_2^2 - S_1^2)$$

In this case $S_1 = 9$ in. $- 6$ in. $= 3$ in. and $S_2 = 11$ in. $- 6$ in. $= 5$ in.

$$\text{Total work} = \tfrac{1}{2}k(S_2^2 - S_1^2) = \tfrac{1}{2}(\tfrac{20}{6})(5^2 - 3^2) = 26.7 \text{ in.-lb}$$

Answer is (E)

PROBLEM 6-28

A 3-in. diameter solid steel shaft 10 ft long is subjected to a constant torque of 100,000 in.-lb at each end, together with an axial tensile load of 70,000 lb also applied at each end. For a circular cross section the polar moment of inertia $J = \pi d^4/32$. The maximum compressive stress, in psi, in the shaft under this loading is nearest to

(A)   0
(B) 10,000

(C) 15,000

(D) 18,800

(E) 19,500

Solution

$$\sigma_x = \sigma_{\text{tension}} = \frac{P}{A} = \frac{70,000}{\frac{\pi}{4}\,3^2} = 9900\text{ psi}$$

$$\sigma_y = 0$$

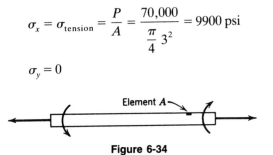

Figure 6-34

Torsion

$$\tau_{xy} = \frac{Tr}{J} = \frac{T(d/2)}{\pi(d^4/32)} = \frac{16T}{\pi d^3} = \frac{16(100,000)}{\pi(27)} = 18,860\text{ psi}$$

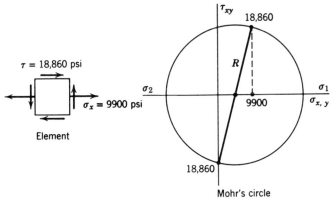

Element

Mohr's circle

Figure 6-35

$$R = (18,860^2 + 4950^2)^{1/2} = 19,500$$

$$(\tau_{xy})_{\text{max}} = R = 19,500\text{ psi}$$

$$\sigma_1 = 4950 + R = 24,450\text{ psi tension}$$

$$\sigma_2 = 4950 - R = -14,550\text{ psi compression}$$

Answer is (C)

PROBLEM 6-29

The *least radius of gyration* is required in the design of

(A) shaft couplings
(B) columns
(C) helical springs
(D) cantilevered beams
(E) riveted joints

Solution

The ratio L/r of the column length to the least radius of gyration is the column slenderness ratio.

Answer is (B)

PROBLEM 6-30

A rectangular shape has a cross section 30 cm wide and 60 cm in height. The least radius of gyration, in cm, for this shape is closest to

(A) 10
(B) 15
(C) 20
(D) 25
(E) 30

Solution

The radius of gyration $r = (I/A)^{1/2}$ where I is the moment of inertia about a centroidal axis and A is the cross-sectional area.

When the least moment of inertia is used, the computation gives the least radius of gyration. For a rectangular cross section,

$$I = \frac{bh^3}{12} = \frac{60 \times 30^3}{12} = 135{,}000 \text{ cm}^4$$

$$A = 30 \times 60 = 1800 \text{ cm}^2$$

$$\text{Least } r = \left(\frac{135{,}000}{1800}\right)^{1/2} = 8.66 \text{ cm}$$

Answer is (A)

PROBLEM 6-31

A beam has a cross-sectional area of 72 cm² and a radius of gyration of 3 cm. The corresponding value of the moment of inertia for the beam is nearest to

(A)   24
(B)   216
(C)   650
(D) 1,300
(E) 15,000

Solution

Radius of gyration $r = (I/A)^{1/2}$. Here $I = r^2 A = 3^2 \times 72 = 648 \text{ cm}^4$.

$$\text{Answer is (C)}$$

PROBLEM 6-32

The *slenderness ratio* of a column is generally defined as the ratio of its

(A) length to its minimum width
(B) unsupported length to its maximum radius of gyration
(C) length to its moment of inertia
(D) unsupported length to its least radius of gyration
(E) unsupported length to its minimum cross-sectional area

Solution

Euler's formula

$$P_{cr} = \frac{\pi^2 EI}{L^2}$$

may be rewritten to compute the critical stress.

$$I = r^2 A$$

$$\sigma_{cr} = \frac{P_{cr}}{A} = \frac{\pi^2 E r^2 A}{AL^2} = \frac{\pi^2 E}{(L/r)^2}$$

where L is the unsupported length and r is the least radius of gyration. The ratio L/r is called the column *slenderness ratio*.

$$\text{Answer is (D)}$$

PROBLEM 6-33

The section modulus, in cm^3, about the central axis of a 10-cm × 25-cm beam on edge is closest to

(A)   250
(B) 1,000
(C) 6,000
(D) 13,000
(E) 15,600

Solution

$$\text{Section modulus } Z = \frac{I}{c}$$

where I is the moment of inertia and c is the distance from the neutral axis to the extreme fiber.

For a rectangular cross section,

$$I = \frac{bh^3}{12} \qquad c = \frac{h}{2}$$

$$Z = \frac{I}{c} = \frac{bh^3/12}{h/2} = \frac{bh^2}{6}$$

In this case

$$Z = \frac{10(25)^2}{6} = 1042 \text{ cm}^3$$

Answer is (B)

PROBLEM 6-34

A cantilevered beam 20 in. long and of square cross section, 1 in. on a side, is loaded at its end, through the centroid of the cross section by a vertical force of magnitude 150 lb. The magnitude of the maximum bending stress, in psi, is nearest to which value?

(A) 3,000
(B) 6,000
(C) 12,000
(D) 15,000
(E) 18,000

Solution

The   stress $\sigma = Mc/I$ with $I = bh^3/12 = 1(1)^3/12 = \frac{1}{12}$ in.^{4} and
$M = (150 \text{ lb})(20 \text{ in.}) = 3000$ in.-lb.

$$\sigma = \frac{(3000)(\frac{1}{2})}{\frac{1}{12}} = 18,000 \text{ psi}$$

Answer is (E)

PROBLEM 6-35

The maximum unit fiber stress at any vertical section in a beam is obtained by dividing the moment at that section by

(A) the section modulus
(B) the cross-sectional area
(C) one-half the distance to the point where the shear is zero
(D) the radius of gyration
(E) the moment of inertia

Solution

The basic formula for maximum fiber stress in a beam is

$$\text{Unit stress} = \frac{Mc}{I}$$

The section modulus Z is defined as I/c, so the basic formula reduces to

$$\text{Unit stress} = \frac{M}{Z} = \frac{\text{moment}}{\text{section modulus}}$$

Answer is (A)

PROBLEM 6-36

Consider the reinforced concrete beam shown in Fig. 6-36. Assume a balanced design, using straight-line theory.

Data:
f_c = maximum allowable compressive unit stress = 1350 psi
f_s = maximum allowable tensile unit stress in the longitudinal reinforcement = 20,000 psi
A_s = area of longitudinal steel reinforcement = 2.18 in.^{2}
n = ratio of modulus of elasticity of steel to that of concrete = 10

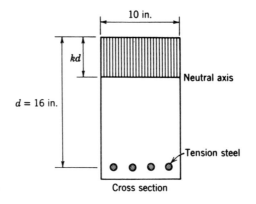

Figure 6-36   Cross section

The distance kd, in inches, is closest to

(A) 4.3
(B) 6.5
(C) 7.1
(D) 8.0
(E) 8.6

Solution

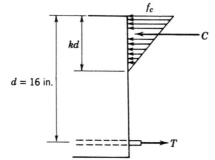

Figure 6-37

$$\text{Width } b = 10 \text{ in.}$$
$$C = \tfrac{1}{2} f_c kdb = \tfrac{1}{2}(1350)(kd)(10) = 6750kd$$
$$T = A_s f_s = 2.18(20,000) = 43,600 \text{ lb}$$

In a *balanced design* the maximum allowable concrete stress f_c and allowable steel stress f_s are simultaneously developed.

$$\sum F_x = 0 \qquad T = C$$

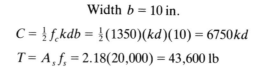

$$43,600 = 6750kd \qquad kd = \frac{43,600}{6750} = 6.46 \text{ in.}$$

Answer is (B)

PROBLEM 6-37

In the design of a reinforced concrete beam, it is commonly assumed that

  (A) the concrete has one-fourth of the tensile strength of the steel
  (B) the modulus of elasticity for concrete is the same as the modulus of elasticity for steel
  (C) the steel has a tensile stress of 1800 psi induced by shrinkage of the concrete
  (D) the dead load of the beam may be neglected for long spans
  (E) the tensile strength of the concrete is nearly zero

Solution

Conventional concrete design is done by a cracked section analysis; that is, it is assumed that the portion of the beam in tension will crack with the result that the effective concrete tensile strength is zero.

Answer is (E)

PROBLEM 6-38

Stirrups in a reinforced concrete beam are designed primarily to resist stresses caused by

  (A) diagonal tension
  (B) axial tension
  (C) horizontal shear
  (D) axial compression
  (E) web crippling

Solution

At each point in a loaded concrete beam there are vertical or transverse shearing forces and, for equilibrium, longitudinal shearing forces of equal intensity. These forces tend to distort the beam so that tension exists along a diagonal plane. This diagonal tension acts along a line of action 45° from the beam's axis. Stirrups or other web reinforcement are designed to resist this diagonal tension.

Answer is (A)

7

FLUID MECHANICS

Fluid mechanics studies fluids at rest and in motion by the application of basic principles of mechanics. The discipline covers a broad field since it endeavors to study both liquids and gases under all conditions. In this chapter we restrict our attention almost entirely to a study of the mechanical behavior of incompressible liquids such as water. Topics related to the internal energy changes and the compressibility and chemical behavior of gases may be found in Chapters 8 and 9.

Paralleling the study of the mechanics of solid bodies, fluid mechanics is normally divided into fluid statics and fluid dynamics. Hydrostatics studies the variation of pressure in a stationary fluid body and leads to an understanding of such subjects as buoyancy and manometry; one also learns how to compute the forces exerted on submerged bodies because of this pressure variation. The fundamental principles underlying fluid dynamics are contained in three conservation laws: those expressing the conservation of mass, momentum, and energy. In connection with these principles the calculation of power, head loss, and flow rate is considered. In almost all the dynamics problems, the flow is assumed to be steady and also one-dimensional; that is, the flow variables are only a function of the distance along the conveyance structure and do not vary across the flow cross section.

HYDROSTATICS

The fundamental equation of fluid statics indicates that the rate of change of the pressure p is directly proportional to the rate of change of the depth z, or

$$\frac{dp}{dz} = -\gamma = -\rho g \tag{7-1}$$

where γ is the unit weight (also called specific weight or weight density) of the fluid (for water $\gamma_w = 62.4 \, \text{lb/ft}^3$) and z is positive upward. The fluid density (or mass density) is ρ, and g is the acceleration of gravity. In some cases fluid properties are reported in terms of relative density $S = \rho/\rho_r$, which is the ratio of one fluid density to that of some chosen reference fluid. The ratio could also be written in terms of unit weights as γ/γ_r; if water is chosen as the reference fluid, S is then often called the specific gravity of the fluid.

For incompressible fluids γ is constant and Eq. (7-1) may be integrated to give

$$p_2 = p_1 + \gamma h \tag{7-2}$$

Here point 2 is located a distance h below point 1. In most problems in fluid mechanics one may work in either absolute or gage pressures if proper care is taken to be consistent. Gage pressure is equal to the difference between the absolute pressure and atmospheric pressure; when it is negative, it is often called a vacuum. In English units lb/in.^2 is often written psi, and absolute and gage pressures are distinguished by writing psia and psig.

Manometers are devices that measure pressure differences by a direct application of the hydrostatic principle given in Eq. (7-2). Pressure differences in manometers can easily be computed by systematically applying the equation to each manometer limb.

Example 1

Mercury (Hg) is poured into a U-tube. Then 18 in. of oil of specific gravity $S_0 = 0.90$ is poured into one leg on top of the mercury (Fig. 7-1). What suction, in pounds per square inch, applied to the leg containing only mercury will bring the upper surfaces of the oil and mercury in the two legs to the same level?

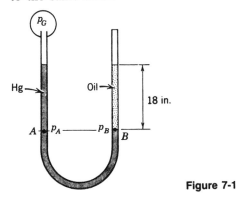

Figure 7-1

Solution

For one fluid, Eq. (7-2) shows that the pressures at two points are equal when the elevation difference between the points is zero so that $p_A = p_B$. But Eq. (7-2) also shows that

$$p_A = p_G + \gamma_M h$$

and

$$p_B = \gamma_0 h$$

Hence

$$p_G = h(\gamma_0 - \gamma_M) = h\gamma_w(S_0 - S_M)$$

where the specific gravity of mercury $S_M = 13.55$.

$$p_G = \tfrac{18}{12}(62.4)(0.90 - 13.55) = -1184 \text{ psfg}$$

$$p_G = \frac{-1184}{144} = -8.22 \text{ psig} = 8.22 \text{ psi vacuum}$$

The buoyant force F_B exerted upward on a floating or submerged object is equal to the weight of the fluid displaced by that object, as can be shown by properly integrating Eq. (7-2) over the surface of the body. Restated,

$$F_B = \gamma_f(\text{volume}) \tag{7-3}$$

where γ_f is the unit weight of the fluid, and the volume is the amount of fluid displaced.

Example 2

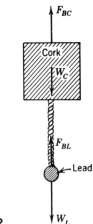

Figure 7-2

A piece of lead (specific gravity $S_L = 11.3$) is tied to $8\,\text{in.}^3$ of cork whose specific gravity is $S_C = 0.25$. They float submerged in water (Fig. 7-2). What is the weight of the lead?

Solution

The net upward force on the cork is $F_{BC} - W_C$, which is just balanced by the net downward force on the lead $W_L - F_{BL}$, or

$$F_{BC} - W_C = W_L - F_{BL}$$

Thus

$$(1 - S_C)\gamma_w\left(\frac{8}{12^3}\right) = (S_L - 1)\gamma_w V_L$$

where V_L is the unknown volume of the lead. Solving first for V_L and then for the weight W_L, we have

$$W_L = S_L \gamma_w V_L = \gamma_w \left(\frac{S_L}{S_L - 1}\right)(1 - S_C)\frac{8}{12^3}$$

$$W_L = 62.4\left(\frac{11.3}{10.3}\right)(0.75)\frac{8}{12^3}$$

$$W_L = 0.238\,\text{lb lead}$$

Distributed fluid pressures cause hydrostatic forces to act on submerged surfaces. It is important to be able to compute the direction, magnitude, and location of these resultant forces. The force on any submerged surface is determinable if one knows how to compute the force acting on a submerged plane area and the centroid or center of gravity of areas and volumes.

The magnitude of the force F on a submerged plane surface is

$$F = \int_A p\,dA = p_C A = \gamma h_C A \tag{7-4}$$

where A is the surface area, γ is the fluid unit weight, and p_C and h_C are, respectively, the pressure and submerged depth of the centroid of the area. This compressive force acts normal to the plane area. Simple formulas can be derived for the location of this force, but they are easily misused. A more direct approach is to compute the first moment of the pressure distribution about some convenient axis, according to the principles of statics, and equate this to the product of F and the distance to the line of action of F.

An extension of these principles allows the computation of the magnitude and location of a hydrostatic force which acts on a submerged curved surface. The horizontal component of this force is computed from Eq. (7-4), where the area A is now the vertical projection of the curved surface. The vertical force component is basically equal to the weight of the fluid which lies vertically above the curved surface, but some care is needed in applying this principle. When all the fluid lies above the surface, the principle applies exactly. If some fluid also lies beneath a portion of the curved surface, then the *net* upward vertical force is a buoyant force equal to the weight of fluid displaced by the presence of the curved surface. The line of action of each component force and the magnitude and direction of the resultant force are then found by direct application of the principles of statics.

Example 3

A closed circular tank 2 ft in diameter and 6 ft deep, with its axis vertical, contains 4 ft of water. Air at a pressure of 5 psig is pumped into the cylinder. Find the normal force per inch of circumference on the vertical wall of the tank and the distance to the center of pressure from the base of the tank.

Solution

First we plot a diagram, Fig. 7-3, of the gage pressure exerted on a unit width of the wall. The hydrostatic pressure p_1 at the base of the tank is

$$p_1 = \gamma h = \frac{(62.4)(4)}{144} = 1.73 \text{ psi}$$

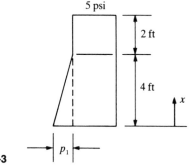

Figure 7-3

The force per inch of circumference equals the area of the pressure diagram, or

$$F = \int p\, dA = 5(6)(12) + \tfrac{1}{2}(1.73)(4)(12)$$

$$= 360 + 41.5 = 401.5 \text{ lb/in.}$$

The distance to the center of pressure x_{cp}, by statics, is

$$x_{cp} = \frac{1}{F} \int px \, dA$$

$$= \frac{1}{401.5} [360(3) + 41.5(\tfrac{4}{3})] = 2.83 \text{ ft} = 34 \text{ in.}$$

CONTINUITY

The law of conservation of mass states that matter is neither created nor destroyed. Applied to a streamtube in incompressible flow, the law requires the flow to vary in a continuous way from cross section to cross section along the streamtube so that at any section

$$Q = \int_A V \, dA = \text{constant} \qquad (7\text{-}5)$$

where Q is the volume rate of flow and V is the velocity at a point in the cross-sectional streamtube area A. When the velocity is assumed to be constant across the section, we obtain the familiar continuity equation

$$Q = A_1 V_1 = A_2 V_2 \qquad (7\text{-}6)$$

This equation is widely used in combination with the energy or momentum equations to solve fluid flow problems. In the United States volume rate of flow may commonly be expressed not only in ft³/sec (cfs) but also in gallons per minute (gpm) and millions of gallons per day (mgd).

Example 4

A fluid flowing steadily through a constant-diameter pipeline has a velocity profile $u(r)$ which varies parabolically across the pipe. Specifically,

$$u = m(1 - s^2)$$

where $s = r/R$. Here m is a constant, r is the distance from the pipe centerline, and R is the pipe radius. What fraction of the total flow in the pipe is flowing between the pipe wall and a distance of 10% of the pipe radius from the wall?

Solution

For the integration of Eq. (7-5) we choose the differential area shown in Fig. 7-4. Then the total volume rate of flow in the pipe is

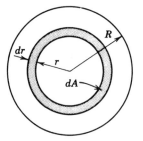

Figure 7-4

$$Q_T = \int_A V \, dA = \int_0^R m(1 - s^2)(2\pi r \, dr)$$

$$= 2\pi m R^2 \int_0^1 (1 - s^2)s \, ds$$

$$= 2\pi m R^2 \left[\frac{s^2}{2} - \frac{s^4}{4} \right]_0^1$$

$$Q_T = \tfrac{1}{2}\pi m R^2$$

The volume rate of flow near the wall is

$$Q = \int_{0.9R}^R m(1 - s^2)(2\pi r \, dr)$$

$$= 2\pi m R^2 \int_{0.9}^1 (1 - s^2)s \, ds$$

$$= 2\pi m R^2 \left[\frac{s^2}{2} - \frac{s^4}{4} \right]_{0.9}^1$$

$$Q = 0.018\pi m R^2$$

Thus the flow fraction is

$$\frac{Q}{Q_T} = \frac{0.018}{0.5} = 0.036$$

ENERGY EQUATION

For one-dimensional, steady incompressible flow a general energy equation that expresses the changes in energy, *per unit weight* of flowing fluid, between points 1 and 2 is

$$\frac{V_1^2}{2g} + \frac{p_1}{\gamma} + z_1 = \frac{V_2^2}{2g} + \frac{p_2}{\gamma} + z_2 + h_L - E_m \qquad (7\text{-}7)$$

In this equation $V^2/2g$ is the kinetic energy or velocity head, p/γ is the pressure energy or pressure head, and z is the potential energy or elevation head. The head loss h_L represents the loss in energy between points 1 and 2 for any of a variety of causes. The last term E_m is the mechanical energy added to the fluid between the two points; this is accomplished by hydraulic machinery. For pumps, energy is added to the flow and E_m is positive. For turbines, E_m is negative since energy is then extracted from the fluid. When E_m and h_L are both zero, Eq. (7-7) is the classic Bernoulli equation.

The head loss term h_L represents a loss in energy that is due primarily to viscosity and the turbulence in the flow. It may be written as

$$h_L = \Sigma K \frac{V^2}{2g}$$

where V is a representative velocity for the point where the loss occurs, K is a loss coefficient related to the nature of the loss-producing element, and one sums the effects of all loss elements between the two points. Particularly important is the head loss caused by pipe friction in a pipe of length L and diameter d. In this case

$$h_L = f \frac{L}{d} \frac{V^2}{2g} \tag{7-8}$$

The equation is the Darcy-Weisbach equation. The friction factor $f = f(\mathbf{R})$, and the Reynolds number is $\mathbf{R} = Vd\rho/\mu$. For laminar flow in pipes $\mathbf{R} < 2300$ and $f = 64/\mathbf{R}$, but in turbulent flow, when $\mathbf{R} > 4000$, f is also a function of the relative roughness e/d for rough pipes. The height of a representative roughness projection is e. Between the laminar and turbulent zones lies an unpredictable transition zone. Most losses other than that due to pipe friction are termed minor losses, since their magnitude is usually small in comparison to the pipe friction loss in practical problems. A different K value is needed for each type of loss element.

The power produced or expended in a given situation is directly related to the change in energy in the flow. If the weight rate of flow is $Q\gamma$ and the energy change per unit weight of fluid is E, then the power gained or expended is the product $Q\gamma E$. In terms of horsepower (1 hp = 550 ft-lb/sec) the relation is

$$\text{Horsepower} = \frac{Q\gamma E}{550} \tag{7-9}$$

For turbines or pumps E is the same as E_m.

Example 5

A Venturi meter with a throat diameter of 6 in. is placed in a 12-in.-diameter pipeline. It meters the flow of oil having a specific gravity $S_0 = 0.80$. A differential manometer containing a fluid of specific gravity $S_M = 3.20$ is connected between the pipe and the throat section and shows a deflection of 2 ft. Above the manometer liquid the tubes are filled with oil (Fig. 7-5). Neglecting energy losses, what is the indicated flow through the meter?

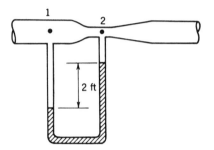

Figure 7-5

Solution

Here we apply the Bernoulli and continuity equations in combination. The Bernoulli equation is

$$\frac{V_1^2}{2g} + \frac{p_1}{\gamma_0} + z_1 = \frac{V_2^2}{2g} + \frac{p_2}{\gamma_0} + z_2$$

Since $z_1 = z_2$,

$$\frac{p_1 - p_2}{\gamma_0} = \frac{V_2^2 - V_1^2}{2g}$$

Considering the manometer,

$$p_1 + S_0 \gamma_w(2) = p_2 + S_M \gamma_w(2)$$

and by noting that $S_0 \gamma_w = \gamma_0$, we obtain

$$\frac{p_1 - p_2}{\gamma_0} = 2\left(\frac{S_M}{S_0} - 1\right)$$

From Eq. (7-6), the continuity equation is $V_1 A_1 = V_2 A_2$, or

$$V_2 = \frac{A_1}{A_2} V_1 = \left(\frac{d_1}{d_2}\right)^2 V_1 = \left(\tfrac{12}{6}\right)^2 V_1 = 4V_1$$

Therefore

$$2\left(\frac{S_M}{S_0} - 1\right) = \frac{(4V_1)^2 - V_1^2}{2g} = \frac{15V_1^2}{2g}$$

$$V_1^2 = \frac{1}{15}\left[2(32.2)(2)\left(\frac{3.20}{0.80} - 1\right)\right] = 25.8$$

$$V_1 = 5.08 \text{ ft/sec}$$

Finally

$$Q = A_1V_1 = \frac{\pi}{4} (1)^2(5.08) = 4.0 \text{ cfs}$$

Example 6

A water main with a 24 in. inside diameter carries a flow of 20 cfs. If the friction factor is 0.02 and the pump is 85% efficient, how much horsepower is required to pump the water through 10,000 ft of pipeline?

Solution

The velocity in the main is

$$V = \frac{Q}{A} = \frac{20}{(\pi/4)(2)^2} = 6.37 \text{ ft/sec}$$

Using Eq. (7-8), the Darcy-Weisbach formula for head loss caused by friction, we obtain

$$h_L = f \frac{L}{d} \frac{V^2}{2g}$$

$$h_L = (0.02)\left(\frac{10,000}{2}\right)\frac{(6.37)^2}{2(32.2)}$$

$$h_L = 63.0 \text{ ft}$$

This head loss is equal to the amount of energy E that must be added per pound of fluid. The net power requirement is

$$\frac{Q\gamma h_L}{550} = \frac{(20)(62.4)(63.0)}{550} = 143 \text{ hp}$$

Since the pump efficiency is only 85%, the total power requirement is $143/0.85 = 168$ hp.

Pipelines of different sizes and lengths commonly occur in combination. When pipes are connected end-to-end in series, the head loss of the combination is the sum of the head losses in the individual pipes, while the flow rate is the same for each element. The situation is reversed for a parallel combination of pipes, that is, a case where two or more pipes are connected between the same two end points. Then the total flow rate is the sum of the individual flow rates, while the head lost is identical for each pipe.

MOMENTUM EQUATION

The steady-flow momentum conservation equation takes the form

$$\mathbf{F}_S + \mathbf{F}_B = \int_S \rho \mathbf{V} V_n \, dA \tag{7-10}$$

This is a vector equation. The first term represents the net external surface force acting *on* a fluid volume which is enclosed by the surface S; this term includes all contributions arising from the surface pressure distribution and from surface viscous stresses acting on S. The second term is the net body force within the control volume; usually this is just the weight of the enclosed fluid acting in the direction of gravity. The right term is the net flux of momentum flowing out of the region enclosed by S. V_n is the fluid velocity component exiting normal (perpendicular) to S. For uniform flow across an entrance section 1 and an exit section 2, we may write

$$\begin{aligned} \Sigma F_x &= \rho Q(V_{2x} - V_{1x}) \\ \Sigma F_y &= \rho Q(V_{2y} - V_{1y}) \end{aligned} \tag{7-11}$$

for flow in two dimensions. The volume flow rate is Q. Note that V_x and V_y may be either positive or negative.

Example 7

The 18-in. to 12-in. reducing bend shown in Fig. 7-6 is in a horizontal pipeline conveying oil (of specific gravity $S_0 = 0.90$) at the rate of $10 \, \text{ft}^3/\text{sec}$. The pressure in the line at the 18-in. section is 50 psia when the atmospheric pressure $p_a = 14.7 \, \text{psia}$. Compute the x and y components of thrust caused by the flowing fluid.

Solution

The force components we seek are equal and opposite to the external forces X and Y shown in Fig. 7-6.

The reducing bend and the fluid within it are chosen as a control volume. Atmospheric pressure acts on the entire control volume surface and thus produces no net force. Employing Eq. (7-11), we find

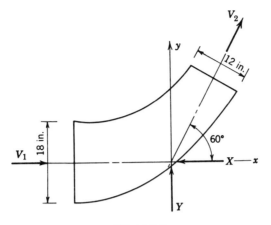

Figure 7-6

$$\Sigma F_x = (p_1 - p_a)A_1 - (p_2 - p_a)A_2 \cos 60° - X$$

and

$$\rho Q(V_{2x} - V_{1x}) = \rho Q(V_2 \cos 60° - V_1)$$

Here

$$Q = 10 \text{ cfs}$$

$$V_1 = \frac{Q}{A_1} = \frac{10}{(\pi/4)(\frac{18}{12})^2} = 5.66 \text{ ft/sec}$$

$$V_2 = \frac{Q}{A_2} = \frac{10}{(\pi/4)(\frac{12}{12})^2} = 12.73 \text{ ft/sec}$$

Assuming no energy losses occur in the bend, we apply the Bernoulli equation between the entrance and exit to find p_2:

$$\frac{V_1^2}{2g} + \frac{p_1}{\rho g} + z_1 = \frac{V_2^2}{2g} + \frac{p_2}{\rho g} + z_2$$

Since there is no elevation change $z_1 = z_2$, and

$$p_2 = p_1 + \frac{\rho}{2}(V_1^2 - V_2^2)$$

$$p_2 = 50 + \tfrac{1}{2}(0.90)(1.94)[(5.66)^2 - (12.73)^2]\tfrac{1}{144}$$

$$p_2 = 50 - 0.79 = 49.2 \text{ psia}$$

The momentum equation now becomes

$$(50 - 14.7)(144) \frac{\pi}{4} \left(\frac{18}{12}\right)^2 - (49.2 - 14.7)(144) \frac{\pi}{4} \left(\frac{12}{12}\right)^2 \cos 60° - X$$
$$= (0.90)(1.94)(10)(12.73 \cos 60° - 5.66)$$
$$X = 7020 \text{ lb}$$

The momentum equation in the y direction is

$$\Sigma F_y = \rho Q(V_{2y} - V_{1y})$$
$$Y - (p_2 - p_a)A_2 \sin 60° = \rho Q(V_2 \sin 60° - 0)$$
$$Y - (49.2 - 14.7)(144) \frac{\pi}{4} \left(\frac{12}{12}\right)^2 \sin 60° = (0.90)(1.94)(10)(12.73 \sin 60°)$$
$$Y = 3570 \text{ lb}$$

The thrust components of the fluid are equal and opposite to X and Y, so the x component is 7020 lb to the right and the y component is 3570 lb downward.

OPEN CHANNEL FLOW RATE

The fundamental equation for the flow rate Q in uniform flow as a function of depth of flow and channel characteristics is the Manning equation. In English units it is

$$Q = \frac{1.49}{n} AR^{2/3}S^{1/2} \tag{7-12}$$

To use metric units, replace 1.49 by 1.0. The roughness coefficient n may vary from 0.01 for smooth uniform channels to 0.03 or higher for irregular natural river channels, and S is the channel bottom slope. The cross-sectional area of flow is A, and R is a shape parameter for the channel section. Called the hydraulic radius, $R = A/P$, where P is the wetted perimeter of the cross section. Using this equation to find Q when the other factors are known is straightforward, but solving for the flow depth y, when Q is known, usually requires trial-and-error computations.

Example 8

The trapezoidal channel shown in Fig. 7-7, with $S = 0.0009$ and $n = 0.025$, carries a discharge $Q = 300$ cfs. Compute the flow depth y and the average velocity V.

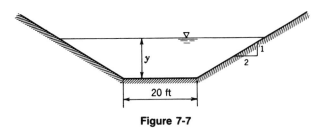

Figure 7-7

Solution

In terms of y, the area A and hydraulic radius R are

$$A = y(20 + 2y)$$

$$R = \frac{A}{P} = \frac{y(20 + 2y)}{20 + 2y\sqrt{5}}$$

The Manning equation then gives

$$300 = \frac{1.49}{0.025} [y(20 + 2y)] \left[\frac{y(20 + 2y)}{20 + 2y\sqrt{5}} \right]^{2/3} (0.0009)^{1/2}$$

or

$$7680 + 1720y = [y(10 + y)]^{5/2}$$

This equation must be solved by trial and error.

Trial y	$7680 + 1720y$		$[y(10 + y)]^{5/2}$
3.00	12,840		9,500
3.30	13,356		12,760
3.40	13,528		14,010
3.36	13,460	$\approx$	13,500

The depth of flow is thus $y = 3.36$ ft.

For this depth the area is $A = 3.36[20 + 2(3.36)] = 89.7$ ft^2 and the average velocity is

$$V = \frac{Q}{A} = \frac{300}{89.7} = 3.34 \text{ ft/sec}$$

PROBLEM 7-1

The anchor block at a bend in a pipeline must be designed primarily to resist forces caused by

(A) friction and acceleration

(B) friction and pressure

(C) pressure and acceleration caused by gravity

(D) static head

(E) pressure and velocity

Solution

The anchor block at a bend in a pipeline must be designed primarily to resist forces caused by pressure and velocity.

Answer is (E)

PROBLEM 7-2

The line showing the pressure head plus the potential head plus the velocity head at any section of a pipe is called the

(A) total head line

(B) energy gradient

(C) energy head line

(D) hydraulic gradient

(E) combined head line

Solution

The line showing the sum of the pressure, potential, and velocity heads is the total head line; it graphically shows the sum of the three terms in the Bernoulli equation.

Answer is (A)

PROBLEM 7-3

Cavitation results from

(A) laminar flow

(B) low soil permeability

(C) turbulent water flow

(D) rough, irregular flow boundaries

(E) excessive pore water pressure in soil

Solution

Cavitation is a high-speed water phenomenon that occurs when the fluid pressure approaches vapor pressure. This can only occur in turbulent flow.

Answer is (C) ok

PROBLEM 7-4

Capillarity results from

- (A) excess pore water pressure
- (B) surface tension
- (C) low fluid pressures
- (D) inadequate compaction
- (E) turbulent flow

Solution

Capillarity results from surface tension.   *ok*

Answer is (B)

PROBLEM 7-5

The hydraulic jump is utilized for

- (A) energy dissipation
- (B) pressure regulation
- (C) lifting of water
- (D) transport of sediment
- (E) evaporation rate increase

Solution

In the hydraulic jump energy is dissipated.

Answer is (A)

PROBLEM 7-6

The primary function of a weir is

    (A) wildlife conservation
    (B) channel diversion
    (C) measurement of discharge
    (D) prevention of scour
    (E) energy dissipation

Solution

Weirs are used primarily to measure discharge in channels.

$$\text{Answer is (C)}$$

PROBLEM 7-7

If the pressure in a plenum is noted as 6 in. of water column, this is most nearly equivalent to

    (A) $6\,\text{lb/ft}^2$
    (B) $392\,\text{lb/ft}^2$
    (C) $9\,\text{lb/ft}^2$
    (D) $31\,\text{lb/ft}^2$
    (E) $39\,\text{lb/ft}^2$

Solution

$$p = \gamma z = 62.4 \text{ lb/ft}^3 \times 0.5 \text{ ft} = 31.2 \text{ lb/ft}^2$$

$$\text{Answer is (D)}$$

PROBLEM 7-8

The value of the coefficient of viscosity of air at 19.2°C is 1.828×10^{-4} poises. The equivalent value, expressed in pounds per foot-second, is closest to

(A) 1.23×10^{-2}
(B) 5.57×10^{-3}
(C) 1.02×10^{-3}
(D) 1.23×10^{-5}
(E) 1.02×10^{-6}

Solution

The poise is equal to 1 g/cm-sec and is named after Poiseuille. Application of conversion factors gives

$$1.828 \times 10^{-4} \; \frac{g}{cm\text{-}sec} \times \frac{1\;lb}{454\;g} \times \frac{2.54 \times 12\;cm}{1\;ft} = 1.227 \times 10^{-5} \; \frac{lb}{ft\text{-}sec}$$

Answer is (D)

PROBLEM 7-9

A fluid flows at a constant velocity in a pipe. The fluid completely fills the pipe, and the Reynolds number is such that the flow is just subcritical and laminar. If all other parameters remain unchanged and the viscosity of the fluid is decreased a significant amount, one would generally expect the flow to

(A) not change
(B) become turbulent
(C) become more laminar
(D) increase
(E) temporarily increase

Solution

Considering the Reynolds number,

$$\mathbf{R} = \frac{VD\rho}{\mu}$$

where V is the average velocity, D is the pipe diameter, ρ is the density of the fluid, and μ is the viscosity of the fluid, we see that a decrease in μ will increase $\mathbf{R}$ so that it will be greater than the Reynolds number for the transition to turbulent flow, and the flow will become turbulent.

Answer is (B)

PROBLEM 7-10

The transition between laminar and turbulent flow in a pipe usually occurs at a Reynolds number of approximately

(A) 350
(B) 900
(C) 1800
(D) 2300
(E) 3850

Solution

The lower critical Reynolds number, below which laminar flow always occurs in a pipe, is approximately 2300.

Answer is (D)

PROBLEM 7-11

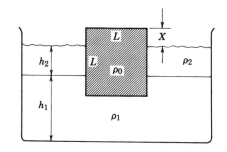

Figure 7-8

A cube of wood with sides of length $L = 24$ in. and of density $\rho_0 = 1.55$ slug/ft^3 floats in two fluids, as shown in Fig. 7-8. The heavier fluid is water of density $\rho_1 = 1.94$ slug/ft^3 and depth $h_1 = 36$ in. The lighter fluid is oil of density $\rho_2 = 1.65$ slug/ft^3 and forms a layer of thickness $h_2 = 6$ in. Neglecting surface tension, the distance X, in inches, that the cube projects above the surface of the lighter fluid is nearest to

(A) 3.9
(B) 4.2
(C) 4.5
(D) 4.8
(E) 5.1

Solution

Archimedes' principle tells us that the buoyant force is equal to the weight of the liquid displaced. The weight of the displaced liquid is

$$\text{Volume} \times \text{unit weight} = L^2 h_2 \rho_2 g + L^2 (L - h_2 - X) \rho_1 g$$

for the two liquid layers. The weight of the block $(L^3 \rho_0 g)$ is equal to this buoyant force

$$L^3 \rho_0 g = L^2 g [h_2 \rho_2 + \rho_1 (L - h_2 - X)]$$
$$L \rho_0 = h_2 \rho_2 + \rho_1 L - \rho_1 h_2 - X \rho_1$$
$$X \rho_1 = h_2 \rho_2 + \rho_1 L - \rho_1 h_2 - L \rho_0$$
$$X = h_2 \left(\frac{\rho_2}{\rho_1} - 1 \right) - L \left(\frac{\rho_0}{\rho_1} - 1 \right)$$
$$X = 6 \left(\frac{1.65}{1.94} - 1 \right) - 24 \left(\frac{1.55}{1.94} - 1 \right)$$
$$X = 3.93 \text{ in.}$$

$$\text{Answer is (A)}$$

PROBLEM 7-12

An "overflow" can brimful of water is suspended from the hook of a spring balance. After a small block of wood is placed in the water, the balance reading is

(A) increased by the weight of the wood block
(B) decreased by the weight of the wood block
(C) decreased by the weight of the water displaced
(D) increased by the weight of the wood above the water surface
(E) unaffected

Solution

When a floating body is placed in a liquid, it sinks until it displaces its own weight of liquid.

Assuming that the wood floats, the weight of water that will overflow is equal to the weight of the wood block. The balance reading will be unchanged.

$$\text{Answer is (E)} \checkmark$$

PROBLEM 7-13

An object weights 100 lb in air and 25 lb in fresh water.

Question 1. Its volume in cubic feet is closest to

(A) 0.75
(B) 1.1
(C) 1.2
(D) 1.3
(E) 1.5

Solution

$$\text{Buoyant force} = (\text{weight in air}) - (\text{weight in water})$$

$$= 100 - 25 = 75 \text{ lb}$$

Since the buoyant force is equal to the weight of the water displaced by the object,

$$\text{Volume} = \frac{\text{weight of displaced water}}{\text{unit weight of water}} = \frac{75 \text{ lb}}{62.4 \text{ lb}/\text{ft}^3} = 1.2 \text{ ft}^3$$

Answer is (C)

Question 2. Its specific gravity (relative density) is most nearly

(A) 0.75
(B) 1.1
(C) 1.2
(D) 1.3
(E) 1.5

Solution

The buoyant force is also the weight of a volume of water equal to the volume of the object. By definition,

$$\text{Specific gravity } S = \frac{\text{weight in air}}{\text{weight of equal volume of water}} = \frac{100}{75} = 1.33$$

Answer is (D)

PROBLEM 7-14

It is said that Archimedes discovered his principle while seeking to detect a suspected fraud in the construction of a crown. The crown was thought to have been made from an alloy of gold and silver instead of pure gold. The crown had a mass of 1000 g in air and 940 g in pure water. The volume of the alloy equaled the combined volume of the components. If the density of gold is 19.3 g/cm^3 and the density of silver is 10.5 g/cm^3, the mass of gold, in g, in the crown is most nearly

(A) 780
(B) 810
(C) 860
(D) 910
(E) 940

Solution

Archimedes' principle says that an object immersed in a fluid will undergo an apparent loss of weight equal to the weight of the displaced fluid. Since the weight of an object is a nearly constant multiple of its mass on earth, the principle can also be applied directly to masses.

The mass of displaced water is $1000 - 940$ or 60 g. As the density of water is 1 g/cm^3, the volume of displaced water equals the volume of the crown, or 60 cm^3. Let

$$x = \text{mass of gold in the crown}$$
$$1000 - x = \text{mass of silver in the crown}$$

Then the volume of gold plus the volume of silver equals the volume of the crown, or

$$\frac{x}{19.3} + \frac{1000 - x}{10.5} = 60 \qquad 0.0518x + 95.24 - 0.0952x = 60$$

$$0.0434x = 35.24$$

$$\text{Mass of gold} = x = \frac{35.24}{0.0434} = 812 \text{ g}$$

Answer is (B)

PROBLEM 7-15

A uniform solid rod 24 in. long is supported at one end by a string 6 in. above the water. The specific gravity of the rod is $\frac{5}{9}$. Assume the cross section of the rod is small. The length of the immersed section of the rod, in inches, is most nearly

(A) 8
(B) 10
(C) 12
(D) 14
(E) 16

Solution

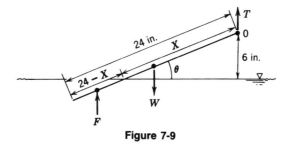

Figure 7-9

There are three forces acting on the rod:

(1) the weight of the rod $W = \gamma V = \frac{5}{9}\gamma_w(24)(A)$,
    where A = cross-sectional area of the rod
(2) the buoyant force $F = \gamma_w(24 - X)(A)$
(3) the tensile force T

The rod will be in equilibrium if $\Sigma M_0 = 0$.

$$W(12 \cos \theta) - F\left(24 - \frac{24 - X}{2}\right)\cos \theta = 0$$

Substituting the values of W and F in this equation, we have

$$12(\tfrac{5}{9}\gamma_w)(24A) - \gamma_w(24 - X)(A)\left(24 - \frac{24 - X}{2}\right) = 0$$

$$\frac{12(5)(24)}{9} - (24 - X)\left(\frac{24 + X}{2}\right) = 0$$

$$160 - \frac{24^2 - X^2}{2} = 0 \qquad X^2 = 256 \qquad X = 16 \text{ in.}$$

Therefore the immersed portion of the rod is 24 in. $-$ 16 in. $= 8$ in.

Answer is (A)

PROBLEM 7-16

A brass weight of density $8.4\ \text{g/cm}^3$ is dropped from the water surface in a tank. The water in the tank is 8.0 m deep. Neglecting water viscosity, the time, in sec, it takes for the weight to reach the bottom of the tank is closest to

(A) 0.90
(B) 0.96
(C) 1.12
(D) 1.28
(E) 1.36

Solution

Since viscosity directly or indirectly causes all fluid drag, both friction drag (shear) and pressure drag (fore and aft), neglecting viscosity implies neglecting drag forces. Thus the only forces acting on the brass weight are gravity and buoyant forces, both of which are constant. Therefore the acceleration will be constant and the following formula applies:

$$s = \tfrac{1}{2}at^2 \tag{1}$$

or, solving for the time t to cover the vertical distance s at constant acceleration a,

$$t = \left(\frac{2s}{a}\right)^{1/2} \tag{2}$$

Using $\Sigma F = ma$, we have

$$a = \frac{\Sigma F}{m} = \frac{\gamma_{\text{brass}}V - \gamma_{\text{water}}V}{\rho_{\text{brass}}V} \tag{3}$$

where γ = specific weight, ρ = mass density, and V = volume of the brass weight. Since $\gamma = \rho g$, Eq. (3) becomes

$$a = \frac{g(\rho_{\text{brass}} - \rho_{\text{water}})V}{\rho_{\text{brass}}V} = g\left(1 - \frac{\rho_{\text{water}}}{\rho_{\text{brass}}}\right) = 9.81\ \text{m/sec}^2\left(1 - \frac{1}{8.4}\right)$$

$$a = 9.81(0.881) = 8.64\ \text{m/sec}^2$$

Substituting back into Eq. (2),

$$t = \left(\frac{2 \times 8}{8.64}\right)^{1/2} = (1.85)^{1/2} = 1.36\ \text{sec}$$

Answer is (E)

PROBLEM 7-17

A rectangular barge 25 ft wide × 46 ft long × 8 ft deep floats in a canal lock which is 32 ft wide × 60 ft long × 12 ft deep. With no load on the barge other than its own weight, the bottom of the barge is 3 ft beneath the water surface, and the depth of the water in the lock is 7 ft. If a load of steel that weighs 75 tons is added to the barge, the new depth of water, in ft, in the lock is most nearly

(A) 7.0
(B) 7.5
(C) 8.2
(D) 9.0
(E) 10.1

Solution

The barge displacement with no load is $25 \times 46 \times 3 = 3450 \text{ ft}^3$. According to Archimedes, the weight of the steel will equal the weight of the additional water that is displaced. The additional volume of water to be displaced is then $75 \times 2000/62.4 = 2400 \text{ ft}^3$, and the new barge displacement is therefore $3450 + 2400 = 5850 \text{ ft}^3$.

The actual volume of water in the lock is $(32 \times 60 \times 7) - 3450 = 13{,}440 - 3450 = 9990 \text{ ft}^3$. The new *apparent* volume of water in the lock is equal to the actual volume of water plus the displacement of the loaded barge or $9990 + 5850 = 15{,}840 \text{ ft}^3$, which is equivalent to a new water depth in the lock of $15{,}840/(32 \times 60) = 8.25 \text{ ft}$.

Answer is (C)

PROBLEMS 7-18

A barge loaded with rocks floats in a canal lock with both the upstream and the downstream gates closed. If the rocks are dumped into the canal lock water with both gates still in the closed position, the water level in the lock will theoretically

(A) rise
(B) rise and then return to original level
(C) fall
(D) fall and then return to original level
(E) remain the same

Solution

With the rocks in the barge, the barge displaces additional water equal to the weight of the rocks. When the rocks are thrown into the canal, they displace their own volume of water. Thus with rocks of specific gravity greater than 1, the water level in the lock will fall.

Answer is (C)

PROBLEM 7-19

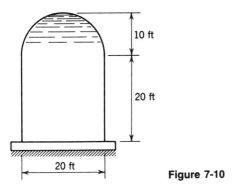

20 ft

Figure 7-10

A cylindrical water tank with a hemispherical dome has the dimensions shown. The tank is full. The total force, in lb, exerted by the water on the base of the tank is most nearly

(A) 500,000
(B) 520,000
(C) 550,000
(D) 590,000
(E) 640,000

Solution

The pressure on the bottom of the tank is due to the total head of 30 ft of water. The force is equal to the pressure times the area.

$$p = 30(62.4) = 1872 \text{ lb/ft}^2$$

$$F = pA = 1872\left(\frac{\pi}{4}\right)20^2 = 588,000 \text{ lb}$$

Answer is (D)

PROBLEM 7-20

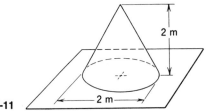

Figure 7-11

A hollow steel cone, with internal dimensions as shown, has a small hole at the apex. The cone is filled with water of weight 9800 N/m³. The minimum weight, in N, of the cone W_c that will prevent the water from uplifting the cone and flowing out is closest to

(A) 61,600
(B) 41,100
(C) 46,200
(D) 20,500
(E) 30,800

Solution

The pressure at the base of the cone is $p = \gamma h = (9800 \text{ N/m}^3)(2 \text{ m}) = 19,600 \text{ N/m}^2$(pascals). The uplift force is $R = pA = (19,600)[(\pi/4)2^2] = 61,600$ N. For vertical equilibrium of the cone and its enclosed water,

$$W_w + W_c = R$$

where W_w = weight of the enclosed water. Here

$$W_w = \gamma_w V_c = \gamma_w \times \tfrac{1}{3}(\text{area of base}) \times h$$
$$= (9800)(\tfrac{1}{3})\pi(2) = 20,500 \text{ N}$$

Hence

$$W_c = 61,600 - 20,500 = 41,100 \text{ N}$$

Answer is (B)

PROBLEM 7-21

In this manometer the unit weight for the first fluid is $\gamma_1 = 56 \text{ lb/ft}^3$, and for the second $\gamma_2 = 99 \text{ lb/ft}^3$. The pressure difference $p_A - p_B$, in lb/ft², is most nearly

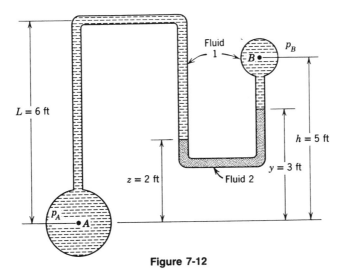

Figure 7-12

(A) 125
(B) 180
(C) 280
(D) 323
(E) 380

Solution

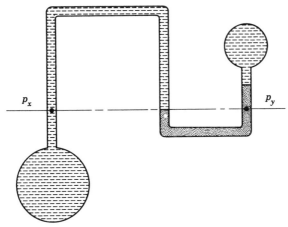

Figure 7-13

$$\Delta p = p_A - p_B$$
$$p_B = p_y - (y - z)\gamma_2 - (h - y)\gamma_1$$
$$p_A = p_x + z\gamma_1 \quad \text{or} \quad p_A - z\gamma_1 = p_x$$

Since $p_x = p_y$,

$$p_A - z\gamma_1 = p_B + (y - z)\gamma_2 + (h - y)\gamma_1$$
$$\Delta p = (h - y + z)\gamma_1 + (y - z)\gamma_2$$
$$\Delta p = (5 - 3 + 2)(56) + (3 - 2)(99)$$
$$\Delta p = 323 \text{ lb/ft}^2$$

Answer is (D)

PROBLEM 7-22

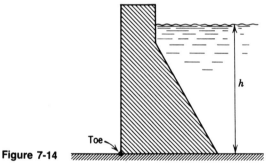

Toe

Figure 7-14

The moment tending to overturn the dam about the toe will increase in proportion to

(A) $h^{1/2}$
(B) h
(C) $h^{3/2}$
(D) h^2
(E) h^3

Solution

$$\text{Moment} = \frac{\gamma h^2}{2} \times \frac{h}{3} = \frac{\gamma h^3}{6}$$

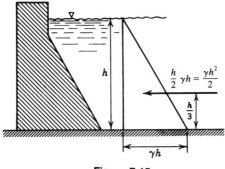

Figure 7-15

The moment will increase in proportion to h^3.

Answer is (E)

PROBLEM 7-23

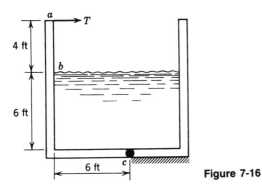

Figure 7-16

The figure shows a cross-sectional view of a 10-ft long rectangular water tank. The wall of the tank, abc, is hinged at c and supported by a horizontal tie rod at a. The force T, in lb, in the tie rod is most nearly

(A) 1000
(B) 2000
(C) 7000
(D) 8000
(E) 9000

Solution

A free-body diagram of the hinged wall is shown in Fig. 7-17; it includes the hydrostatic forces R_1 and R_2 and the hinge reactions C_x and C_y. The forces

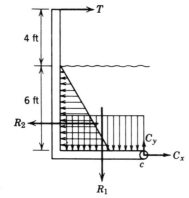

Figure 7-17

R_1 and R_2 are the concentrated forces equivalent to the shaded pressure distributions. By use of Eq. (7-4),

$$R_1 = 10(6)(62.4)(6) = 22,460 \text{ lb}$$
$$R_2 = 10(\tfrac{1}{2})(6)(62.4)(6) = 11,230 \text{ lb}$$

Now apply $\Sigma\, M_c = 0$ to obtain

$$2R_2 + 3R_1 - 10T = 0$$
$$T = \frac{2(11,230) + 3(22,460)}{10} = \frac{89,840}{10}$$
$$T = 8984 \text{ lb}$$

Answer is (E)

PROBLEM 7-24

In Fig. 7-18 the gate AB rotates about an axis through B. The gate width is 4 ft. A torque T is applied to the shaft through B. The torque T, in lb-ft, to keep the gate closed is closest to

(A) 5,000
(B) 10,000
(C) 20,000
(D) 30,000
(E) 40,000

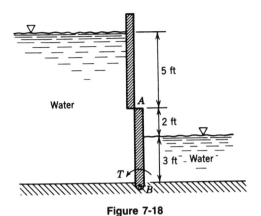

Figure 7-18

Solution

A cross-sectional view of the pressure distribution on gate AB is shown in Fig. 7-19. Also shown are three equivalent concentrated forces acting on the

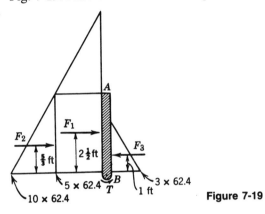

Figure 7-19

4-ft long gate, their locations, and the magnitude of the pressure (lb/ft^2) at three points. The magnitudes of the forces are

$$F_1 = 5(62.4)(5)(4) \quad = 6240 \text{ lb}$$
$$F_2 = \tfrac{1}{2}(5)(62.4)(5)(4) = 3120 \text{ lb}$$
$$F_3 = \tfrac{1}{2}(3)(62.4)(3)(4) = 1123 \text{ lb}$$

Equilibrium around B is ensured by summing moments at B:

$$\sum M_B = 0 = T - 2.5F_1 - \tfrac{5}{3}F_2 + 1F_3$$
$$T = 2.5(6240) + \tfrac{5}{3}(3120) - 1123 = 19{,}677 \text{ lb-ft}$$

Answer is (C)

PROBLEM 7-25

A rectangular gate 1.5 m wide and 2.5 m high is hinged at the top and can be opened at the bottom by pulling on a cable. The water level is 0.6 m over

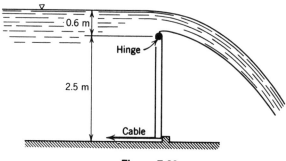

Figure 7-20

the top of the gate. Water weighs $9800\ \text{N/m}^3$. The tension in the cable, in newtons, which is required to open the gate is most nearly

   (A) 8,500
   (B) 10,500
   (C) 35,000
   (D) 41,000
   (E) 70,000

Solution

First construct a free-body diagram of the gate. We use the pressure distribution shown in Fig. 7-21 as a reasonable approximation to the true

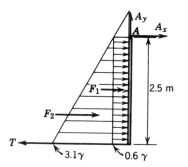

Figure 7-21

distribution. Actually the pressure is locally zero at A. Using $\gamma = 9800\ \text{N/m}^3$ for water, the equivalent concentrated forces are

$$F_1 = (0.6\gamma)(2.5)(1.5) = 2.25\gamma$$
$$F_2 = \tfrac{1}{2}(2.5\gamma)(2.5)(1.5) = 4.69\gamma$$

Now sum moments around point A.

$$\sum M_A = 0 = \tfrac{1}{2}(2.5)F_1 + \tfrac{2}{3}(2.5)F_2 - 2.5T$$
$$0 = \tfrac{1}{2}(2.25\gamma) + \tfrac{2}{3}(4.69\gamma) - T$$
$$T = 41{,}700 \text{ N}$$

Since the assumed pressure distribution slightly overestimates the true pressure distribution, this force is slightly larger than the true result.

Answer is (D)

PROBLEM 7-26

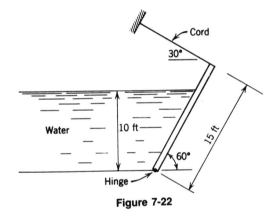

Figure 7-22

The gate is 10 ft wide. The tension in the cord, in lb, is closest to

(A) 7,000
(B) 8,000
(C) 9,000
(D) 10,000
(E) 12,000

Solution

First a free-body diagram of the gate is drawn. The hydrostatic force F acts perpendicularly to the face of the gate at a distance $d/3$ from point O. From

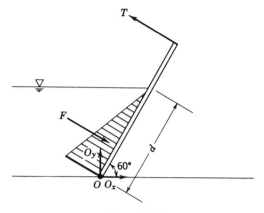

Figure 7-23

trigonometry $\sin 60° = 10/d$, and $d = 11.55$ ft. The centroid of the sub-merged portion of the gate is 5 ft below the water surface. Using Eq. (7-4),

$$F = \gamma h_c A = \gamma(5)(10d)$$

Summing moments around O for equilibrium,

$$\Sigma M_O = 0 = 15T - \frac{d}{3} F$$

$$T = \frac{dF}{3(15)} = \frac{50\gamma d^2}{45} = \frac{50}{45}(62.4)(11.55)^2$$

$$T = 9250 \text{ lb}$$

Answer is (C)

PROBLEM 7-27

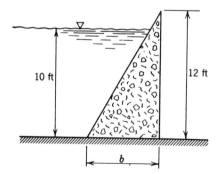

Figure 7-24

Masonry weighs 150 lb/ft^3 and the coefficient of friction between the bottom of the dam and the stream bed is 0.4. The required value of b, in ft, to prevent overturning of the masonry dam is most nearly

(A) 4.6
(B) 4.7
(C) 5.1
(D) 5.5
(E) 5.9

Solution

The problem can be readily solved by first calculating the horizontal and vertical components of the hydrostatic force.

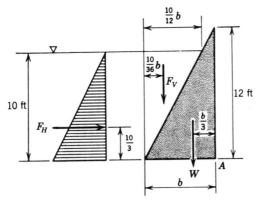

Figure 7-25

$$F_H = \frac{\gamma h}{2} h = \frac{(62.4)(10)}{2} (10) = 3120 \text{ lb}$$

$$F_V = \frac{1}{2} \gamma h \left(\frac{10}{12} b \right) = \frac{(62.4)(10)}{2} \left(\tfrac{10}{12} b \right) = 260b \text{ lb}$$

$\Sigma M_A = 0$

$$-\tfrac{10}{3} F_H + \tfrac{26}{36} b F_V + \frac{b}{3} W = 0$$

$$-\tfrac{10}{3} (3120) + \tfrac{26}{36} b (260b) + \frac{b}{3} (900b) = 0$$

$$-10,400 + 188b^2 + 300b^2 = 0$$

$$b = \left(\frac{10,400}{488} \right)^{1/2} = (21.3)^{1/2} = 4.62 \text{ ft}$$

Although a friction factor for sliding is given, the problem asks only for the base width necessary to prevent overturning; hence a check on sliding is not needed.

Answer is (A)

PROBLEM 7-28

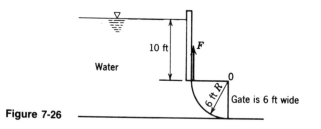

Figure 7-26

Question 1. The horizontal component of the forces acting on the radial gate, and its line of action above the base, are most nearly

   (A) 6,800 lb and 2.00 ft
   (B) 6,800 lb and 3.00 ft
   (C) 22,500 lb and 3.00 ft
   (D) 29,230 lb and 2.77 ft
   (E) 29,230 lb and 3.00 ft

Solution

First construct a free-body diagram of the gate and some of the water outside the gate:

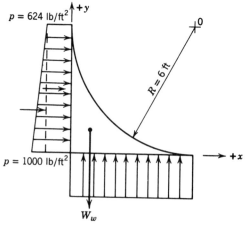

Figure 7-27

At the top of the gate the pressure is $p = 10\gamma = 624$ lb/ft^2. At the base of the gate the pressure is $p = 16\gamma = 1000$ lb/ft^2.

The horizontal force is

$$F_h = 624(6)(6) + \tfrac{376}{2}(6)(6) = 22{,}460 + 6770 = 29{,}230 \text{ lb}$$

Its line of action is

$$\bar{y} = \frac{\Sigma\, yF}{\Sigma\, F} = \frac{3(22{,}460) + 2(6770)}{29{,}230} = 2.77 \text{ ft above base}$$

$$\text{Answer is (D)}$$

Question 2. The vertical component of the forces acting on the gate, and its line of action from point 0, are most nearly

(A) 30,200 lb and 2.55 ft
(B) 33,100 lb and 2.85 ft
(C) 36,000 lb and 3.00 ft
(D) 33,100 lb and 3.15 ft
(E) 30,200 lb and 3.45 ft

Solution

The net vertical force is

$$F_v = pA - W_w$$

$$F_v = 1000(36) - (62.4)\left(36 - \frac{36\pi}{4}\right)(6) = 36{,}000 - 2900 = 33{,}100 \text{ lb}$$

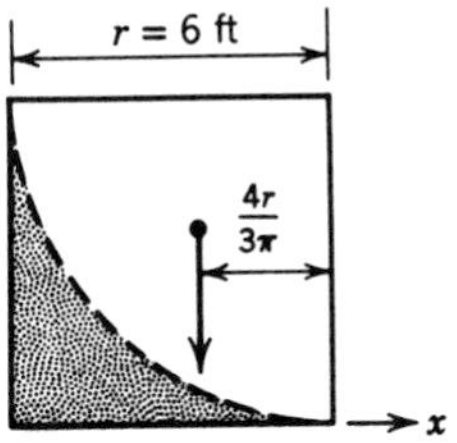

Figure 7-28

The centroid of the water alone, measured from the left edge, is

$$\bar{x} = \frac{\Sigma\, Ax}{A} = \frac{6(6)(3) - \frac{\pi}{4}(6)^2\left[6 - \frac{4(6)}{3\pi}\right]}{6(6) - \frac{\pi}{4}(6)^2}$$

$$\bar{x} = \frac{108 - 97.6}{7.73} = \frac{10.4}{7.73} = 1.35 \text{ ft}$$

The line of action for F_v is therefore

$$\bar{x} = \frac{36,000(3) - 2900(1.35)}{33,100} = \frac{108,000 - 3900}{33,100}$$

$$\bar{x} = \frac{104,100}{33,100} = 3.15 \text{ ft}$$

The distance from point 0 is $6.0 - 3.15 = 2.85$ ft.

Answer is (B)

Question 3. Neglecting the gate's weight, the force F, in lb, required to open the gate is most nearly

(A) 18,000
(B) 13,600
(C) 10,000
(D) 4,500
(E) 0

Solution

The force to open the gate is found by using $\Sigma M_0 = 0$.

$$6F = -33,100(2.85) + 29,230(3.23)$$

$$= -94,400 + 94,400$$

Therefore $F = 0$. This is to be expected since the pressure forces acting directly on the cylinder have lines of action that pass through the center of rotation of the gate.

Answer is (E)

PROBLEM 7-29

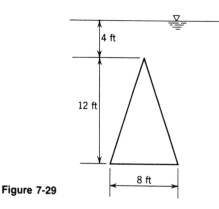

Figure 7-29

Hydrostatic pressure acts on the triangular gate shown in Fig. 7-29. Use the water surface as the reference plane in locating the resultant force.

Question 1. The resultant force, in lb, on the gate is closest to

(A) 24,000
(B) 30,000
(C) 32,000
(D) 36,000
(E) 48,000

Solution

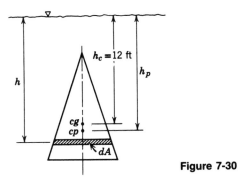

Figure 7-30

Integrate the pressure distribution over the gate to find the force. The force on the differential area dA is

$$dF = \gamma h \, dA$$

$$F = \int dF = \gamma \int h \, dA$$

But $\int h \, dA$ is the moment of the area or $h_c A$; hence

$$F = \gamma h_c A = 62.4(12)(\tfrac{8}{2})(12) = 36{,}000 \text{ lb}$$

Answer is (D)

Question 2. The distance, in ft, from the water surface to the point of application of the resultant is most nearly

(A)   7.3
(B) 10.0

(C) 12.0
(D) 12.6
(E) 13.0

Solution

The point of application of the force is located on the vertical centerline of the gate since it is symmetrical. With the water surface as the reference plane, h_p is equal to the moment of the force divided by the force: $h_p = M/F$. $dM = h\,dF$ with $dF = \gamma h\,dA$, so $dM = \gamma h^2\,dA$ and $M = \gamma \int h^2\,dA$. Since $\int h^2\,dA$ is the moment of inertia of the area with respect to the water surface, $M = \gamma I_{ws}$.

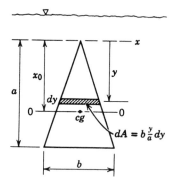

Figure 7-31

We must now determine the moment of inertia of the triangle. Using the transfer equation $I_{ws} = I_0 + Ah_c^2$, we find

$$I_x = \int y^2\,dA = \frac{b}{a}\int_0^a y^3\,dy = \frac{b}{a}\left.\frac{y^4}{4}\right|_0^a = \frac{ba^4}{4a} = \frac{ba^3}{4}$$

$$I_0 = I_x - Ax_0^2$$

$$I_0 = \frac{8(12)^3}{4} - \frac{8}{2}(12)(8)^2 = 3456 - 3072 = 384 \text{ ft}^4$$

$$I_{ws} = 384 + \tfrac{8}{2}(12)(12)^2$$

$$= 384 + 6912 = 7296 \text{ ft}^4$$

$$h_p = \frac{M}{F} = \frac{\gamma I_{ws}}{F}$$

$$= \frac{(62.4)(7296)}{36,000} = 12.65 \text{ ft}$$

Answer is (D)

PROBLEM 7-30

A vertical water softener is to operate under the following conditions:

1. Water flow of 307 gpm
2. Maximum flow rate of 8.0 gpm/ft^2 of area
3. Supply water with a hardness of 12.0 grains/gal
4. Softener contains 95.0 ft^3 of exchange resin
5. Exchange value of resin is 24,000 grains/ft^3

Question 1. The required diameter, in ft, of the softener is nearest to

(A) 8
(B) 7
(C) 6
(D) 5
(E) 4

Solution

$$Q = AV$$

$$307 \text{ gpm} = \frac{\pi}{4} D^2 \times 8 \text{ gpm/ft}^2 \qquad D = \left[\left(\frac{307}{8} \right) \left(\frac{4}{\pi} \right) \right]^{1/2} = \sqrt{48.9} = 7 \text{ ft}$$

Answer is (B)

Question 2. The number of gallons of water softened between regenerations is nearest to

(A) 70,000
(B) 95,000
(C) 140,000
(D) 165,000
(E) 190,000

Solution

$$\text{Capacity between regenerations} = \frac{\text{total exchange capacity}}{\text{supply water hardness}}$$

$$= \frac{24,000 \text{ grains/ft}^3 \times 95 \text{ ft}^3}{12 \text{ grains/gal}}$$

$$= 190,000 \text{ gal}$$

Answer is (E)

PROBLEM 7-31

The theoretical head, in m, required to push water through a 3-cm round orifice at 15 m/sec is closest to

(A) 9.8
(B) 11.5
(C) 12.7
(D) 13.4
(E) 22.9

Solution

$$h = \frac{v^2}{2g} = \frac{(15)^2}{2(9.81)} = 11.47 \text{ m}$$

Answer is (B)

PROBLEM 7-32

Water is flowing through a pipe. The following data are known:

$$D = 2 \text{ in.} \qquad h_f = 20 \text{ ft}$$

$$p = 70 \text{ lb/in.}^2 \qquad R = 4 \times 10^5$$

$$n = 0.015 \qquad V = 25 \text{ ft/sec}$$

The rate of flow, in U.S. gallons per minute, is closest to

(A) 200
(B) 245
(C) 310
(D) 770
(E) 980

Solution

$$Q = VA = V\left(\frac{\pi}{4}\right)D^2 = (25 \text{ ft/sec})\left(\frac{\pi}{4}\right)\left(\frac{2}{12} \text{ ft}\right)^2 = 0.545 \text{ ft}^3/\text{sec}$$

$$= 0.545 \times 60 \text{ sec/min} \times 7.48 \text{ gal/ft}^3 = 245 \text{ gpm}$$

Answer is (B)

PROBLEM 7-33

Water flows through two orifices in the side of a large water tank. The water surface in the tank is constant. The upper orifice is 16 ft above the ground surface, and this stream strikes the ground 8 ft from the base of the tank. The stream from the lower orifice strikes the ground 10 ft from the base of the tank. For convenience, assume $g = 32$ ft/sec^2.

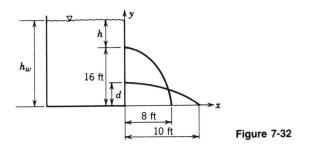

Figure 7-32

Question 1. The height h_w, in ft, of the water surface above the ground is

(A) 20
(B) 19
(C) 18
(D) 17
(E) 16

Solution

For the upper stream, $X = vt$ and $Y = \frac{1}{2}gt^2$ on the trajectory of the stream. Eliminating t, $X^2 = (2v^2/g)Y$, where v is the velocity at the vena contracta of the stream:

$$v^2 = \frac{X^2 g}{2Y} = \frac{(8^2)(32)}{(2)(16)} = 64 \qquad v = 8 \text{ ft/sec}$$

Neglecting friction, the Bernoulli equation may be written as

$$\frac{V_1^2}{2g} + \frac{p_1}{\gamma} + h = \frac{V_2^2}{2g} + \frac{p_2}{\gamma}$$

where subscript 2 refers to the vena contracta and subscript 1 refers to the surface of the water in the tank.

$$V_2 = v = \sqrt{2g}\left(h + \frac{p_1 - p_2}{\gamma} + \frac{V_1^2}{2g} \right)^{1/2}$$

where $p_1 = p_2 =$ atmospheric pressure and $V_1 = 0$. Therefore

$$V_2 = v = \sqrt{2gh}$$

Using the velocity at the vena contracta, we find h is

$$8 = \sqrt{2(32)h} \qquad h = 1 \text{ ft}$$
$$h_w = 16 + h = 17 \text{ ft}$$

Answer is (D)

Question 2. If the two orifices are known to be more than 2 ft apart, the height d, in ft, of the lower orifice above the ground is closest to

(A) 1.6
(B) 3.8
(C) 8.5
(D) 13.2
(E) 15.4

Solution

The velocity at the vena contracta of the lower orifice will be

$$v = \sqrt{2(32)(17 - d)}$$

Using the equation $X^2 = (2v^2/g)Y$, where now $Y = d$,

$$10^2 = \frac{2[2(32)(17 - d)]}{32} \text{ (d)} \quad \text{or} \quad 100 = 4(17 - d)(d)$$

Solving for d,

$$d^2 - 17d + 25 = 0$$
$$d = \frac{17 \pm \sqrt{17^2 - 4(25)}}{2} = \frac{17 \pm \sqrt{189}}{2} = \frac{17 \pm 13.8}{2}$$
$$d = 15.4 \text{ ft} \quad \text{or} \quad d = 1.6 \text{ ft}$$

The first answer must be rejected because it is too near to the upper orifice.

Answer is (A)

PROBLEM 7-34

In a horizontal Venturi meter water flows through a 4-in.-diameter constriction in an 8-in.-diameter pipe. Simple piezometer columns indicate a 10-ft difference in head h between the upstream and constricted sections. Neglecting energy losses, the discharge through the pipe, in ft³/sec, is closest to

(A) 2.3
(B) 3.7
(C) 5.1
(D) 7.5
(E) 9.1

Solution

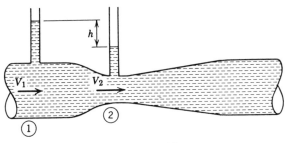

Figure 7-33

The continuity equation is

$$Q = A_1 V_1 = A_2 V_2 \tag{1}$$

The Bernoulli equation is

$$\frac{p_1}{\gamma} + \frac{V_1^2}{2g} + z_1 = \frac{p_2}{\gamma} + \frac{V_2^2}{2g} + z_2 \tag{2}$$

In our case, $z_1 - z_2 = 0$ and $(p_1/\gamma) - (p_2/\gamma) = h$, so the equation reduces to

$$\frac{V_1^2}{2g} + h = \frac{V_2^2}{2g} \tag{3}$$

Squaring Eq. (1),

$$Q^2 = A_1^2 V_1^2 = A_2^2 V_2^2 \qquad V_1^2 = \frac{Q^2}{A_1^2} \qquad V_2^2 = \frac{Q^2}{A_2^2}$$

Substituting these values into Eq. (3),

$$\frac{Q^2}{A_1^2 2g} + h = \frac{Q^2}{A_2^2 2g}$$

Factoring,

$$Q^2\left(\frac{1}{A_1^2} - \frac{1}{A_2^2}\right) = -2gh$$

$$Q^2\left[\left(\frac{A_1}{A_2}\right)^2 - 1\right] = 2ghA_1^2 \qquad Q = A_1\left[\frac{2gh}{(A_1/A_2)^2 - 1}\right]^{1/2}$$

Finally, recalling that the area ratio varies as the square of the diameters, we compute

$$Q = \frac{\pi}{4}\left(\frac{8}{12}\right)^2\left[\frac{2(32.2)(10)}{(8/4)^4 - 1}\right]^{1/2} = 2.29 \text{ ft}^3/\text{sec}$$

Answer is (A)

PROBLEM 7-35

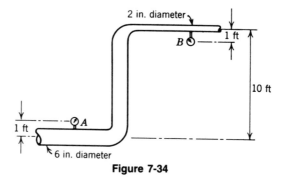

Figure 7-34

Gage A indicates a pressure of 25 psi, and Gage B indicates a pressure of 15 psi. Assume that losses are negligible in the transition from the 6-in. pipe to the 2-in. pipe. The rate of flow in gallons per minute for this system is closest to

(A) 1000
(B) 650
(C) 400
(D) 310
(E) 290

Solution

The Bernoulli equation is

$$\frac{p_1}{\gamma} + \frac{V_1^2}{2g} + z_1 = \frac{p_2}{\gamma} + \frac{V_2^2}{2g} + z_2 \qquad (1)$$

Also, 1 ft of water $= 0.433$ psi. Each of the terms can be evaluated with reference to a datum plane passing through the 6-in.-diameter pipe.

$$p_6 = 25 \text{ psig} + 1 \text{ ft water} = 25 + 0.433 = 25.4 \text{ psig} \qquad z_6 = 0$$

$$p_2 = 15 \text{ psig} - 1 \text{ ft water} = 15 - 0.433 = 14.6 \text{ psig} \qquad z_2 = 10 \text{ ft}$$

$$\frac{25.4(144)}{62.4} + \frac{V_6^2}{(2)(32.2)} + 0 = \frac{14.6(144)}{62.4} + \frac{V_2^2}{(2)(32.2)} + 10 \qquad (2)$$

$$58.7 + 0.0155 V_6^2 = 33.7 + 0.0155 V_2^2 + 10$$

$$V_2^2 - V_6^2 = \frac{58.7 - 43.7}{0.0155} = 968 \text{ ft}^2/\text{sec}^2 \qquad (3)$$

The continuity equation shows

$$Q = A_6 V_6 = A_2 V_2 \qquad (4)$$

Using

$$A_2 = \frac{\pi}{4}\left(\frac{2}{12}\right)^2 = \frac{\pi}{144} = 0.0218 \text{ ft}^2$$

and

$$A_6 = \frac{\pi}{4}\left(\frac{6}{12}\right)^2 = \frac{9\pi}{144} = 0.1963 \text{ ft}^2$$

$$V_6 = \frac{A_2}{A_6} V_2 = \frac{0.0218}{0.1963} V_2 = 0.111 V_2 \qquad V_6^2 = 0.0123 V_2^2$$

Substituting this relation into Eq. (3),

$$V_2^2 - 0.0123 V_2^2 = 968 \text{ ft}^2/\text{sec}^2$$

$$V_2 = \left(\frac{968}{0.9877}\right)^{1/2} = (980)^{1/2} = 31.3 \text{ ft/sec}$$

Now using Eq. (4),

$$Q = A_2 V_2 = (0.0218)(31.3) = 0.68 \text{ ft}^3/\text{sec}$$

But the problem asks for the rate of flow in gallons per minute. Here

$$Q = 0.68 \text{ cfs} \times 7.48 \text{ gal/ft}^3 \times 60 \text{ sec/min} = 305 \text{ gpm}$$

$$\text{Answer is (D)}$$

PROBLEM 7-36

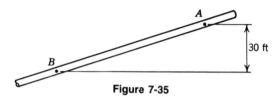

Figure 7-35

In Fig. 7-35 the pipe is of uniform diameter. The gage pressure at A is 20 psi and at B is 30 psi. If the liquid has a specific weight (weight density) of 30 lb/ft^3, the head loss and direction of flow are

(A) 18 ft, from A to B
(B) 29 ft, from A to B
(C) 18 ft, from B to A
(D) 29 ft, from B to A
(E) 7 ft, from A to B

Solution

Assume the direction of flow is from B to A. The Bernoulli equation is

$$\frac{p_B}{\gamma} + \frac{V_B^2}{2g} + z_B = \frac{p_A}{\gamma} + \frac{V_A^2}{2g} + z_A + h_L$$

Since the diameter is uniform, the continuity equation $Q = AV$ tells us that $V_B = V_A$. Therefore the Bernoulli equation reduces to

$$\frac{p_B}{\gamma} + z_B = \frac{p_A}{\gamma} + z_A + h_L$$

$$\frac{30(144)}{30} + 0 = \frac{20(144)}{30} + 30 \text{ ft} + h_L$$

$$144 = 96 + 30 + h_L \qquad \text{head loss } h_L = 18 \text{ ft}$$

Since h_L is positive, the assumed direction of flow (from B to A) is correct.

Answer is (C)

PROBLEM 7-37

A 15-cm ID pipe discharges water at an elevation 10 m below the surface of a reservoir. If the total head loss to the point of discharge is 5 m, the discharge, in m^3/sec, is most nearly

(A) 0.18
(B) 0.22
(C) 0.25
(D) 0.32
(E) 0.70

Solution

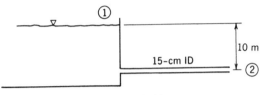

Figure 7-36

The Bernoulli equation is

$$\frac{p_1}{\gamma} + \frac{V_1^2}{2g} + z_1 = \frac{p_2}{\gamma} + \frac{V_2^2}{2g} + z_2 + h_L$$

p_1 and V_1 are zero at the free surface of the water. p_2 is zero since the water discharges at atmospheric pressure. Using point 2 as the reference elevation, $z_2 = 0$. Thus the Bernoulli equation reduces to

$$z_1 = \frac{V_2^2}{2g} + h_L$$

$$10 = \frac{V_2^2}{2g} + 5 \qquad V^2 = 10g$$

$$V = [10(9.81)]^{1/2} = 9.90 \text{ m/sec}$$

$$Q = AV = \frac{\pi}{4}(0.15)^2(9.90) = 0.175 \text{ m}^3/\text{sec}$$

Answer is (A)

PROBLEM 7-38

Friction head in a pipe carrying water varies

(A) inversely with gravity squared
(B) directly with diameter
(C) inversely with diameter

(D) directly with velocity

(E) inversely with the coefficient of friction f

Solution

The formula for friction losses in a pipe is

$$h_L = f \frac{L}{D} \frac{V^2}{2g}$$

From the formula we see that h_L varies inversely with diameter.

Answer is (C)

PROBLEM 7-39

Figure 7-37 shows the plan view of a horizontal parallel pipe section. The main and its branches are circular pipes in which the flow is laminar. The main delivers fluid at a rate Q_0 to the branch lines. These lines have diameters and effective lengths of d_1, d_2, L_1, and L_2, respectively. The only head loss is due to pipe friction. The friction factor for laminar flow in a circular pipe is inversely proportional to the Reynolds number. For the conditions $Q_0 = 9$, $d_2 = 2d_1$, and $L_2 = 2L_1$, the flow rate Q_1 is nearest to

(A) 1

(B) 2

(C) 3

(D) 4

(E) 5

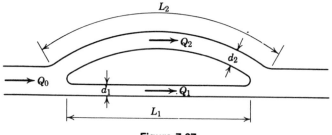

Figure 7-37

Solution

Head loss varies directly with velocity head and pipe length and inversely with pipe diameter. With a coefficient of proportionality f (the friction factor) the equation becomes

$$h_L = f \frac{L}{d} \frac{V^2}{2g}$$

The problem indicates that $f \propto 1/$Reynolds number. Since the Reynolds number for a given fluid is directly proportional to Vd,

$$h_L \propto \frac{1}{Vd} \frac{L}{d} \frac{V^2}{2g} = \frac{LV}{2d^2g} \propto \frac{LV}{d^2}$$

The head loss in each branch must be identical, so

$$\frac{L_1 V_1}{d_1^2} = \frac{L_2 V_2}{d_2^2}$$

Substituting $d_2 = 2d_1$ and $L_2 = 2L_1$

$$\frac{L_1 V_1}{d_1^2} = \frac{2L_1 V_2}{(2d_1)^2} \qquad \frac{V_1}{V_2} = \frac{2L_1 d_1^2}{4L_1 d_1^2} = \frac{1}{2}$$

$$Q = AV \qquad Q_0 = Q_1 + Q_2$$

Therefore

$$Q_0 = Q_1 + Q_2 = \frac{\pi}{4} d_1^2 V_1 + \frac{\pi}{4} (2d_1)^2 (2V_1) = 9$$

$$Q_0 = \frac{\pi}{4} d_1^2 V_1 + \frac{8\pi}{4} d_1^2 V_1 = 9 \qquad Q_1 + 8Q_1 = 9 \qquad Q_1 = 1$$

Answer is (A)

PROBLEM 7-40

A 10-in. pipeline which carries a flow of 5 ft^3/sec branches into a 6-in. line 500 ft long and an 8-in. line 1000 ft long. The 6-in. and 8-in. pipes rejoin and continue as a 10-in. line. The Darcy friction factor is 0.022 for both pipes. The flow rate, in ft^3/sec, in the 8-in. line is closest to

(A) 1.5
(B) 2.0
(C) 2.5
(D) 3.0
(E) 3.5

Solution

Where a pipeline branches and then rejoins, the head loss in each branch must be equal.

$$h_6 = \left[f \frac{L}{d} \frac{V^2}{2g} \right]_6, \qquad h_8 = \left[f \frac{L}{d} \frac{V^2}{2g} \right]_8$$

Since f and $2g$ are constants and $h_6 = h_8$,

$$\left[\frac{L}{d} V^2 \right]_6 = \left[\frac{L}{d} V^2 \right]_8 \quad \text{or} \quad \frac{500}{\frac{1}{2}} V_6^2 = \frac{1000}{\frac{2}{3}} V_8^2$$

$$V_6^2 = \frac{(1000)(\frac{1}{2})}{(500)(\frac{2}{3})} V_8^2 = 1.5 V_8^2 \qquad V_6 = (1.5 V_8^2)^{1/2} = 1.22 V_8$$

Continuity states that $Q_{10} = Q_6 + Q_8 = 5$ ft^3/sec and also $Q = AV$.

$$5 = \frac{\pi}{4} (\tfrac{1}{2})^2 V_6 + \frac{\pi}{4} (\tfrac{2}{3})^2 V_8 \qquad 5 = 0.196 V_6 + 0.349 V_8$$

Using $V_6 = 1.22 V_8$,

$$5 = 0.196(1.22 V_8) + 0.349 V_8 = 0.588 V_8$$

$$V_8 = \frac{5}{0.588} = 8.50 \text{ ft/sec}$$

$$Q_8 = A_8 V_8 = (0.349)(8.50) = 2.97 \text{ ft}^3/\text{sec}$$

Answer is (D)

PROBLEM 7-41

A pump requires 100 hp to pump water (with a specific gravity of 1.0) at a certain capacity to a given elevation. What horsepower is required if the capacity and elevation conditions are the same but the fluid pumped has a specific gravity of 0.8?

(A) 60
(B) 80
(C) 100
(D) 125
(E) 130

Solution

$$\text{Power} = \text{work per unit time} = \frac{\text{weight} \times \text{height}}{\text{time}}$$

If the specific gravity of the fluid is reduced from 1.0 to 0.8, then the weight is reduced by this ratio. Consequently, the horsepower will be reduced similarly to 80 hp.

Answer is (B)

PROBLEM 7-42

The head loss through a 24-in. butterfly valve is approximately 0.3 of a velocity head. If water is being pumped through the valve at a rate of 30 ft³/sec for 6 months out of the year, what is the annual energy charge for pumping through the valve if the cost of energy is $0.11 per kilowatt hour?

(A) $410
(B) $440
(C) $520
(D) $600
(E) $660

Solution

$$Q = AV \qquad V = \frac{Q}{A} = \frac{30}{(\pi/4)2^2} = 9.55 \text{ ft/sec}$$

$$h_L = 0.3 \frac{V^2}{2g} = 0.3 \frac{9.55^2}{(2)(32.2)}$$

$$= 0.425 \text{ ft}$$

$$\text{Lost hp} = \frac{Q\gamma h_L}{550} = \frac{(30)(62.4)(0.425)}{550} = 1.45 \text{ hp}$$

Since 1 hp = 0.746 kW, the cost per year is

$$(1.45)(0.746)(\tfrac{365}{2})(24)(0.11) = \$520$$

Answer is (C)

PROBLEM 7-43

Water is supplied by pumping for a certain industrial application. Owing to fluctuations in the water demand, two identical pumps are connected in parallel. A plot of their performance is given in Fig. 7-38.

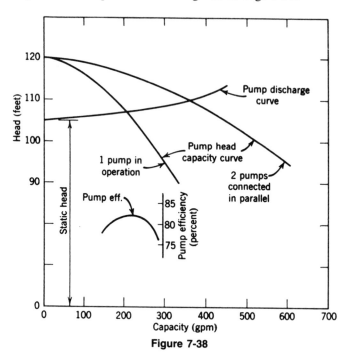

Figure 7-38

Question 1. The maximum combined pump capacity in gpm when both pumps are working is closest to

(A) 180
(B) 210
(C) 285
(D) 360
(E) 600

Solution

The maximum pump capacity is given by the intersection of the pump head and pump discharge curves:

One pump: Range of capacity 0–210 gpm
Two pumps: Range of capacity 0–360 gpm

Answer is (D)

Question 2. The maximum horsepower required when both pumps are working is closest to

(A) 6.2
(B) 6.9
(C) 12.3
(D) 15.4
(E) 18.2

Solution

$$\text{Work rate} = \frac{110 \text{ ft} \times 360 \text{ gpm} \times 8.33 \text{ lb/gal}}{0.81 \text{ eff} \times 33{,}000 \text{ ft-lb/min}}$$

$$= 12.34 \text{ hp total for both pumps}$$

Answer is (C)

PROBLEM 7-44

Water falling from a height of 120 ft at the rate of 1000 ft³/min drives a turbine that is directly connected to an electric generator. The generator rotates at 120 rpm. If the total resisting torque due to friction and other losses is 250 lb-ft and the water leaves the turbine blades with a velocity of 15 ft/sec, the horsepower developed by the generator is nearest to

(A) 207
(B) 214
(C) 221
(D) 228
(E) 235

Solution

The exit velocity head is $V^2/2g = 15^2/(2)(32.2) = 3.5$ ft, so the net head is $120 - 3.5 = 116.5$ ft.

$$\text{Power} = Q\gamma h = (1000)(62.4)(116.5) = (7.27)10^6 \text{ ft-lb/min}$$

From this we must deduct the friction and other losses to obtain the generator output.

$$\text{Lost power} = 2\pi NT = (2\pi)(120)(250) = 188{,}500 \text{ ft-lb/min}$$

$$\text{Generator horsepower} = \frac{7{,}270{,}000 - 188{,}500}{33{,}000} = 215 \text{ hp}$$

Answer is (B)

PROBLEM 7-45

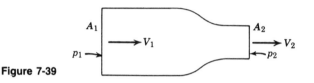

Figure 7-39

The nozzle shown has a large cross-sectional area $A_1 = 4 \text{ ft}^2$ acted on by pressure $p_1 = 23 \text{ lb/in.}^2$ and a small area $A_2 = 1 \text{ ft}^2$ acted on by $p_2 = 10 \text{ lb/in.}^2$. With water flowing into the nozzle at velocity $V_1 = 8 \text{ ft/sec}$ and out at $V_2 = 32 \text{ ft/sec}$, the external force, in lb, required to hold the nozzle stationary is nearest to

(A) 10,300
(B) 11,000
(C) 11,800
(D) 12,500
(E) 13,200

Solution

The force can be found directly by applying the momentum principle, Eq. (7-11), in the direction of flow. Upon recognizing that F_1 and F_2 are simply

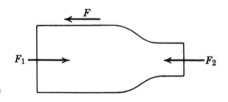

Figure 7-40

the product of the appropriate pressures and areas, and that the discharge Q can be written as $A_1 V_1$, the result from Eq. (7-11) is

$$p_1 A_1 - p_2 A_2 - F = \rho A_1 V_1 (V_2 - V_1)$$

or

$$F = p_1 A_1 - p_2 A_2 - \rho A_1 V_1 (V_2 - V_1)$$
$$F = 23(144)(4) - 10(144)(1) - 1.94(4)(8)(32 - 8)$$
$$F = 10,300 \, \text{lb}$$

Answer is (A)

PROBLEM 7-46

The most efficient shape for a prescribed cross-sectional area in open channel flow is

(A) a square
(B) a parabolic section
(C) a rectangle with the width twice the depth
(D) a semicircle
(E) a half hexagon

Solution

From Manning's equation for a fixed value of the cross-sectional area, the most efficient section is the one with the largest hydraulic radius $R = A/P$, or equivalently the shape with the smallest wetted perimeter P. Consequently the semicircle is most efficient, followed closely by the half hexagon. The rectangle with the proportions listed is the most efficient rectangular shape.

Answer is (D)

PROBLEM 7-47

A 6-ft-diameter pipe has a depth of flow of 5.6 ft. The hydraulic radius, in ft, of the pipe for this depth of flow is nearest to

(A) 1.40
(B) 1.50
(C) 1.75
(D) 2.80
(E) 3.00

Solution

By definition,

$$\text{Hydraulic radius} = \frac{\text{flow area}}{\text{wetted perimeter}} = \frac{A}{P}$$

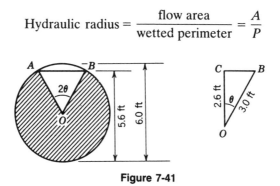

Figure 7-41

From Fig. 7-41 we can see that the determination of A and P both depend on θ. From the figure

$$\theta = \cos^{-1}\left(\frac{2.6}{3.0}\right) = 30°$$

and therefore $BC = 3.0 \sin \theta = 1.5$ ft

$$\text{Area of } \triangle OAB = 2(2.6)(1.5)\tfrac{1}{2} = 3.9 \text{ ft}^2$$

$$\text{Shaded area} = \pi R^2 \left(\tfrac{300}{360}\right) = \pi(3)^2 \left(\tfrac{5}{6}\right)$$

$$= 23.6 \text{ ft}^2$$

$$A = \text{Area of wetted section} = 3.9 + 23.6 = 27.5 \text{ ft}^2$$

$$P = \text{Wetted perimeter} = \pi D \left(\tfrac{300}{360}\right) = \pi(6.0)\tfrac{5}{6}$$

$$= 15.71 \text{ ft}$$

$$\text{Hydraulic radius } R = \frac{A}{P} = \frac{27.5}{15.71} = 1.75 \text{ ft}$$

Answer is (C)

PROBLEM 7-48

A rectangular concrete-lined channel ($n = 0.013$) 30 ft wide flows 4 ft deep on a slope of 0.0009. The discharge, in ft^3/sec, is nearest to

(A)   870
(B)   890

(C) 1040

(D) 1300

(E) 2800

Solution

The Manning equation for open channel flow is

$$Q = \frac{1.49}{n} AR^{2/3}S^{1/2}$$

Here $n = 0.013$, $S = 0.0009$, and $A = (30)(4) = 120 \text{ ft}^2$. The hydraulic radius $R = A/P$, and since the wetted perimeter $P = 30 + 2(4) = 38 \text{ ft}$, $R = 120/38 = 3.16 \text{ ft}$. Hence

$$Q = \frac{1.49}{0.013}(120)(3.16)^{2/3}(0.0009)^{1/2}$$

$$Q = \frac{1.49}{0.013}(120)(2.15)(0.03)$$

$$Q = 887 \text{ ft}^3/\text{sec}$$

Answer is (B)

PROBLEM 7-49

The tendency of a free liquid surface to contract is called

(A) elasticity

(B) adhesion

(C) cohesion

(D) capillarity

(E) surface tension

Solution

This phenomenon is called surface tension.

Answer is (E)

8

THERMODYNAMICS

Thermodynamics studies energy; it normally concentrates primarily on thermal and mechanical energy transfer. Engineering thermodynamics applies this knowledge to the analysis and design of a myriad of engineering devices, including engines of all kinds. Being both broadly based and practically useful, thermodynamics is founded on a small number of fundamental laws. We shall first review some of these basic principles and then look into the application of thermodynamics to some engineering problems of general interest.

We begin by reviewing the concepts of absolute temperature and absolute pressure which are used in gas law calculations both in thermodynamics and in chemistry. The first and second laws of thermodynamics are then examined. Some properties of gases and that important working fluid, steam, are reviewed. This is followed by an examination of some elements of compression processes and cycles. The chapter concludes with a brief review of some heat transfer principles.

In this field, quantities are routinely expressed in a diversity of unit systems, and more than normal care must be exercised to ensure that the units used within an equation are consistent.

ABSOLUTE TEMPERATURE AND PRESSURE

The temperature scales in common use around the world are relative temperature scales. Two such scales, called the Celsius or Centigrade and the Fahrenheit scales, have in the past been constructed by assuming a linear temperature variation between arbitrarily chosen numerical values assigned to the freezing and boiling points for water. The Celsius scale was divided into 100 parts between 0°C (freezing) and 100°C (boiling); the Fahrenheit scale had 180 divisions between 32°F and 212°F.

An absolute temperature scale can be associated with each of the two foregoing relative temperature scales. The lower end of each absolute scale is called absolute zero and is the lowest temperature that can exist. One of these absolute scales, the Kelvin scale, assigns the value 273°K to the freezing point of water; the other, the Rankine scale, assigns the value of 492°R to this point. Also 1 K° = 1 C° and 1 R° = 1 F°. The conversion from one temperature scale to another is easily accomplished by deriving the relation from Fig. 8-1.

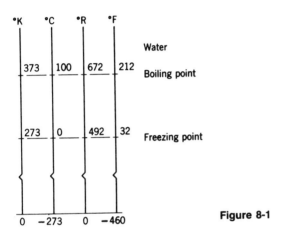

Figure 8-1

Many pressure gages register differential pressures, usually the difference in pressure between the point of interest and the surrounding atmosphere; this is a gage pressure. In many problems in thermodynamics, however, it is important to use absolute pressures that are measured relative to a perfect vacuum. Assuming that atmospheric pressure is known, we can easily convert values from one system to the other since the absolute pressure is equal to the sum of the gage pressure and the local atmospheric pressure. At sea level, atmospheric pressure is normally 14.7 psia.

THERMAL EQUILIBRIUM

A body is in thermal equilibrium when it is not exchanging thermal energy in the form of heat with another body. Directly related to this is the zeroth law of thermodynamics: two bodies each in thermal equilibrium with a third body will also be in thermal equilibrium with each other. On the other hand, if two bodies are placed together when they are not in thermal equilibrium, heat transfer will take place from one body to the other so that equilibrium will eventually be established. The amount of heat transfer Q is

$$Q = mc(T_2 - T_1) \tag{8-1}$$

where m is the mass of the body, T_2 and T_1 are, respectively, the final and initial temperatures of the body, and c is the specific heat (or specific heat capacity) of the body (the amount of heat transfer per unit mass per degree). For nongases c is relatively independent of the nature of the heat transfer process. For gases this is not true; the specific heats c_p and c_v are normally given for a constant-pressure and a constant-volume process for a gas. In some cases a change of phase occurs during the heat transfer process. Then the constant that is characteristic of the phase change is the latent heat L or specific latent heat l, and the heat transfer is $Q = L = ml$.

Example 1

In a thermos flask are 90 g of water and 10 g of ice in equilibrium at a temperature of 0°C. A 100-g piece of metal with a specific heat of 0.40 cal/g-°C and a temperature of 100°C is dropped into the flask. What is the final equilibrium temperature, assuming no heat loss or gain to or from the surroundings? (The latent heat of fusion of water is 80 cal/g.)

Solution

Since the system within the thermos is thermally insulated from its surroundings, the heat lost by the metal must equal the heat gained by the ice and water. Also, the ice and water are initially in equilibrium so that all the ice will melt before the water temperature is raised. Stated mathematically with T_e as the unknown equilibrium temperature, Eq. (8-1) gives

$$(100 \text{ g})(0.40 \text{ cal/g-°C})(100°C - T_e)$$
$$= (10 \text{ g})(80 \text{ cal/g}) + (90 \text{ g} + 10 \text{ g})(1.0 \text{ cal/g-°C})T_e$$

or

$$4000 - 40T_e = 800 + 100T_e$$
$$140T_e = 3200$$
$$T_e = 22.9°C$$

FIRST LAW OF THERMODYNAMICS

This law is a statement that energy is conserved. It is a very general law. For different classes of problems the law may be written in varying mathematical forms for ease of application. In all cases, however, the equations state that during a given process the net amount of heat Q transferred into a system is equal to the net work output W of the system plus the change in the system internal energy ΔE, that is,

$$Q = W + \Delta E \qquad (8\text{-}2)$$

Other definitions or sign conventions may also be selected here; what is important is that the net energy change of the system is the difference between the amount of energy entering and leaving the system. In this equation the internal energy is a property of the system, while heat and work generally are not properties. One exception, however, is the adiabatic process $(Q = 0)$ where the net work is identified with the change in internal energy.

Most constant-mass or nonflow processes can be analyzed directly by Eq. (8-2), but for flow processes the terms are usually modified to make more explicit the different contributing energy terms. The foregoing work term for a flow process represents both shaft work and the flow work pv, where p is the fluid pressure and v is the fluid specific volume (reciprocal of the density ρ); let W now represent shaft work only. We also split the internal energy per unit mass into kinetic energy per unit mass $V^2/2$, potential energy per unit mass gz and specific internal energy u and at this time introduce the enthalpy $h = u + pv$. Since these terms are often not all in the same units, care in converting to common units is necessary, as the problems will show. The steady flow energy equation, written on a unit time basis, is then

$$\left(\frac{V^2}{2} + gz + h\right)_1 \dot{m} + Q = \left(\frac{V^2}{2} + gz + h\right)_2 \dot{m} + W \qquad (8\text{-}3)$$

The mass rate of flow $\dot{m}$ is the same in steady flow for the entering and exiting stream. The subscript 1 denotes terms related to the entering fluid stream and the subscript 2 refers to the exiting stream. For additional fluid streams, additional groups of terms must be added. For only one stream entering and exiting, Eq. (8-3) may be divided by $\dot{m}$ to express the first law on a unit mass basis.

SECOND LAW OF THERMODYNAMICS

The second law of thermodynamics, unlike the first law, is not a conservation law. One statement of the second law, following Kelvin and Planck, says that a system cannot operate cyclically and produce a net work output while exchanging heat only at one fixed temperature. Systems whose operations violate the second law are impossible. Related to the second law are the concepts of reversibility, entropy, and thermal efficiency.

A reversible process is one where *both* the system *and* its surroundings can be restored exactly to a prior state; all other processes are irreversible. Strictly speaking, no real process is reversible, but it is a useful concept and many actual processes closely approximate this condition. One system property, the entropy S, is defined in terms of a reversible heat transfer process as

$$S_B - S_A = \Delta S_{AB} = \int_A^B dS = \int_A^B \frac{dQ_{rev}}{T} \tag{8-4}$$

between any system states A and B. This definition of entropy in terms of the reversible heat transfer dQ_{rev} applies regardless of whether the actual process is reversible or irreversible. From Eq. (8-4) we note that a reversible, adiabatic ($dQ = 0$) process is isentropic ($S = $ constant). In general, any process satisfying two conditions, those of being reversible and adiabatic or isentropic, also satisfies the third condition. Finally, in all nonisentropic processes the overall entropy of the process must increase, according to the second law.

The efficiency η of a cyclical process is the ratio of the net work output to the total heat input. If Q_1 is the heat input and Q_2 is the heat output of a cycle, then the first law shows the thermal efficiency of the cycle is

$$\eta = 1 - \frac{Q_2}{Q_1} \tag{8-5}$$

The maximum thermal efficiency occurs for a reversible engine; the second law then shows the thermal efficiency to be $\eta = 1 - T_2/T_1$, where T_1 and T_2 are, respectively, the (absolute) temperature of the heat received and rejected.

Example 2

A mass flow of 200 lbm of air per minute is passing through a steady flow machine. Entrance conditions are: pressure $= 400$ psia; specific volume $= 1.387$ ft^3/lbm; temperature $= 1040°F$; velocity $= 100$ ft/sec. Exit conditions are: pressure $= 20$ psia; specific volume $= 9.25$ ft^3/lbm; temperature $= 40°F$; velocity $= 110$ ft/sec. The transferred heat given up by the air is 40 Btu/lbm. The entrance and exit connections are at the same elevation. For air, $c_p = 0.24$ Btu/lbm-°R.

Find the shaft horsepower. Is the work done on or by the air?

Solution

We apply the steady flow energy equation, Eq. (8-3), to this process. Since the equation is written on a unit time basis, Q and W represent heat transfer and work per unit time. The entrance and exit elevations are identical; their net effect is therefore zero. Hence Eq. (8-3) is

$$W = Q + \dot{m}\left(h_1 - h_2 + \frac{V_1^2 - V_2^2}{2}\right)$$

The entrance and exit temperatures are $T_1 = 1040 + 460 = 1500°R$, $T_2 = 40 + 460 = 500°R$. It can be shown for perfect gases that $h = c_p T$, so that

$$h_1 - h_2 = c_p(T_1 - T_2)$$
$$h_1 - h_2 = (0.24 \text{ Btu/lbm-°R})(1500 - 500)°R = 240 \text{ Btu/lbm}$$

In computing the velocity terms, we must convert from mechanical to thermal units:

$$\tfrac{1}{2}(V_1^2 - V_2^2) = \frac{\tfrac{1}{2}(100^2 - 110^2) \text{ ft}^2/\text{sec}^2}{(32.2 \text{ lbm-ft/lbf-sec}^2)(778 \text{ ft-lbf/Btu})} = -0.042 \frac{\text{Btu}}{\text{lbm}}$$

Hence

$$W = \left(-40 \frac{\text{Btu}}{\text{lbm}}\right)\left(200 \frac{\text{lbm}}{\text{min}}\right) + \left(200 \frac{\text{lbm}}{\text{min}}\right)(240 - 0.042) \frac{\text{Btu}}{\text{lbm}}$$

$$W = 40{,}000 \frac{\text{Btu}}{\text{min}}$$

$$W = \frac{\left(40{,}000 \frac{\text{Btu}}{\text{min}}\right)\left(\frac{1}{60} \frac{\text{min}}{\text{sec}}\right)\left(778 \frac{\text{ft-lbf}}{\text{Btu}}\right)}{\left(550 \frac{\text{ft-lbf}}{\text{sec-hp}}\right)}$$

$$W = 943 \text{ hp done by the air}$$

STEAM TABLES

In thermodynamic analyses it is regularly necessary to know the properties of fluids which do not obey a simple equation of state. This is the case for water in the liquid and vapor (steam) states, since it is so commonly used as the working fluid in machines. Steam tables provide these data in tabular form;* much of the same data is alternately displayed in thermodynamic charts such as the Mollier diagram, which is an enthalpy-entropy chart for steam.

For liquid-vapor mixtures of water, called unsaturated steam, the value of any extensive property of the mixture, such as the specific volume v, enthalpy h, entropy s, or internal energy u, is the sum of individual property

* J. H. Keenan et al., *Steam Tables* (New York: Wiley, 1985). Abridged versions of such tables appear in many thermodynamics textbooks.

values for the liquid and vapor phases. As an example, consider specific volume. Let the subscript f indicate the saturated liquid state and the subscript g denote the saturated vapor state. The difference in specific volume between the two saturated states is v_{fg}, that is,

$$v_{fg} = v_g - v_f \tag{8-6}$$

For a mass fraction x (called quality) of vapor in a mixture, the mixture specific volume v_x is found by summing the liquid and vapor fractional components; thus

$$v_x = (1 - x)v_f + xv_g$$

or

$$\tag{8-7}$$

$$v_x = v_f + xv_{fg}$$

In many problems the steam quality x is not given directly but instead is determinable from two stated properties. Once x is found, the other fluid properties can also be computed in the same manner as v_x.

Example 3

One pound of a mixture of steam and water at 160 psia is contained in a rigid vessel. Heat is added to the vessel until the contents are at 560 psia and 600°F. Determine the quantity of heat, in Btu, added to the tank contents.

Solution

Using the tables for steam at 560 psia and 600°F, we find the final enthalpy to be $h_2 = 1293.4$ Btu/lbm, and the vapor is in a superheated state. The associated specific volume is $v_2 = 1.0224$ ft³/lbm. Since the containing vessel is rigid, this is also the specific volume at 160 psia when the process began. At 160 psia, $v_f = 0.01815$ ft³/lbm and $v_g = 2.834$ ft³/lbm. Hence

$$v_1 = (1 - x)v_f + xv_g$$
$$1.0224 = (1 - x)(0.0182) + x(2.834)$$

The steam quality is

$$x = 0.357$$

Then the original enthalpy was

$$h_1 = (1 - x)h_f + xh_g$$
$$h_1 = (1 - 0.357)(335.93) + (0.357)(1195.1)$$
$$h_1 = 642.7 \text{ Btu/lbm}$$

Since no work can be done on a fluid in a rigid container, the heat added, according to the first law, is

$$Q = E_2 - E_1 = u_2 - u_1$$

Internal energy can be found from enthalpy $h = u + pv$, and the result is

$$Q = h_2 - h_1 - (p_2 - p_1)v = 1293.4 - 642.7 - (560 - 160)(144)(1.0224)/778$$

$$Q = 575 \text{ Btu}$$

GAS LAW RELATIONS

The behavior of many common gases is reasonably well described by a set of simple equations. These equations are called gas laws, and the gases that are assumed to obey these laws exactly are called perfect (or ideal) gases.

The equation of state for a mass m of a gas is

$$pV = mRT \qquad (8\text{-}8)$$

Here V is the volume occupied by the gas, and p and T are the absolute pressure and temperature. R is a gas constant related to the universal gas constant $\bar{R}$ by the relation $R = \bar{R}/M$, M being the molecular weight of the gas. Numerically, $\bar{R} = 1545$ ft-lbf/lbm-mole-°R, and for air $R = 53.35$ ft-lbf/lbm-°R. From the state equation we obtain Charles' law $p/T = \text{constant}$ for a constant-volume process and Boyle's law $pV = \text{constant}$ for a constant-temperature process. For a given mass of gas they may be combined to give $pV/T = \text{constant}$, which is sometimes called the universal gas law. (Additional problems involving gas laws are presented in Chapter 9.)

Some other simple relations are useful in describing further the behavior of perfect gases. For a perfect gas the specific heats c_p and c_v are each constant, and it can be shown that $c_p - c_v = R$. Also, the internal energy and enthalpy are each directly proportional to the absolute temperature; specifically $u = c_v T$ and $h = c_p T$. By evaluating Eq. (8-4) for different paths, several different expressions for the entropy change of a perfect gas between two points can be derived. One relation is

$$s_2 - s_1 = c_p \ln \left(\frac{v_2}{v_1} \right) + c_v \ln \left(\frac{p_2}{p_1} \right) \qquad (8\text{-}9)$$

Equation (8-9) is useful in learning about compression processes. Using the specific heat ratio $k = c_p/c_v$, Eq. (8-9) can be rearranged to show that an isentropic or reversible adiabatic compression process satisfies the relation

$$pV^k = \text{constant} \qquad (8\text{-}10)$$

By using the equation of state, the alternative relations $TV^{k-1} = \text{constant}$ and $Tp^{(1-k)/k} = \text{constant}$ can be derived. The isentropic compression process is actually a special case of the more general polytropic process described by $pV^n = \text{constant}$, where n is not equal to k. The polytropic process is generally not isentropic.

Example 4

The air pressure in an automobile tire was checked at a service station and found to be 30 psig when the temperature was 65°F. Later the same tire was checked again, and the pressure gage read 35 psi. Assuming that the atmospheric pressure of 14.7 psi did not change, what was the new temperature of the air in the tire?

Solution

If we assume that the volume of the tire remained constant, a special case of the universal gas law, called Charles' law, may be applied. The law shows $p/T = \text{constant}$ for a constant-volume process. Here

$$p_1 = 30 + 14.7 = 44.7 \text{ psia}$$

$$p_2 = 35 + 14.7 = 49.7 \text{ psia}$$

$$T_1 = 65 + 460 = 525°\text{R}$$

and T_2 is to be found. Thus

$$\frac{p_1}{T_1} = \frac{p_2}{T_2}$$

$$T_2 = T_1\left(\frac{p_2}{p_1}\right) = 525\left(\frac{49.7}{44.7}\right) = 584°\text{R}$$

$$T_2 = 584 - 460 = 124°\text{F}$$

Example 5

Five pounds of a gas initially at 0.0 psig and 60°F are compressed. Later it is found that the increase in pressure was 1900% and the decrease in volume was 90%. The barometric pressure was 24.44 in. of mercury during the compression process. Find

(a) the value of n in $pV^n = \text{constant}$

(b) the final gage pressure and final volume if the initial volume was 10 ft^3

Solution

(a) From $pV^n = \text{constant}$,

$$\frac{p_2}{p_1} = \left(\frac{V_1}{V_2}\right)^n$$

Here $p_2 = 19p_1$ and $V_2 = 0.10V_1$, so that

$$19 = \left(\frac{1}{0.10}\right)^n = (10)^n$$

Taking logarithms,

$$\log_{10}(19) = n \log_{10}(10) = n$$

Hence

$$n = 1.279$$

(b) $$V_2 = 0.10V_1 = 0.10(10) = 1.0 \text{ ft}^3$$

The initial pressure $p_1 = 0.0 \text{ psig} = 24.44 \text{ in.}$ of mercury on the absolute scale.

$$24.44 \text{ in. of mercury} = \frac{24.44}{12} \cdot \frac{(62.4)(13.55)}{144} = 12.0 \text{ psia}$$

The final pressure is therefore $p_2 = 19p_1 = 19(12.0) = 228 \text{ psia}$, or

$$p_2 = 228 - 12 = 216 \text{ psig}$$

CYCLES

Cyclical processes play a significant role in engineering thermodynamics. The characteristics of several basic cycles are briefly reviewed here. Each is an idealized cycle that is composed entirely of reversible processes.

The Carnot cycle is a reversible four-process cycle consisting of alternating isothermal heat transfer processes and adiabatic compression or expansion processes. Since the second law shows that a reversible cycle is the most efficient of all cycles operating between two given temperature levels, the simple Carnot cycle is often used as a standard against which the performance characteristics of all other cycles may be measured.

The Otto, Diesel, and Brayton cycles are four-process gas cycles. If air is the working fluid, they are called air-standard cycles. In all three cycles every second process is an isentropic expansion or compression process. The Otto cycle is used for reciprocating engines; the other two processes in this cycle are constant-volume heat transfer processes. The Diesel cycle is thermodynamically like the Otto cycle except that the idealized heat supply process is presumed to occur at constant pressure. In the Brayton cycle both heat transfer processes occur at constant pressure; it is the standard cycle for gas turbines.

The idealized cycle which describes the operation of steam power plants is the Rankine cycle, which, of course, is a vapor cycle. The cycle consists, in turn, of an isentropic expansion process for the steam, a constant-pressure heat transfer process, an isentropic compression of the liquid, and a constant-pressure heat transfer process to the liquid. The first process could occur at any of several states (see Fig. 8-2).

State diagrams for these basic cycles are given in Fig. 8-2. A variable listed beside a process line indicates that the variable is constant during the process. The extra line on the Rankine cycle plots divides the vapor and liquid phases.

HEAT TRANSFER

Engineers often want to know not only the amount but also the rate of heat transfer during a process. For this reason the various heat transmission modes are outlined here. All heat transfer occurs by some combination of the three mechanisms of conduction, convection, and radiation.

Conductive heat transfer occurs in the absence of mass transfer. The rate of heat transfer q, in Btu/hour, is given in this case by Fourier's law

$$q = -kA \frac{dT}{dx} \tag{8-11}$$

where k is the thermal conductivity in Btu/hr-ft-°F, A is the cross-sectional area in square feet, and dT/dx is the temperature gradient in the direction of heat flow.

Convective heat transfer involves the mass transfer of fluid from regions of high temperature to regions of lower temperature. The process is described by Newton's cooling law as

$$q = hA \, \Delta T \tag{8-12}$$

Here h is called the convective heat transfer coefficient or film coefficient and is measured in Btu/hr-ft^2-°F. ΔT is the temperature difference causing the process.

Radiation depends on the transmission of electromagnetic waves to achieve heat transfer. The rate of emission of heat is governed by the Stefan-Boltzmann law for black-body radiation, which may be written

$$q = A\sigma(T_1^4 - T_2^4) \tag{8-13}$$

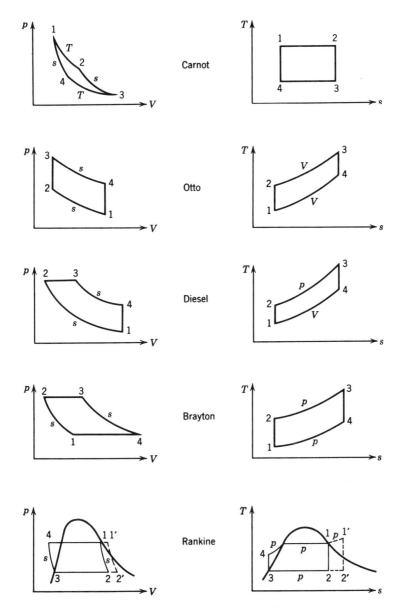

Figure 8-2 Thermodynamic cycles

This expression applies for a small body having an area A which emits radiation at a temperature T_1 (absolute) to some point which is at temperature T_2. The Stefan-Boltzmann constant is σ. For nonblack bodies this constant is replaced by $e\sigma$, e being the dimensionless emissivity of the particular body.

Example 6

A masonry wall has a 4-in. thick facing wall bonded to an 8-in. thick concrete backing. On a day when the room temperature is 68°F and the outside temperature is 12°F, the inner surface temperature of the concrete is 57°F and the outer surface temperature of the brick is 19°F. See Fig. 8-3.

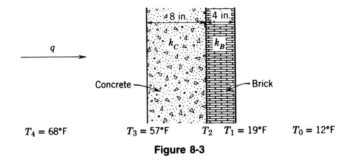

$T_4 = 68°F$ $T_3 = 57°F$ T_2 $T_1 = 19°F$ $T_0 = 12°F$

Figure 8-3

The thermal conductivities for brick and concrete are 0.36 and 0.68 Btu/hr-ft-°F, respectively. Determine

(a) the overall heat transfer coefficient
(b) the convective heat transfer coefficients for the concrete and brick walls

Solution

(a) Here we assume a steady-state heat flow. A combination of conduction and convection is occurring. By analogy to Eqs. (8-11) and (8-12), an overall heat transfer coefficient U may be defined by the equation

$$q = UA \, \Delta T = UA(T_4 - T_0)$$

From Eqs. (8-11) and (8-12) we may write several expressions for the ratio q/A:

$$\frac{q}{A} = h_C(T_4 - T_3) = \frac{k_C}{L_C}(T_3 - T_2) = \frac{k_B}{L_B}(T_2 - T_1) = h_B(T_1 - T_0)$$

Using the two conduction expressions, we can solve for the unknown temperature T_2 and then determine q/A. Then U can be computed. Thus

$$\frac{q}{A} = \frac{0.68}{\frac{8}{12}}(57 - T_2) = \frac{0.36}{\frac{4}{12}}(T_2 - 19)$$

$$1.02(57 - T_2) = 1.08(T_2 - 19)$$

$$T_2 = 37.4°F$$

and $$\frac{q}{A} = 1.02(57 - T_2) = 1.02(57 - 37.4)$$

$$\frac{q}{A} = 20 \text{ Btu/hr-ft}^2$$

Therefore

$$U = \frac{q}{A} \cdot \frac{1}{\Delta T} = 20 \frac{1}{(68 - 12)}$$

$$U = 0.357 \text{ Btu/hr-ft}^2\text{-°F}$$

(b) The convective heat transfer coefficients can now be found.

$$\frac{q}{A} = h_C(T_4 - T_3) = h_B(T_1 - T_0)$$

$$20 = h_C(68 - 57) = h_B(19 - 12)$$

For the concrete,

$$h_C = \tfrac{20}{11} = 1.82 \text{ Btu/hr-ft}^2\text{-°F}$$

For the brick,

$$h_B = \tfrac{20}{7} = 2.86 \text{ Btu/hr-ft}^2\text{-°F}$$

PROBLEM 8-1

The temperature 45°C is equal to

(A)   45°F
(B)   57°F
(C)   113°F
(D)   81°F
(E)   25°F

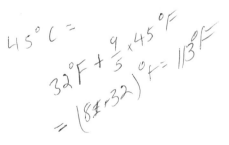

Solution

The melting point of ice and the boiling point of water are, respectively, 0°C = 32°F and 100°C = 212°F. By direct proportion,

$$45°C = 32°F + 45\left(\frac{212 - 32}{100}\right)$$

$$= 113°F$$

The solution is equivalent to using the formula

$$°C = \tfrac{5}{9}(°F - 32)$$

Answer is (C)

PROBLEM 8-2

How much heat is required to raise 10 g of water from 0°C to 1°C?

(A) 10 Btu
(B) 1 Btu
(C) 1 cal
(D) 5 J
(E) 10 cal ✓

$10 \ cal$

Solution

One cal will raise the temperature of 1 g of water 1°C. Hence 10 g of water will require 10 cal.

Answer is (E)

PROBLEM 8-3

The amount of heat, in J, required to raise the temperature of 1 ft³ of water by 1°F is nearest to

(A) 65,800
(B) 48,500
(C) 28,300
(D) 15,700
(E)   62

Solution

One Btu will raise the temperature of 1 lb of water by 1°F, and water weighs 62.4 lbf/ft³; hence 62.4 Btu are required. The SI equivalent is

$$(62.4 \text{ Btu})\left(778 \ \frac{\text{ft-lbf}}{\text{Btu}}\right)\left(0.3048 \ \frac{\text{m}}{\text{ft}}\right)\left(4.45 \ \frac{\text{N}}{\text{lbf}}\right) = 65{,}800 \text{ N-m}$$

or 65,800 joules (J).

Answer is (A)

PROBLEM 8-4

An adiabatic process is one in which

    (A) the pressure is constant
    (B) internal energy is constant
    (C) no work is done
    (D) no heat is transferred
    (E) friction is not considered

Solution

By definition, an adiabatic process is one in which no heat is transferred into or out of the system.

$$\text{Answer is (D)}$$

PROBLEM 8-5

How much boiling water, in g, is required to melt 1000 g of ice at 0°C and produce a mixture at 20°C?

    (A)   200
    (B)   250
    (C)   800
    (D) 1000
    (E) 1250

Solution

The amount of heat lost by the boiling water must equal the heat gained by the ice and cold water.

Let l = heat of fusion of ice = 80 cal/g

      c = heat capacity of water = 1 cal/g-°C

     M_w = mass of boiling water at 100°C

      M_i = mass of ice = 1000 g

    ΔT_w = decrease in hot water temperature

     ΔT_i = increase in temperature for ice

Then

$$M_w c \, \Delta T_w = lM_i + M_i c \, \Delta T_i$$

$$M_w(1)(100-20) = 80(1000) + (1000)(1)(20-0)$$

$$80 M_w = 100{,}000$$

$$M_w = 1250 \text{ g of boiling water}$$

Answer is (E)

PROBLEM 8-6

A mass of 0.2 kg of a metal having a temperature of 100°C is plunged into 0.04 kg of water at 20°C. The temperature of the water and metal becomes 48°C. The latent heat of ice at 0°C is 335 kJ/kg, and the specific heat capacity of water is 4.19 kJ/kg-°C.

Question 1. The specific heat capacity of the metal, in kJ/kg-°C, is closest to

(A) 0.25
(B) 0.30
(C) 0.40
(D) 0.45
(E) 0.75

Solution

Assuming no heat loss to the surroundings, we equate the amount of heat lost by the metal and the amount of heat gained by the water. We employ the formula $Q = Mc \, \Delta T$ and we use the subscripts m for metal and w for water. Then

$$M_m c_m \, \Delta T_m = M_w c_w \, \Delta T_w$$

$$0.2 c_m (100 - 48) = 0.04(4.19)(48 - 20)$$

The specific heat capacity of the metal is therefore

$$c_m = 0.451 \text{ kJ/kg-°C}$$

Answer is (D)

Question 2. The number of kg of ice at 0°C that can be melted by 0.2 kg of this metal at 100°C is nearest to

(A) 0.015
(B) 0.024
(C) 0.027
(D) 0.036
(E) 0.045

Solution

Again we equate the heat lost by the metal and the heat gained by the ice. Let l = latent heat of ice. We obtain

$$M_m c_m \Delta T_m = l M_i$$

$$0.2(0.451)(100 - 0) = 335 M_i$$

Hence

$$M_i = 0.027 \text{ kg}$$

Answer is (C)

PROBLEM 8-7

The melting rate of snow will normally be accelerated when warm rain falls. The following data are provided:

Snow	Rain
Depth 50 in.	Amount 2 in.
Water content 40%	Temperature 68°F
Temperature 32°F (melting point)	
Heat of fusion 144.0 Btu/lb of water	

Assume the weight of melted snow and rain water to be 62.4 lb/ft³. The percentage of the snow pack that will be melted by the rain is nearest to

(A) 1.0
(B) 2.5
(C) 4.0
(D) 5.0
(E) 10.0

Solution

We first calculate the amount of heat available in the 68°F rain. Then we calculate the amount of snow at 32°F that will be converted to water at 32°F in absorbing this amount of heat.

The quantity of heat available above 32°F, per square foot of surface, is

$$\tfrac{2}{12}(62.4)(68 - 32) = 374.4 \, \text{Btu/ft}^2$$

This amount of heat will melt an amount of snow equivalent to $374.4/144.0 = 2.6 \, \text{lb}$ of water, or

$$\left(\frac{2.6}{62.4}\right)\left(\frac{1}{0.4}\right) = 0.104 \, \text{ft}^3 \text{ of snow/ft}^2 \text{ of surface}$$

This equals $0.104 \times 12 = 1.25$ in. of snow. The percentage of the 50-in. snow pack that is melted is

$$\frac{1.25}{50}(100) = 2.5\%$$

Answer is (B)

PROBLEM 8-8

The first law of thermodynamics may be referred to as the principle of the conservation of

(A) mass
(B) momentum
(C) heat
(D) energy
(E) enthalpy

Solution

The first law of thermodynamics is a statement of the principle of energy conservation which relates the net heat Q and net work W to the internal energy change ΔE of a system.

Answer is (D)

PROBLEM 8-9

A perfect gas is contained in a piston-and-cylinder machine. Within the machine the pressure of the gas is always directly proportional to its volume. Initially the gas is at a pressure of 15 psia and a volume of 1 ft³. Heat is transferred reversibly to the gas until its pressure is 150 psia. If the movement of the piston is frictionless, the work done by the gas, expressed in Btu, is most nearly

(A) 25
(B) 80
(C) 135
(D) 200
(E) 275

Solution

The gas pressure p is directly proportional to its volume V, or $p = KV$. The initial pressure p_0 is $p_0 = 15(144)$ lb/ft^2 absolute when the initial volume $V_0 = 1$ ft^3. Hence

$$15(144) = K(1) \qquad K = 2160 \text{ lb/ft}^5$$

When the final pressure is $p_1 = 150(144)$ lb/ft^2 absolute, the volume is

$$V_1 = \frac{p_1}{K} = \frac{150(144)}{2160} = 10 \text{ ft}^3$$

The work done is

$$W = \int_{V_0}^{V_1} p \, dV = \int_1^{10} KV \, dV = 2160 \left[\frac{V^2}{2} \right]_1^{10}$$
$$W = 2160(\tfrac{1}{2})(10^2 - 1) = 106{,}900 \text{ ft-lb}$$

The work done by the gas, in Btu, is

$$W = \frac{106{,}900}{778} = 137.4 \text{ Btu}$$

Answer is (C)

PROBLEM 8-10

Steam enters a turbine at a velocity of 100 ft/sec and an enthalpy of 1410 Btu/lbm; it leaves the turbine at 390 ft/sec and an enthalpy of 990 Btu/lbm. Heat is lost to the surroundings at a rate of 22 Btu/lbm of steam. If the steam flow rate is 75,000 lbm/hr, the power output in kilowatts is closest to

(A) 6,400
(B) 6,500
(C) 6,800
(D) 8,700
(E) 11,800

Solution

The turbine work output can be found by using the steady flow energy equation written on a unit mass basis and neglecting changes in potential energy. Using the subscript notation i = inlet, e = exit, the equation is

$$W = (h_i - h_e) + \frac{V_i^2 - V_e^2}{2} + Q$$

$$W = (1410 - 990)\,\frac{\text{Btu}}{\text{lbm}} + \frac{\left(100^2 - 390^2\right)\dfrac{\text{ft}^2}{\text{sec}^2}}{2\left(32.2\,\dfrac{\text{lbm-ft}}{\text{lbf-sec}^2}\right)\left(778\,\dfrac{\text{ft-lbf}}{\text{Btu}}\right)} - 22\,\frac{\text{Btu}}{\text{lbm}}$$

$$W = 420 - 2.84 - 22 = 395.2 \text{ Btu/lbm}$$

The heat term is negative because it is a loss. The power output P is equal to work times mass rate of flow. Since the equivalence between kilowatts and Btu/hr is

$$1\text{ kW} = \frac{\left(1000\,\dfrac{\text{N-m}}{\text{sec}}\right)(60^2)\dfrac{\text{sec}}{\text{hr}}}{\left(4.448\,\dfrac{\text{N}}{\text{lbf}}\right)\left(0.3048\,\dfrac{\text{m}}{\text{ft}}\right)\left(778\,\dfrac{\text{ft-lbf}}{\text{Btu}}\right)} = 3413\,\frac{\text{Btu}}{\text{hr}}$$

$$P = \frac{(395.2)(75,000)}{3413} = 8680 \text{ kW}$$

Answer is (D)

PROBLEM 8-11

Fluid enters a turbine with a velocity of 2 m/sec and an enthalpy of 900 Btu/lbm and leaves with an enthalpy of 850 Btu/lbm and a velocity of 100 m/sec. Heat losses are 1200 W, and the mass rate of flow is 1 kg/sec. The inlet to the turbine is 3 m higher than the outlet. The maximum theoretical power, in kilowatts, that can be developed by the turbine is nearest to

(A) 48
(B) 53
(C) 80
(D) 105
(E) 116

Solution

Only the enthalpies are not already in metric units; let us first convert from Btu/lbm to J/kg:

$$1\,\frac{\text{Btu}}{\text{lbm}} = \left(1\,\frac{\text{Btu}}{\text{lbm}}\right)\left(778\,\frac{\text{ft-lbf}}{\text{Btu}}\right)\left(\frac{0.3048\,\text{m}}{\text{ft}}\right)\left(\frac{4.45\,\text{N}}{\text{lbf}}\right)\left(\frac{2.20\,\text{lbm}}{\text{kg}}\right)$$

$$= 2320\,\text{J/kg}$$

Now we may employ the steady flow energy equation, written on a unit mass basis, which is

$$\frac{V_1^2}{2} + gz_1 + h_1 + \frac{dQ}{dm} = \frac{V_2^2}{2} + gz_2 + h_2 + \frac{dW}{dm}$$

Here $z =$ elevation above a chosen datum, and the subscripts 1 and 2 represent the turbine entrance and exit, respectively.

$$\frac{dQ}{dm} = \frac{dQ/dt}{dm/dt} = \frac{1200\,\text{W}}{1\,\text{kg/sec}} = 1200\,\text{J/kg}$$

$$\frac{(\frac{1}{2})(2^2 - 100^2)\left(\frac{\text{m}}{\text{sec}}\right)^2}{(1\,\text{kg-m/N-sec}^2)} + \frac{(9.81\,\text{m/sec}^2)(3\,\text{m})}{(1\,\text{kg-m/N-sec}^2)}$$

$$+ (900 - 850)\,\frac{\text{Btu}}{\text{lbm}}\left(2320\,\frac{\text{J/kg}}{\text{Btu/lbm}}\right) - 1200\,\frac{\text{J}}{\text{kg}} = \frac{dW}{dm}$$

$$\frac{dW}{dm} = 109.8\,\text{kJ/kg}$$

The work per unit time, or power, is therefore

$$\frac{dW}{dt} = \left(\frac{dW}{dm}\right)\left(\frac{dm}{dt}\right) = \left(109.8\,\frac{\text{kJ}}{\text{kg}}\right)\left(1\,\frac{\text{kg}}{\text{sec}}\right) = 109.8\,\text{kW}$$

Answer is (D)

PROBLEM 8-12

Entropy

(A) remains constant during an irreversible process
(B) is independent of temperature
(C) is a maximum at absolute zero
(D) is a measure of unavailable energy
(E) is the reciprocal of enthalpy

Solution

Entropy is a measure of unavailable energy.

Answer is (D)

PROBLEM 8-13

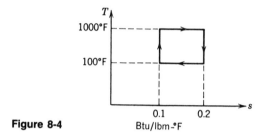

Figure 8-4

In the temperature-entropy diagram in Fig 8-4, the thermal efficiency of the process is nearest to

(A) 40%
(B) 60%
(C) 80%
(D) 90%
(E) 93%

Solution

Temperatures in this problem must be expressed in an absolute system, in this case degrees Rankine.

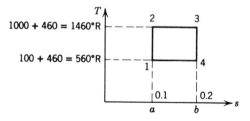

Figure 8-5

Thermal efficiency is defined as

$$\eta = \frac{W}{Q} = \frac{\text{net work}}{\text{heat added}} = \frac{\text{Area 1-2-3-4}}{\text{Area } a\text{-2-3-}b} = \frac{(0.2 - 0.1)(1460 - 560)}{(0.2 - 0.1)(1460)}$$

$$\eta = \tfrac{900}{1460} = 0.616 = 61.6\%$$

Answer is (B)

PROBLEM 8-14

A Carnot engine uses steam (a vapor) as the thermodynamic medium. One thousand Btu/min is supplied by a source at 500°F. The temperature of the refrigerator is 120°F. The output of the engine, in horsepower, is closest to

(A) 6.6
(B) 9.3
(C) 12.6
(D) 14.2
(E) 17.9

Solution

Converting to degrees Rankine,

$$T_1 = 500 + 460 = 960°\text{R} \qquad T_2 = 120 + 460 = 580°R$$

Figure 8-6 is a temperature-entropy diagram for this Carnot cycle. The

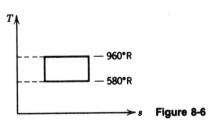

Figure 8-6

thermal efficiency η is

$$\eta = \frac{T_1 - T_2}{T_1} = \frac{960 - 580}{960}$$

$$\eta = 0.396$$

The net work output is then

$$\eta Q_1 = (0.396)\left(1000 \ \frac{Btu}{min}\right) = 396 \ \frac{Btu}{min}$$

The equivalent horsepower is

$$hp = \frac{\left(396 \ \frac{Btu}{min}\right)\left(778 \ \frac{ft\text{-}lbf}{Btu}\right)}{\left(60 \ \frac{sec}{min}\right)\left(550 \ \frac{ft\text{-}lbf}{sec\text{-}hp}\right)} = 9.34 \ hp$$

Answer is (B)

PROBLEM 8-15

Which one of the following is an extensive property?

(A) temperature
(B) pressure
(C) mass
(D) specific heat capacity
(E) specific volume

Solution

Intensive properties of a thermodynamic system are independent of the mass of the system; extensive properties depend on the amount of mass present. Clearly, mass itself is an extensive property.

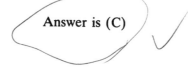

Answer is (C)

PROBLEM 8-16

For a single component system, the number of properties required to define a phase uniquely is

(A) 1
(B) 2
(C) 3
(D) 4
(E) 5

Solution

The required number is 2. As an example, the state of an ideal gas is completely defined if two of the three quantities (p, V, T) appearing in the gas law are known, for then the third variable and other quantities can be found.

Answer is (B)

PROBLEM 8-17

A room experiences a heat gain of 100,000 Btu/hr and must be maintained at 80°F. If $c_p = 0.24$ Btu/lbm-°R and $R = 53.3$ ft-lbf/lbm-°R for air, the number of cubic feet per minute of 64°F air required to maintain the desired temperature is nearest to

(A) 26,000

(B) 8,200

(C) 5,800

(D) 4,100

(E)   450

Solution

Entering the room ($p = 14.7$ psi, $T = 64°F = 524°R$) the mass of 1 ft^3 of air is, according to the gas law, approximately

$$m = \frac{pV}{RT} = \frac{\left(14.7\ \frac{lbf}{in^2}\right)\left(144\ \frac{in.^2}{ft^2}\right)(1\ ft^3)}{\left(53.3\ \frac{ft\text{-}lbf}{lbm\text{-}°R}\right)(524°R)} = 0.0758\ lbm$$

The heat capacity C of 1 ft^3 of air is therefore

$$C = (0.24\ Btu/lbm\text{-}°R)(0.0758\ lbm) = 0.018\ Btu/°R$$

Here $\Delta T = 80 - 64 = 16°F$, and thus the amount of heat to be removed is $100,000/60 = 1667$ Btu/min. Hence the required volume of air is

$$V = \frac{1667}{16(0.018)} = 5800\ ft^3/min$$

Answer is (C)

PROBLEM 8-18

At the place where Piccard started his ascent in the stratosphere balloon, the temperature was 17°C and the pressure 640 mm Hg. At the highest altitude reached, the temperature was −48°C and the pressure 310 mm Hg. None of the gas was vented. The fractional part of its total capacity to which the balloon was filled before ascending so that it would be fully expanded at the highest altitude reached is most nearly

(A) 0.56
(B) 0.62
(C) 0.68
(D) 0.74
(E) 0.80

Solution

Assuming air to be a perfect gas, here we can apply the universal gas law

$$\frac{p_1 V_1}{T_1} = \frac{p_2 V_2}{T_2}$$

In this equation absolute temperatures and pressures must be used. Initially, $T_1 = 17°C = 17 + 273 = 290°K$, $p_1 = 640$ mm Hg, and V_1 is unknown. At the highest altitude, $T_2 = -48°C = -48 + 273 = 225°K$, $p_2 = 310$ mm Hg, and $V_2 = $ total capacity of balloon. Hence

$$\frac{V_1}{V_2} = \frac{p_2}{p_1}\frac{T_1}{T_2} = \left(\frac{310}{640}\right)\left(\frac{290}{225}\right) = 0.624$$

and the balloon therefore was filled to 0.624 or five-eighths of its capacity before it was launched.

Answer is (B)

PROBLEM 8-19

An automobile tire registered a gage pressure of 28 psi when the temperature was 70°F. After driving for a while, the gage pressure was found to be 31 psi. Assume that the barometric pressure was constant at 14.3 psi and that the tire volume was constant. The temperature of the tire, in °F, at the time of the second reading was most nearly

(A) 75
(B) 78

(C) 90
(D) 108
(E) 127

Solution

Assuming air to behave as a perfect gas while undergoing a constant-volume process, we may apply Charles' law $p_1/T_1 = p_2/T_2$. In this equation we must use absolute pressures and temperatures.

$$p_1 = 28 + 14.3 = 42.3 \text{ psia}$$
$$p_2 = 31 + 14.3 = 45.3 \text{ psia}$$
$$T_1 = 70 + 460 = 530°R$$

Thus

$$T_2 = T_1 \frac{p_2}{p_1} = 530\left(\frac{45.3}{42.3}\right) = 568°R$$
$$T_2 = 568 - 460 = 108°F$$

Answer is (D)

PROBLEM 8-20

Nitrogen is pumped into a 10-ft^3 tank until the pressure gage reads 185.3 psi and the temperature of the gas is 200°F. The tank is cooled until the temperature of the nitrogen is 80°F. Assume atmospheric pressure of 14.7 psia. The following data are also available:

Gas	Chemical Formula	Molecular Weight	R ft-lbf lbm-°R	c_p Btu lbm-°R	c_v Btu lbm-°R	k $\frac{c_p}{c_v}$
Nitrogen	N_2	28.0	55.1	0.248	0.177	1.40

Question 1. The amount of heat removed, in Btu, from the nitrogen gas is nearest to

(A) 156
(B) 168
(C) 219
(D) 236
(E) 555

Solution

We first determine the mass of the nitrogen in the tank by using the equation of state for a perfect gas $pV = mRT$, where p and T must be in absolute units.

$$m = \frac{pV}{RT} = \frac{(185.3 + 14.7)(12^2)(10)}{(55.1)(200 + 460)} = 7.92 \text{ lbm}$$

We can find the heat removed from the equation $Q = mc(T_2 - T_1)$. For this constant-volume process, the specific heat $c = c_v$, the specific heat at constant volume. Hence

$$Q = mc_v(T_2 - T_1) = (7.92)(0.177)(80 - 200) = -168.2 \text{ Btu}$$

The minus sign indicates heat removed.

Answer is (B)

Question 2. The final gage pressure, in psi, is nearest to

- (A) 65
- (B) 74
- (C) 149
- (D) 152
- (E) 164

Solution

Using the gas law at constant volume (Charles' law) $V_1 = V_2$,

$$\frac{p_2}{p_1} = \frac{T_2}{T_1}$$

or

$$p_2 = (185.3 + 14.7)\left(\frac{80 + 460}{200 + 460}\right) = 200\left(\frac{540}{660}\right) = 163.6 \text{ psia}$$

$$p_2 = 163.6 - 14.7 = 148.9 \text{ psig}$$

Answer is (C)

PROBLEM 8-21

Two moles of oxygen at 50 psia and 40°F are in a container that is connected by a valve to a second container filled with 5 moles of nitrogen at 30 psia and

140°F. The valve is opened, and adiabatic mixing occurs. The following data are also supplied:

Gas	Chemical Formula	Molecular Weight	R $\frac{\text{ft-lbf}}{\text{lbm-}°R}$	c_p $\frac{\text{Btu}}{\text{lbm-}°R}$	c_v $\frac{\text{Btu}}{\text{lbm-}°R}$	k $\frac{c_p}{c_v}$
Nitrogen	N_2	28.0	55.1	0.248	0.177	1.40
Oxygen	O_2	32.0	48.3	0.219	0.157	1.39

Question 1. The equilibrium temperature, in °F, of the mixture is closest to

(A) 90
(B) 95
(C) 100
(D) 105
(E) 110

Solution

First we determine the initial volume of each gas using the state equation $pV = mRT$.

$$2 \text{ lb-moles of oxygen} = 2(32) = 64 \text{ lbm oxygen}$$

$$5 \text{ lb-moles of nitrogen} = 5(28) = 140 \text{ lbm nitrogen}$$

$$N_2: \quad V = \frac{mRT}{p} = \frac{(140)(55.1)(140 + 460)}{(30)(12^2)} = 1071 \text{ ft}^3$$

$$O_2: \quad V = \frac{mRT}{p} = \frac{(64)(48.3)(40 + 460)}{(50)(12^2)} = 215 \text{ ft}^3$$

An adiabatic process is one in which no net heat transfer to the surroundings occurs, so the heat lost by the nitrogen in mixing is gained by the oxygen. The mixing is a constant-volume process so that c_v, the specific heat at constant volume, is the proper specific heat value to use. The heat transfer Q in this process is $Q = mc_v(T_2 - T_1)$. Let T be the final, or equilibrium, temperature of the mixture. Then

$$\text{Heat lost by the nitrogen} = 140(0.177)(140 - T) = 3470 - 24.8T$$

$$\text{Heat gained by the oxygen} = 64(0.157)(T - 40) = 10.0T - 402$$

$$3470 - 24.8T = 10.0T - 402$$

$$34.8T = 3870$$

The equilibrium temperature is

$$T = 111°F$$

Answer is (E)

Question 2. The equilibrium pressure, in psia, of the mixture is nearest to

(A) 30.0
(B) 33.3
(C) 36.7
(D) 40.0
(E) 43.3

Solution

For a mixture of gases the pressure P of the mixture is equal to the sum of the partial pressures of the constituent gases (Dalton's law). Since T and V are now common to both gases and $p = mRT/V$,

$$P = p_{N_2} + p_{O_2} = [140(55.1) + 64(48.3)]\left(\frac{111 + 460}{1071 + 215}\right)$$

$$= 4800 \text{ lb/ft}^2 \text{ abs} = 4800/144 = 33.3 \text{ psia}$$

Answer is (B)

PROBLEM 8-22

The temperature at which the vapor in a mixture starts to condense when the mixture is cooled at constant pressure is called the

(A) dry-bulb temperature
(B) wet-bulb temperature
(C) dew point
(D) relative humidity
(E) specific humidity

Solution

The dry-bulb temperature is measured with an ordinary thermometer. The wet-bulb temperature is measured by a thermometer covered with a liquid film. Relative humidity is the ratio of the amount of vapor present to the amount of vapor required to produce saturation at the same temperature, whereas specific humidity is normally the mass ratio of water vapor to air in a mixture.

Answer is (C)

PROBLEM 8-23

A closed cylinder contains 3.00 ft^3 of dry air at a temperature of 60°F and a gage pressure of 10 psi. A piston very suddenly reduces the enclosed volume to 2.00 ft^3.

Question 1. The new temperature of the air, in °F, is most nearly

(A) 71
(B) 104
(C) 128
(D) 152
(E) 177

Solution

We assume the air behaves as an ideal gas. From the description of the piston movement we may also reasonably assume the volume change to be adiabatic.

$$p_1 = 10 \text{ psig} = 10 + 14.7 = 24.7 \text{ psia} \qquad p_2 = ?$$
$$T_1 = 60°F = 60 + 460 = 520°R \qquad\qquad T_2 = ?$$
$$V_1 = 3.00 \text{ ft}^3 \qquad\qquad\qquad\qquad V_2 = 2.00 \text{ ft}^3$$

The specific heat ratio for air is $k = 1.4$. From the above assumptions it can be shown that both TV^{k-1} and pV^k are constant during the compression process.

$$\frac{T_1}{T_2} = \left(\frac{V_2}{V_1}\right)^{k-1}$$

$$T_2 = T_1\left(\frac{V_1}{V_2}\right)^{k-1} = 520\left(\frac{3.00}{2.00}\right)^{1.4-1}$$

$$T_2 = 520(1.5)^{0.4} = 520(1.176)$$

$$T_2 = 612°R = 612 - 460 = 152°F$$

Answer is (D)

Question 2. The new gage pressure of the air, in psi, is closest to

(A) 17.6
(B) 18.4
(C) 22.6
(D) 28.9
(E) 30.7

Solution

$$\frac{p_1}{p_2} = \left(\frac{V_2}{V_1}\right)^k$$

$$p_2 = p_1\left(\frac{V_1}{V_2}\right)^k = 24.7\left(\frac{3.00}{2.00}\right)^{1.4} = 24.7(1.764)$$

$$p_2 = 43.6 \text{ psia} = 43.6 - 14.7 = 28.9 \text{ psig}$$

Answer is (D)

PROBLEM 8-24

Air is compressed polytropically so that the quantity $pV^{1.4}$ is constant. If 0.02 m^3 of air at atmospheric pressure (101.3 kN/m^2) and 4°C are compressed to a gage pressure of 405 kN/m^2, the final temperature of the air, in °C, is closest to

(A) 155
(B) 165
(C) 175
(D) 185
(E) 195

Solution

One approach is to use the compression relation to determine the final volume and then employ the gas law to obtain the temperature.

From the compression relation we have

$$(101.3)(0.02)^{1.4} = (405 + 101.3)V^{1.4}$$

$$(101.3)(0.00418) = (506)V^{1.4}$$

$$V^{1.4} = 8.37(10^{-4}) \qquad V = 6.34(10^{-3}) \text{ m}^3$$

Using the gas law,

$$\frac{p_1V_1}{T_1} = \frac{p_2V_2}{T_2}$$

$$\frac{101.3(0.02)}{4+273} = \frac{(506)(0.00634)}{T_2}$$

$$T_2 = 439°K = 439 - 273 = 166°C$$

Answer is (B)

PROBLEM 8-25

The volume in the single cylinder of an air compressor is $0.57\,\text{ft}^3$ at the beginning of the compression stroke with air at atmospheric pressure. The piston compresses the air polytropically to 69.8 psig according to the law $pV^{1.35}$ = constant. At the end of the compression stroke, the volume, in ft^3, is nearest to

(A) 0.10
(B) 0.12
(C) 0.14
(D) 0.16
(E) 0.18

Solution

Assuming atmospheric pressure to be 14.7 psia, we have

$$p_1V_1^{1.35} = p_2V_2^{1.35}$$

or

$$V_2 = V_1\left(\frac{p_1}{p_2}\right)^{1/1.35}$$

$$V_2 = (0.57)\left(\frac{14.7}{14.7+69.8}\right)^{1/1.35}$$

$$V_2 = (0.57)(0.174)^{1/1.35} = (0.57)(0.274)$$

$$V_2 = 0.156\,\text{ft}^3$$

Answer is (D)

PROBLEM 8-26

If an ideal gas is compressed from a lower pressure to a higher pressure at a constant temperature, which of the following is true?

(A) the work required is zero
(B) the volume remains constant
(C) the volume varies inversely with the absolute pressure
(D) heat is being absorbed
(E) none of these

Solution

For an isothermal compression the pressure-volume relation is governed by the universal gas law, which gives $pV =$ constant.

Answer is (C)

PROBLEM 8-27

Atmospheric air at 14.7 psia and 70°F is compressed in a cylinder-piston arrangement until the air is at 100 psia and 300°F. The amount of heat removed per unit mass, in Btu/lbm, from the air during the compression process is closest to

(A) 30
(B) 45
(C) 60
(D) 95
(E) 125

Solution

For a polytropic process it can be shown that

$$\frac{T_2}{T_1} = \left(\frac{p_2}{p_1}\right)^{(n-1)/n}$$

Here

$$p_1 = 14.7 \text{ psia} \qquad\qquad p_2 = 100 \text{ psia}$$

$$T_1 = 70 + 460 = 530°R \qquad T_2 = 300 + 460 = 760°R$$

$$\frac{760}{530} = \left(\frac{100}{14.7}\right)^{1-1/n}$$

$$1.434 = (6.80)^{1-1/n}$$

Taking the logarithm of each term in the equation,

$$\ln(1.434) = \left(1 - \frac{1}{n}\right)\ln(6.80)$$

$$0.360 = \left(1 - \frac{1}{n}\right)(1.92) \qquad n = 1.23$$

The work done is

$$W = \int p \, dV = \frac{p_2 V_2 - p_1 V_1}{1 - n} = \frac{R(T_2 - T_1)}{1 - n} = \frac{(53.3)(760 - 530)}{(1 - 1.23)(778)} = -68.5 \frac{\text{Btu}}{\text{lbm}}$$

The heat transfer Q can now be found by using the first law of thermodynamics. For air, $c_v = 0.171$ Btu/lbm-°F.

$$Q = W + (u_2 - u_1)$$

$$Q = W + c_v(T_2 - T_1)$$

$$Q = -68.5 + 0.171(760 - 530)$$

$$Q = -29.2 \text{ Btu/lbm}$$

The minus sign indicates that the heat is transferred from, not to, the air.

Answer is (A)

PROBLEM 8-28

Air enters an engine in which the expansion ratio is 5 at a temperature of 70°F. The expansion is polytropic in accordance with $pV^{1.36} = C$, and the specific heat capacity for this process is given by $c_n = c_v(k - n)/(1 - n)$. The amount of heat added to the air, in Btu/lbm, during the expansion process is closest to

(A) 0.6

(B) 1.2

(C) 4.5

(D) 10.0

(E) 40.0

Solution

For polytropic compression of a perfect gas

$$\frac{T_2}{T_1} = \left(\frac{V_1}{V_2}\right)^{n-1}$$

or

$$T_2 = (70 + 460)(\tfrac{1}{5})^{1.36-1} = 297°R$$

For air, $c_v = 0.1715$ Btu/lbm-°R and $k = 1.4$. Thus

$$c_n = c_v \frac{k - n}{1 - n} = (0.1715)\frac{1.4 - 1.36}{1 - 1.36} = -0.0191 \frac{Btu}{lbm\text{-}°R}$$

The heat added is therefore

$$Q = c_n(T_2 - T_1) = -0.0191(297 - 530) = 4.45 \text{ Btu/lbm}$$

Answer is (C)

PROBLEM 8-29

Air having an initial pressure $p_i = 14$ psia, temperature $T_i = 80°F$, and volume $V_i = 28.6$ ft^3 is compressed isentropically in a nonflow process to a final pressure $p_f = 120$ psia.

Question 1. The final volume, in ft^3, is nearest to

(A) 15.5

(B) 12.1

(C) 6.2

(D) 3.7

(E) 1.4

Solution

For an isentropic process $pV^k = $ constant. For air, $k = 1.4$. Hence

$$\frac{p_f}{p_i} = \left(\frac{V_i}{V_f}\right)^k$$

$$V_f = V_i\left(\frac{p_i}{p_f}\right)^{1/k}$$

$$V_f = 28.6(\tfrac{14}{120})^{1/1.4} = 6.16 \text{ ft}^3$$

Answer is (C)

Question 2. The final temperature, in °F, is nearest to

(A) 815
(B) 540
(C) 370
(D) 200
(E) 190

Solution

Also for an isentropic process (with $T_i = 80 + 460 = 540°$R)

$$\frac{T_f}{T_i} = \left(\frac{p_f}{p_i}\right)^{(k-1)/k}$$

$$T_f = 540(\tfrac{120}{14})^{(1.4-1)/1.4}$$

$$T_f = 540(8.57)^{0.286}$$

$$T_f = 998°\text{R} = 998 - 460 = 538°\text{F}$$

Answer is (B)

Question 3. The increase in internal energy, in Btu, is nearest to

(A) 80
(B) 85
(C) 110
(D) 155
(E) 220

Solution

To determine the change in internal energy ΔU, we must find the mass of the air in this process. Using the equation of state $pV = mRT$, we have

$$m = \frac{p_i V_i}{RT_i} = \frac{\left(14 \frac{\text{lbf}}{\text{in}^2}\right)\left(144 \frac{\text{in}^2}{\text{ft}^2}\right)(28.6 \text{ ft}^3)}{\left(53.35 \frac{\text{ft-lbf}}{\text{lbm-}^\circ\text{R}}\right)(540^\circ\text{R})}$$

$$m = 2.00 \text{ lbm}$$

Now $\Delta U = mc_v(T_f - T_i)$. Using $c_v = 0.1715 \text{ Btu/lbm-}^\circ\text{R}$ for air,

$$\Delta U = (2.00)(0.1715)(998 - 540) = 157 \text{ Btu}$$

Answer is (D)

PROBLEM 8-30

One hundred gal/min of kerosene is to be heated from 85°F to 195°F (zero vaporization) in an exchanger using 30 psia steam of 96% quality. The heat losses to the surrounding air have been estimated to be 3% of the heat transferred from the condensing steam to the kerosene. For kerosene, specific gravity = 0.82, specific heat = 0.52 Btu/lbm-°F. For saturated water at 30 psia, $h_f = 218.8 \text{ Btu/lbm}$, $v_f = 0.0170 \text{ ft}^3/\text{lbm}$. For saturated steam, $h_g = 1164.0 \text{ Btu/lbm}$, $v_g = 13.76 \text{ ft}^3/\text{lbm}$. If the steam condensate leaves at its saturation point, the amount of steam, in lbm/hr, that is required by the exchanger is nearest to

(A) 2590
(B) 2670
(C) 2700
(D) 3260
(E) 3310

Solution

One gallon of kerosene has a mass of

$$\left(62.4 \frac{\text{lbm}}{\text{ft}^3}\right)\left(\frac{1 \text{ ft}^3}{7.48 \text{ gal}}\right)(0.82) = 6.84 \text{ lbm/gal}$$

The heat from the condensing steam Q_{st} will be equal to the heat gained by the kerosene Q_k plus 3% of Q_k due to the heat losses. The amount of heat gained by the kerosene is

$$Q_k = \dot{m}c_p \, \Delta T = \left(100 \, \frac{\text{gal}}{\text{min}} \times 60 \, \frac{\text{min}}{\text{hr}} \times 6.84 \, \frac{\text{lbm}}{\text{gal}}\right)\left(0.52 \, \frac{\text{Btu}}{\text{lbm-}°\text{F}}\right)(195 - 85)°\text{F}$$

$$= 23.5 \times 10^5 \, \text{Btu/hr}$$

Therefore

$$Q_{st} = 1.03 \, Q_k = 24.2 \times 10^5 \, \text{Btu/hr}$$

We must now calculate the heat transfer Q'_{st} per lbm of steam, which is equal to the difference in enthalpies before and after the process. At 30 psia and a steam quality $x = 96\% = 0.96$, the initial enthalpy is

$$h_x = xh_g + (1 - x)h_f$$
$$= 0.96(1164.0) + (1 - 0.96)(218.8) = 1117 + 8.75$$
$$= 1125.8 \, \text{Btu/lbm}$$

Thus

$$Q'_{st} = (h_x - h_f) = (1125.8 - 218.8) = 907.0 \, \text{Btu/lbm}$$

The required amount of steam in pounds per hour is then

$$\frac{Q_{st}}{Q'_{st}} = \frac{24.2 \times 10^5 \, \text{Btu/hr}}{907.0 \, \text{Btu/lbm}} = 2670 \, \frac{\text{lbm}}{\text{hr}}$$

Answer is (B)

PROBLEM 8-31

Steam enters the blades of a turbine at 300 psia and 1000°F. The discharge is at 3 psia.

Question 1. The work done, in Btu/lbm of steam, for an isentropic expansion through the turbine is nearest to

(A) 430
(B) 445
(C) 460
(D) 475
(E) 490

Solution

The solution of this problem requires the use of either a Mollier diagram or steam tables. Using a Mollier diagram (Fig. 8-7),

$$h_1 = 1524 \text{ Btu/lbm} \quad \text{and} \quad h_2 = 1067 \text{ Btu/lbm}$$

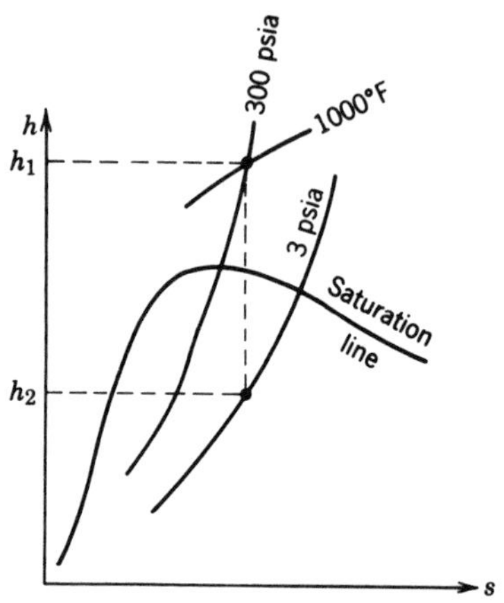

Figure 8-7

for isentropic expansion. The work W is

$$W = h_1 - h_2 = 457 \text{ Btu/lbm}$$

Answer is (C)

Question 2. If the efficiency of the turbine is 90%, the quality of the exhaust is nearest to

(A) 99%
(B) 98%
(C) 97%
(D) 96%
(E) 95%

Solution

If the turbine efficiency is 90%, the actual work done W' is

$$W' = 0.9W = 0.9(457) = 411 \text{ Btu/lbm}$$

Also, $W' = h_1 - h_2' = 1524 - h_2'$, so that

$$h_2' = 1524 - 411 = 1113 \text{ Btu/lbm}$$

On the diagram at the intersection of the lines $h_2' = 1113$ Btu/lbm and $p = 3$ psia, we read the steam quality x as

$$x = 99\%$$

Answer is (A)

PROBLEM 8-32

A small steam turbine is supplied with steam at 1000 psia and 100% quality, and exhausts at 14.7 psia. The turbine uses 40 lbm of steam per hour for each horsepower delivered at the turbine shaft. Heat losses from the turbine to its surroundings are negligible. The entropy per pound, in Btu/lbm-°R, of the exhaust steam is closest to

(A) 1.76
(B) 1.72
(C) 1.68
(D) 1.64
(E) 1.60

Solution

From steam tables it is found that the initial enthalpy of the steam is $h_1 = 1191.8$ Btu/lbm. Since heat losses Q are negligible and the work done per unit mass w is

$$\frac{W}{\dot{m}} = w = \frac{1 \text{ hp}}{40 \text{ lbm steam/hr}} = (1 \text{ hp}) \frac{2545 \text{ Btu/hr-hp}}{40 \text{ lbm/hr}} = 63.6 \frac{\text{Btu}}{\text{lbm}}$$

the first law yields

$$0 = Q = W + \dot{m}(h_2 - h_1)$$

$$h_2 = h_1 - w = 1191.8 - 63.6 = 1128.2 \text{ Btu/lbm}$$

The enthalpy h_2 and the exhaust pressure $p_2 = 14.7$ psia determine the final thermodynamic state of the steam. The steam tables show that the steam quality at this state is less than 100%. Consequently we must interpolate in the tables to find the final entropy.

$$h_g = 1150.4 \qquad s_g = 1.7566$$
$$h_f = 180.1 \qquad s_f = 0.3120$$
$$\overline{h_{fg} = 970.3 \qquad s_{fg} = 1.4446}$$

Interpolating,

$$\frac{s_2 - s_g}{s_{fg}} = \frac{h_2 - h_g}{h_{fg}}$$

$$s_2 = 1.7566 + \frac{1.4446}{970.3}(1128.2 - 1150.4)$$

$$= 1.7566 - 0.0331 = 1.7235 \text{ Btu/lbm-}°R$$

Answer is (B)

PROBLEM 8-33

Three lbm of steam expand isentropically (nonflow) from $p_1 = 300$ psia and $T_1 = 700°F$ to $T_2 = 200°F$. Additional data are as follows:

Initial conditions:

$p_1 = 300$ psia
$T_1 = 700°F$
$v_1 =$ specific volume of steam $= 2.227$ ft^3/lbm
$h_1 =$ specific enthalpy of steam $= 1368.3$ Btu/lbm
$s_1 =$ specific entropy of steam $= 1.6751$ Btu/lbm-$°R$

Final conditions:

$T_2 = 200°F$
$h_f =$ enthalpy of water $= 167.99$ Btu/lbm
$h_{fg} =$ change of enthalpy during evaporation of water $= 977.9$ Btu/lbm
$s_f =$ entropy of saturated water $= 0.2938$ Btu/lbm-$°R$
$s_{fg} =$ change of entropy during evaporation of water
$\qquad = 1.4824$ Btu/lbm-$°R$
$v_g =$ specific volume of steam $= 33.64$ ft^3/lbm
$p_2 =$ final pressure $= 11.53$ psia

Question 1. The final quality of the steam in this process is most nearly

(A) 99%
(B) 97%
(C) 95%
(D) 93%
(E) 91%

Solution

In an isentropic (reversible, adiabatic) process, no heat is transferred $(Q = 0)$, and there is no change in entropy $(\Delta s = 0)$. Since the steam is initially superheated,

$$s_1 = s_2 = s_f + xs_{fg} = 0.2938 + x(1.4824) = 1.6751$$

The final quality of the steam is then

$$x = \frac{1.6751 - 0.2938}{1.4824} = 0.932 = 93.2\%$$

Answer is (D)

Question 2. The work done in this process, in Btu, is most nearly

(A) 230
(B) 290
(C) 460
(D) 700
(E) 870

Solution

The first law states, per pound, $q = \Delta u + w$. Since $q = 0$,

$$w = -\Delta u = u_1 - u_2$$

If h_2 and v_2 are found, then both specific energies can be calculated by the relation $u = h - pv$.

$$h_2 = h_f + xh_{fg} = 167.99 + 0.932(977.9) = 1079 \text{ Btu/lbm}$$
$$v_2 = xv_g + (1 - x)v_f$$

In this case v_f is unknown, but since $1 - x = 0.068$ and v_f is normally much smaller than v_g, it is a good approximation to neglect the term $(1 - x)v_f$ and compute v_2 as

$$v_2 = xv_g = 0.932(33.64) = 31.4 \text{ ft}^3$$

Now we can solve for u_1 and u_2:

$$u_1 = h_1 - p_1v_1 = 1368.3 - \frac{300(144)(2.227)}{778} = 1244.6 \text{ Btu/lbm}$$

$$u_2 = h_2 - p_2 v_2 = 1079 - \frac{11.53(144)(31.4)}{778} = 1012 \text{ Btu/lbm}$$

Hence

$$w = u_1 - u_2 = 1244.6 - 1012 = 232.6 \text{ Btu/lbm}$$

or, for 3 lbm,

$$W = 3(232.6) = 697.8 \text{ Btu}$$

Answer is (D)

PROBLEM 8-34

One lbm of air completes a reversible cycle consisting of the following two processes: (1) From a volume of 2 ft^3 and a temperature of 40°F, the air is compressed adiabatically to half the original volume. (2) Heat is then added at constant pressure until the original volume is reached.

Question 1. The work done in the first process, in ft-lbf, is closest to

(A) 12,600
(B) 21,300
(C) 27,600
(D) 29,800
(E) 38,600

Solution

Plots of the *p-V* and *T-s* planes for this process are shown in Fig. 8-8. For a

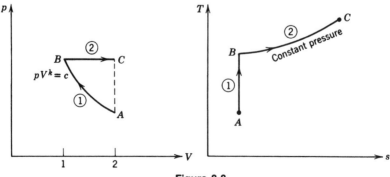

Figure 8-8

reversible adiabatic process the heat transfer $Q = 0$. The temperature change is given by

$$\frac{T_B}{T_A} = \left(\frac{V_A}{V_B}\right)^{k-1} = (2.0)^{0.4} = 1.32$$

since $k = 1.4$ for air. $T_A = 40 + 460 = 500°R$ so that $T_B = 1.32(500) = 660°R$. The work done is (also see Problem 8-27)

$$W = m\,\frac{R}{k-1}\,(T_B - T_A)$$

$$W = (1\;\text{lbm})\left(\frac{53.35\;\dfrac{\text{ft-lbf}}{\text{lbm-°R}}}{1.4 - 1}\right)(660 - 500)°R$$

$$W = 21{,}300\;\text{ft-lbf}$$

Answer is (B)

Question 2. The heat added in the second process, in Btu, is closest to

(A) 80
(B) 113
(C) 143
(D) 160
(E) 170

Solution

At constant pressure we may use the gas law to obtain

$$\frac{V_B}{V_C} = \frac{T_B}{T_C}$$

$$T_C = T_B\,\frac{V_C}{V_B} = 660\left(\frac{2}{1}\right) = 1320°R$$

The heat transfer Q, using $c_p = 0.24$ Btu/lbm-°R for air, is

$$Q = mc_p(T_C - T_B) = (1)(0.24)(1320 - 660)$$

$$Q = 158.4\;\text{Btu}$$

Answer is (D)

PROBLEM 8-35

A mass of 0.5 kg of air, to be considered a perfect gas, is contained in a cylinder and piston machine at an initial volume of 0.03 m^3 and at a pressure of 700 kN/m^2 (state 1). The gas is expanded reversibly at constant temperature to a volume of 0.06 m^3 (state 2). The gas is then compressed reversibly and adiabatically to a volume of 0.03 m^3 (state 3). It is then returned to its initial state (state 1) by a reversible constant-volume process.

Question 1. The net change in entropy, in J/kg-°K, for this cycle is most nearly

   (A) −3.1
   (B) 0
   (C) 1.6
   (D) 3.1
   (E) 4.8

Solution

A *T-s* diagram of the cycle is shown in Fig. 8-9. For this reversible cyclical process, the change in entropy Δ*s* is zero, as is shown in the diagram.

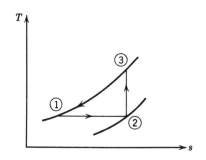

Figure 8-9

Answer is (B)

Question 2. The net heat transferred from the gas, in kJ, is most nearly

   (A) 2.2
   (B) 4.4
   (C) 8.9
   (D) 16.8
   (E) 21.0

Solution

The air mass $m = 0.5$ kg. At state 1, $p_1 = 700$ kN/m^2 and $V_1 = 0.03$ m^3. Also, $V_2 = 0.06$ m^3, $V_3 = 0.03$ m^3. Using the equation of state,

$$T_1 = \frac{p_1 V_1}{mR} = \frac{(7 \times 10^5 \text{ N/m}^2)(0.03 \text{ m}^3)}{(0.5 \text{ kg})(286.8 \text{ N-m/kg-}°\text{K})} = 146.4°\text{K}$$

For a constant-temperature process the gas law gives

$$p_2 = p_1\left(\frac{V_1}{V_2}\right) = (700 \text{ kN/m}^2)\left(\frac{1}{2}\right) = 350 \text{ kN/m}^2$$

Then T_3 can be found from the relation

$$\frac{T_3}{T_2} = \left(\frac{V_2}{V_3}\right)^{k-1} = \left(\frac{2}{1}\right)^{1.4-1} = 1.32$$

and

$$T_3 = 1.32 T_2 = 1.32(146.4) = 193.2°\text{K}$$

The net heat transfer per unit mass is

$$\oint_{\text{cycle}} dQ = {}_1Q_2 + {}_2Q_3 + {}_3Q_1$$

$$= T_1 R \ln\left(\frac{V_2}{V_1}\right) + 0 + c_v(T_1 - T_3)$$

$$= (146.4°\text{K})(286.8 \text{ N-m/kg-}°\text{K}) \ln\left(\frac{0.06}{0.03}\right)$$

$$+ (716 \text{ N-m/kg-}°\text{K})(146.4 - 193.2)°\text{K}$$

$$= 29.1 - 33.5 = -4.4 \text{ kJ/kg}$$

For the 0.5 kg mass, the result is $(-4.4 \text{ kJ/kg})(0.5 \text{ kg}) = -2.2 \text{ kJ}$. The negative sign indicates the heat is transferred away from the gas.

Answer is (A)

PROBLEM 8-36

A gas having a molecular weight of 36 and initially at 50 psia and 1200°F is expanded adiabatically in a turbine to a pressure of 15 psia. The specific heat ratio k of the gas is 1.25, and the isentropic efficiency of the turbine is 0.70.

The gas is expanded at a steady rate of 1 lbm/sec, and the gas velocity is low at both the inlet to and the exhaust from the turbine. The horsepower developed by the turbine is most nearly

- (A) 71
- (B) 84
- (C) 97
- (D) 110
- (E) 137

Solution

The isentropic efficiency $\eta = \dot{W}_A / \dot{W}_s = 0.7$, where $\dot{W}_A$ is the actual power and $\dot{W}_s$ is the power if the process were isentropic. We shall determine $\dot{W}_s$ and thus find $\dot{W}_A$. For an adiabatic expansion we have

$$\frac{T_2}{T_1} = \left(\frac{p_2}{p_1}\right)^{(k-1)/k}$$

$$T_2 = (1200 + 460)(\tfrac{15}{50})^{(1.25-1)/1.25}$$

$$= (1660)(0.3)^{0.2} = 1660(0.786) = 1305°R$$

Neglecting changes in kinetic and potential energy for an ideal gas that undergoes a steady flow, adiabatic process, the isentropic power $\dot{W}_s$ is

$$\dot{W}_s = \dot{m}c_p(T_1 - T_2) = \dot{m}\,\frac{kR}{k-1}(T_1 - T_2)$$

where $R = \bar{R}/M$ with $\bar{R} =$ universal gas constant and $M =$ molecular weight.

$$R = \frac{1545}{36} = 42.9 \ \frac{\text{ft-lbf}}{\text{lbm-}°R}$$

$$\dot{W}_s = (1)\,\frac{(1.25)(42.9)}{1.25 - 1}\,(1660 - 1305)$$

$$\dot{W}_s = 76{,}100 \ \text{ft-lbf/sec}$$

The actual power is

$$\dot{W}_A = \eta\dot{W}_s = \frac{0.7(76{,}100)}{550(\text{ft-lbf/sec-hp})}$$

$$\dot{W}_A = 96.9 \ \text{hp}$$

Answer is (C)

PROBLEM 8-37

A turbine, which is part of a Rankine cycle, receives steam at a pressure of 180 psia and 400°F from the boiler. The turbine exhausts at a pressure of 5 psia. Assume that the turbine operates adiabatically and reversibly. Neglect pump work.

Question 1. The heat efficiency of the cycle is closest to

(A) 0.17
(B) 0.20
(C) 0.23
(D) 0.26
(E) 0.28

Solution

Since the turbine process is both adiabatic and reversible, it is also isentropic, so that $s_1 = s_2$.

From the tables for superheated steam at 180 psia and 400°F we find $h_1 = 1214.0$ Btu/lbm and $s_1 = 1.5745$ Btu/lbm-°R. At 5 psia, the tables give

$$h_f = 130.13 \text{ Btu/lbm} \qquad h_{fg} = 1001.0 \text{ Btu/lbm}$$
$$s_f = 0.2347 \text{ Btu/lbm-°R} \qquad s_{fg} = 1.6094 \text{ Btu/lbm-°R}$$

The steam quality x is determinable from the equation

$$s_1 = s_2 = s_f + x s_{fg}$$
$$1.5745 = 0.2347 + x(1.6094) \quad \text{or} \quad x = 0.832 = 83.2\%$$

Now we can find the final enthalpy h_2:

$$h_2 = h_f + x h_{fg}$$
$$h_2 = 130.13 + (0.832)(1001.0)$$
$$h_2 = 962.96 \text{ Btu/lbm}$$

The heat efficiency $\eta = \dfrac{\text{net work output}}{\text{total heat input}}$

$$\eta = \frac{h_1 - h_2}{h_1 - h_f} = \frac{1214.0 - 962.96}{1214.0 - 130.13} = 0.23 = 23\%$$

Answer is (C)

Question 2. The number of lbm steam per horsepower-hour is closest to

(A) 8
(B) 10
(C) 12
(D) 14
(E) 16

Solution

The steam rate $w = \dfrac{\text{work/hp-hr}}{\text{work/lbm steam}}$

$$w = \frac{2545}{h_1 - h_2} = \frac{2545}{1214.0 - 962.96} = 10.1 \text{ lbm steam/hp-hr}$$

Answer is (B)

PROBLEM 8-38

A nozzle is designed to expand air from 100 psia and 90°F to 20 psia. Assume an isentropic expansion and a zero initial velocity. The air flow rate is 3 lbm/sec, and the molecular weight of air is 29.

Question 1. The exit velocity, in feet per second, is nearest to

(A)   280
(B)   880
(C) 1320
(D) 1560
(E) 2040

Solution

Applying the first law to the nozzle flow,

$$\frac{V_1^2}{2} + h_1 = \frac{V_2^2}{2} + h_2$$

Noting that the initial velocity $V_1 = 0$,

$$V_2 = [2(h_1 - h_2)]^{1/2}$$

which for an ideal gas can be expressed as

$$V_2 = \left[2c_p T_1 \left(1 - \frac{T_2}{T_1} \right) \right]^{1/2}$$

Since the expansion is isentropic and the initial and final pressures are known, we shall use the isentropic relation $Tp^{(1-k)/k} = \text{constant}$ and $k = 1.4$ for air to obtain

$$V_2 = \left\{ 2c_p T_1 \left[1 - \left(\frac{p_2}{p_1} \right)^{(k-1)/k} \right] \right\}^{1/2}$$

With the aid of some conversion factors we find

$$V_2 = \left\{ 2 \left(32.2 \frac{\text{lbm-ft}}{\text{lbf-sec}^2} \right) \left(778 \frac{\text{ft-lbf}}{\text{Btu}} \right) \left(0.24 \frac{\text{Btu}}{\text{lbm-}^\circ\text{R}} \right) (90 + 460)^\circ\text{R} \right.$$

$$\left. \times \left[1 - \left(\frac{20}{100} \right)^{(1.4-1)/1.4} \right] \right\}^{1/2}$$

$$V_2 = \{ 6.61(10^6)[1 - 0.631] \text{ ft}^2/\text{sec}^2 \}^{1/2} = 1562 \text{ ft/sec}$$

Answer is (D)

Question 2. The correct exit cross-sectional area, in in.^{2}, is closest to

(A) 2.00
(B) 1.75
(C) 1.50
(D) 1.25
(E) 1.00

Solution

Since $\dfrac{T_2}{T_1} = \left(\dfrac{p_2}{p_1} \right)^{(k-1)/k} = 0.631$,

$$T_2 = 0.631(90 + 460) = 347^\circ\text{R}$$

By the equation of state, the specific volume v_2 is

$$v_2 = \frac{RT_2}{p_2} = \frac{\bar{R}T_2}{Mp_2} = \frac{(1545)(347)}{(29)(20)(144)}$$

$$v_2 = 6.42 \text{ ft}^3/\text{lbm}$$

The mass rate of flow is 3 lbm/sec; thus conservation of mass requires

$$w = \frac{A_2 V_2}{v_2}$$

or

$$A_2 = \frac{w v_2}{V_2} = \frac{3(6.42)}{1562} = 0.0123 \text{ ft}^2$$

$$A_2 = (0.0123)(144) = 1.77 \text{ in}^2$$

Answer is (B)

PROBLEM 8-39

Figure 8-10 is a schematic diagram of a typical refrigeration system showing the major components, as well as the state of the refrigerant between the components (that is, liquid or vapor). Assume the refrigeration system has a 10-ton capacity while operating with a 40°F suction temperature and a condensing temperature of 102°F. Also assume the heat added to the vapor during compression is 29.6 Btu/min per ton of refrigeration, and a ton of refrigeration equals 12,000 Btu/hr.

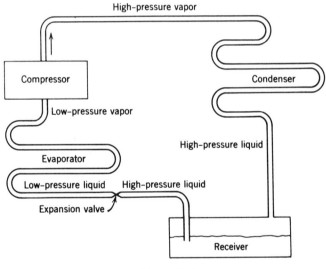

Figure 8-10

The amount of water, in gallons per minute, required for condensing if a 15°F temperature rise is allowed is nearest to

(A) 12
(B) 14
(C) 16
(D) 18
(E) 20

Solution

The heat rejected by the system must equal the heat entering the system plus the work done on the system. The heat in is

$$\left(\frac{12{,}000 \text{ Btu/hr}}{1 \text{ ton}}\right)\left(\frac{1}{60}\frac{\text{hr}}{\text{min}}\right)(10 \text{ ton}) = 2000 \frac{\text{Btu}}{\text{min}}$$

The heat rejected is then $2000 + 29.6(10) = 2296$ Btu/min. This heat must be accepted by the cooling water with a coolant temperature rise of 15°F. Since 1 Btu will raise 1 lb of water 1°F, the flow rate of cooling water must be $2296/15 = 153$ lb/min of water. One gal of water weighs

$$\left(62.4 \frac{\text{lb}}{\text{ft}^3}\right)\left(231 \frac{\text{in.}^3}{\text{gal}}\right)\left(\frac{1}{1728}\frac{\text{ft}^3}{\text{in.}^3}\right) = 8.34 \frac{\text{lb}}{\text{gal}}$$

Hence $153/8.34 = 18.4$ gal/min of cooling water are required.

Answer is (D)

PROBLEM 8-40

A counterflow heat exchanger is operating with the hot liquid entering at 400°F and leaving at 327°F. The cool liquid enters at 100°F and leaves at 283°F. The logarithmic mean temperature difference between the hot and cool liquids is nearest to

(A) 172
(B) 166
(C) 152
(D) 136
(E) 120

Solution

A diagram of the process is shown in Fig. 8-11. The logarithmic mean

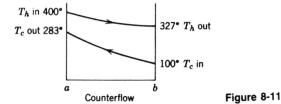

T_h in 400°
T_c out 283°
327° T_h out
100° T_c in
a
b
Counterflow

Figure 8-11

temperature difference ΔT_m, as used in heat exchanger calculations, is defined as

$$\Delta T_m = \frac{\Delta T_a - \Delta T_b}{\ln (\Delta T_a / \Delta T_b)}$$

Here

$$\Delta T_a = 400 - 283 = 117°$$
$$\Delta T_b = 327 - 100 = 227°$$

Hence

$$\Delta T_m = \frac{117 - 227}{\ln (117/227)} = \frac{-110}{-0.663} = 166°F$$

Answer is (B)

PROBLEM 8-41

A windowless wall of a house 8 ft high by 15 ft long is of wood frame construction with a $\frac{3}{4}$-in. stucco exterior, 3 in. of insulation, and a $\frac{1}{2}$-in. plasterboard interior. The outside temperature is 100°F and the inside temperature is 75°F. The conductivities of the stucco and the insulation are 12.00 and 0.27, respectively. The conductivity of the plasterboard is 2.82. (Ignore the effects of the studs and both the inside and outside surface conductances.)

The heat flow through the wall, in Btu/hr, is most nearly

(A) 270
(B) 268
(C) 266
(D) 264
(E) 262

Solution

A cross section of the wall is shown in Fig. 8-12. The basic heat conduction equation is

$$Q = k \frac{A}{L} (T_a - T_b) \quad \frac{\text{Btu}}{\text{hr}}$$

where k = conductivity in Btu-in./ft^2-°F-hr, A = wall area in square feet,

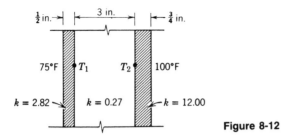

Figure 8-12

L = wall thickness in inches, and $T_a - T_b$ = temperature difference in degrees Fahrenheit. For steady heat flow through the wall, the heat flow Q is identical in each layer of material. Thus

$$100 - T_2 = \frac{0.75}{12} \frac{Q}{A}$$

$$T_2 - T_1 = \frac{3}{0.27} \frac{Q}{A}$$

$$T_1 - 75 = \frac{0.5}{2.82} \frac{Q}{A}$$

Adding these three equations gives

$$100 - 75 = \frac{Q}{A} \left(\frac{0.75}{12} + \frac{3}{0.27} + \frac{0.5}{2.82} \right)$$

Using $A = 8(15) = 120 \text{ ft}^2$, we have

$$(25)(120) = Q(0.06 + 11.11 + 0.18)$$

$$Q = \frac{(25)(120)}{11.35} = 264 \frac{\text{Btu}}{\text{hr}}$$

Answer is (D)

PROBLEM 8-42

One end of a copper bar of cross-sectional area 4 cm^2 and length 80 cm is kept in steam at 1 atm pressure; the other end is in contact with melting ice. The thermal conductivity of copper is $k = 420 \text{ J/m-sec-°C}$, and the specific latent heat of ice is 335 kJ/kg.

The amount of ice, in grams, that will be melted in 10 min if the sides of the copper rod are insulated is closest to

(A) 35.0
(B) 37.5

(C) 40.0

(D) 42.5

(E) 45.0

Solution

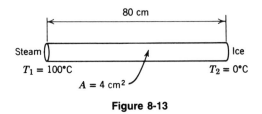

Figure 8-13

The rate at which heat is transferred from the hot end to the cold end of the rod is

$$Q = k \frac{A}{L} (T_1 - T_2)$$

$$Q = \left(\frac{420 \text{ J}}{\text{m-sec-}°\text{C}}\right)\left(\frac{4}{80} \text{ cm}\right)\left(\frac{1 \text{ m}}{100 \text{ cm}}\right)(100 - 0)°\text{C} = 21 \frac{\text{J}}{\text{sec}}$$

In 10 min the total heat transferred is

$$\left(21 \frac{\text{J}}{\text{sec}}\right)\left(60 \frac{\text{sec}}{\text{min}}\right)(10 \text{ min}) = 12,600 \text{ J}$$

The amount of ice melted is therefore

$$\frac{12,600 \text{ J}}{335 \text{ kJ/kg}} = 37.6 \text{ g}$$

Answer is (B)

9

CHEMISTRY

Chemistry, as a branch of science, is primarily concerned with kinds of matter and the changes that occur when they are brought together. Chemistry problems cover a multitude of topics, some of which require specific recall. This collection of chemistry problems will emphasize the review of basic principles rather than the acquisition of specific fragments of information.

GAS LAW RELATIONS

For ease of comparison, gas volumes often are given at a temperature of 0°C and 760 mm of mercury (760 torr or 1 atm or 14.7 psia). The values are referred to as *standard temperature and pressure* (STP) or *standard conditions*.

Boyle's law and Charles' law are often combined to form the universal gas law, which is

$$\frac{p_1 V_1}{T_1} = \frac{p_2 V_2}{T_2}$$

Here the subscripts represent two different states. Values of absolute temperature and pressure are required in this relation.

At the same temperature and pressure, equal volumes of different gases contain equal numbers of molecules. One g-mole (gram molecular weight) of any gas contains Avogadro's number 6.02×10^{23} molecules, and at standard conditions it occupies a volume of 22.4 liters.

If different gases are mixed together, each component of the mixture acts as if it alone were present in the container. The pressure exerted by each component in the mixture is proportional to its mole concentration in the mixture. The total pressure of the mixture of gases is equal to the sum of the pressures of the components. Called Dalton's law, this can be written

$$p_T = p_1 + p_2 + p_3 + \cdots + p_n$$

The partial pressure ratio (p_1/p_T) is equal to the mole fraction (n_1/n_T) of that component, or $p_1/p_T = n_1/n_T$, where $n_1 =$ moles of gas component 1 and $n_T =$ total moles of gas. For additional discussion on gas laws, refer to Chapter 8.

Example 1

A candle is burned under an inverted beaker until the flame dies out. A sample of the mixture of perfect gases in the beaker after the flame has burnt out contains 8.30×10^{20} molecules of nitrogen, 0.70×10^{20} molecules of oxygen, and 0.50×10^{20} molecules of CO_2. If the total pressure of the mixture is 760 mmHg, how many moles of gas are present, and what is the partial pressure of each gas?

Solution

Law of partial pressures: The pressure exerted by each component in a gaseous mixture is proportional to its concentration in the mixture, and the total pressure of the gas is equal to the sum of those of its components. Therefore

$$p = p_{N_2} + p_{O_2} + p_{CO_2} = 760 \text{ mmHg}$$

The number of moles of each gas is equal to the number of molecules divided by Avogadro's number (6.02×10^{23}):

Gas	No. molecules	Divided by	Moles of gas
N_2	8.30×10^{20}	6.02×10^{23}	13.79×10^{-4}
O_2	0.70×10^{20}	6.02×10^{23}	1.16×10^{-4}
CO_2	0.50×10^{20}	6.02×10^{23}	0.83×10^{-4}

Total moles gas $= 15.78 \times 10^{-4}$

$$p_{N_2} = p \frac{\text{moles of } N_2}{\text{total moles of gas}} = (760) \frac{13.79 \times 10^{-4}}{15.78 \times 10^{-4}} = 664 \text{ mmHg}$$

$$p_{O_2} = (760) \frac{1.16 \times 10^{-4}}{15.78 \times 10^{-4}} = 56 \text{ mmHg}$$

$$p_{CO_2} = (760) \frac{0.83 \times 10^{-4}}{15.78 \times 10^{-4}} = 40 \text{ mmHg}$$

CHEMICAL BALANCE CALCULATIONS

Frequently a problem describes a chemical reaction and then the quantities of the various components are to be calculated. The starting point is a balanced chemical equation. Using the atomic weights of the elements, the molecular weights of the components of the chemical equation are determined. The coefficients in front of the formulas provide the relative numbers of each kind of molecule. The relative weights of the various components are thus known. The values may be set in a ratio to determine the quantities of each component in a specified situation.

Example 2

One of the principal scale-forming constituents of water is soluble calcium bicarbonate, $Ca(HCO_3)_2$. This substance may be removed by treating the water with lime, $Ca(OH)_2$, in accordance with the following reaction:

$$Ca(HCO_3)_2 + Ca(OH)_2 \rightarrow 2CaCO_3 + 2H_2O$$

Atomic weights: $Ca = 40$ $H = 1$ $C = 12$ $O = 16$

Determine the kilograms of lime required to remove 1 kg of calcium bicarbonate.

Solution

Molecular weights:

$$Ca(HCO_3)_2 = 40 + (1 + 12 + 3 \times 16)(2) = 162$$
$$Ca(OH)_2 = 40 + (16 + 1)(2) = 74$$
$$CaCO_3 = 40 + 12 + (16)(3) = 100$$
$$H_2O = 2(1) + 16 = 18$$

From the balanced chemical equation we read that 1 mole of $Ca(HCO_3)_2$ plus 1 mole of $Ca(OH)_2$ combines to produce 2 moles of $CaCO_3$ and 2 moles of H_2O. Since a mole is a molecular weight of a substance in any desired weight units, here we will use 1 mole = 1 kg-molecular weight (in other problems 1 mole might be a gram-molecular weight or some other unit).

$$Ca(HCO_3)_2 + Ca(OH)_2 \rightarrow 2CaCO_3 + 2H_2O$$
$$162 \text{ kg} + 74 \text{ kg} \rightarrow 2(100) \text{ kg} + 2(18) \text{ kg}$$

The equation shows that 1 mole of lime combines with 1 mole of calcium bicarbonate. Thus 74 kg of lime combine with 162 kg of calcium bicarbonate. For 1 kg of calcium bicarbonate, 74/162 or 0.457 kg of lime is required.

COMBUSTION

Combustion is a specialized and somewhat more complex example of a chemical balance calculation. In simple complete combustion a hydrocarbon fuel (one containing carbon and hydrogen) is burned in the presence of the theoretically correct amount of oxygen to produce carbon dioxide (CO_2) and water (H_2O) as the combustion products. The source of the oxygen is air. (Air = 21% oxygen and 79% nitrogen by volume.) Often this type of problem is readily solved by a balanced calculation if two additional items are recalled:

1. One g-mole of any gas at standard conditions occupies a volume of 22.4 liters.
2. For 1 liter of oxygen, $100/21 = 4.76$ liters of air are required.

Example 3

Propane (C_3H_8) is completely burned in air with carbon dioxide (CO_2) and water (H_2O) being formed. If 15 lbm of propane are burned per hour, how many cubic feet per hour of dry CO_2 are formed at 70°F and atmospheric pressure?

$$\text{Atomic weights:} \quad C = 12 \quad H = 1 \quad O = 16$$

Solution

The equation for the reaction is

$$C_3H_8 + 5O_2 \rightarrow 3CO_2 + 4H_2O \quad \text{(ignoring the } N_2 \text{ in the air)}$$

Molecular weights:

$$C_3H_8 = 36 + 8 = 44 \quad CO_2 = 12 + 32 = 44$$

Since 44 lbm of propane produce 3(44) lbm of carbon dioxide, 15 lbm of propane produce 45 lbm of carbon dioxide. Using the equation of state

$$pV = mRT$$

where

$$p = 14.7 \times 144 = 2117 \text{ lb/ft}^2 \text{ abs}$$

$$m = 45 \text{ lbm } CO_2$$

$$R = \frac{1545 \text{ ft-lbf}}{\text{lbm-mole }°R} \times \frac{1}{\text{mol wt of } CO_2 = 44} = 35.1 \text{ ft-lbf/lbm-}°R$$

$$T = 70°F + 460 = 530°R$$

$$V = \frac{mRT}{p} = \frac{45 \times 35.1 \times 530}{2117} = 395 \text{ ft}^3/\text{hr } CO_2$$

Example 4

The heating value of natural gas is 1000 Btu/ft^3. If a furnace has an output efficiency of 90%, what should be the supply of air in cubic feet per minute for an output of 100,000 Btu/hr? Assume the oxygen content of air is 21% by volume and that natural gas is essentially methane, which combines with oxygen as follows:

$$CH_4 + 2O_2 \rightarrow CO_2 + 2H_2O$$

Solution

The input of natural gas is $100,000/(0.90 \times 1000) = 111 \text{ ft}^3/\text{hr}$. From the balanced equation we see that 1 mole of CH_4 combines with 2 moles of O_2. For 111 ft^3/hr of natural gas, twice this amount, or 222 ft^3/hr, of pure oxygen is required. Since air is 21% by volume oxygen, the amount required is

$$\frac{222}{0.21} = 1058 \text{ ft}^3/\text{hr} \qquad \frac{1058}{60} = 17.6 \text{ ft}^3/\text{min}$$

MOLAR SOLUTIONS

A molar solution designation is an important method of describing the relative amount of a component of a solution in quantitative form. A molar solution contains 1 g-mole of the solute in 1 liter of solution.

Example 5

How much NaCl is in a 1 M solution? How much in a 0.5 M solution?

$$\text{Atomic weights:} \quad Na = 23 \qquad Cl = 35.5$$

Solution

A 1 M solution of NaCl would contain $23 + 35.5 = 58.5$ g of NaCl in 1 liter of solution. Similarly, a 0.5 M solution of NaCl would contain 29.25 g/liter.

pH VALUE

The pH value of a solution is the negative logarithm of the hydrogen ion concentration

$$pH = -\log_{10}[H^+]$$

where $[H^+]$ means hydrogen ion concentration in g-moles per liter. In neutral solutions the pH is 7. If the pH is less than 7, the solution is acidic; if greater than 7, the solution is basic.

PROBLEM 9-1

If 5 g of mass were entirely converted to energy and the velocity of light equals 3×10^{10} cm/sec, the amount of energy liberated, in joules, would be

(A) 15×10^3
(B) 75×10^3
(C) 22×10^{13}
(D) 45×10^{13}
(E) 45×10^{20}

Solution

The energy liberated in ergs is found from the famous energy equation

$$E = mc^2$$

where m = converted mass in grams, and c = speed of light in centimeters per second.

$$E = 5(3 \times 10^{10})^2 = 45 \times 10^{20} \text{ ergs}$$

$$1 \text{ joule} = 10^7 \text{ erg}$$

$$E = 45 \times 10^{13} \text{ joules}$$

Answer is (D)

PROBLEM 9-2

The halogen group includes the elements

(A) Na, Ca, K
(B) F, Cl, Br, I

(C) Au, Ag, Pt
(D) Mg, Mn, Mo
(E) C, O, H

Solution

The halogen group of elements ("the salt producers") contains fluorine (F), chlorine (Cl), bromine (Br), and iodine (I).

Answer is (B)

PROBLEM 9-3

The group of metals which is comprised of lithium, sodium, potassium, rubidium, and cesium forms a closely related family known as

(A) the rare earth group
(B) the metals of the fourth outer group
(C) the alkali metals
(D) the elements of the inner group
(E) the metals of Group VIII

Solution

These metals are all members of Group I_A—the alkali metals.

Answer is (C)

PROBLEM 9-4

The substance that does not belong in the following group is

(A) lye
(B) digestive fluid
(C) sour milk
(D) sulfuric acid
(E) HCl

Solution

Lye is a base, NaOH. The other four are acids.

Answer is (A)

PROBLEM 9-5

The number that expresses the oxidation state of an atom of an element or of a group of atoms is called the

    (A) indicator
    (B) displacement factor
    (C) electrolyte
    (D) valence
    (E) conductance

Solution

Valence is the number of electron pair bonds which an atom shares with other atoms. Valence often is used synonymously with oxidation state of an atom or ion in inorganic chemistry.

Answer is (D)

PROBLEM 9-6

Which of the following statements is false?

    (A) The atomic weight of oxygen is 16.
    (B) The *pascal* is the pressure or stress of one newton per square meter.
    (C) Zero on the absolute temperature scale is approximately $-273°C$.
    (D) The *coulomb* is the quantity of electricity transported in one second by a current of one ampere.
    (E) Alcohol has a freezing point of $-32°C$.

Solution

The freezing point of methyl alcohol is $-94°C$; the freezing point of ethyl alcohol is even lower.

Answer is (E)

PROBLEM 9-7

The chemical formula for ethyl alcohol (ethanol) is

    (A) C_2H_6
    (B) CH_4

(C) CH_3OH
(D) C_2H_5OH
(E) C_3H_7OH

Solution

The formulas given are

(A) ethane
(B) methane
(C) methyl alcohol
(D) ethyl alcohol
(E) propyl alcohol

Answer is (D)

PROBLEM 9-8

The ion responsible for the chemistry of bases in water is

(A) H^+
(B) Na^+
(C) H_3O^+
(D) OH^-
(E) HOH^-

Solution

Hydroxyl ion (OH^-) is the characteristic base of aqueous solutions.

Answer is (D)

PROBLEM 9-9

In the earth's crust (which includes a 10-mile shell and the atmosphere above it), aside from oxygen and silicon, what is the most abundant element found?

(A) aluminum
(B) iron
(C) calcium
(D) sodium
(E) hydrogen

Solution

	%		%
Oxygen	49.2	Calcium	3.4
Silicon	25.7	Sodium	2.4
Aluminum	7.4	Hydrogen	1.0
Iron	4.7		

(Although different sources show slightly different percentages, they agree with this ranking.)

Answer is (A) *0 K*

PROBLEM 9-10

Aside from nitrogen and oxygen, which of the following gases is the most plentiful in dry air?

(A) argon
(B) carbon dioxide *✓ Wrong*
(C) neon
(D) helium
(E) krypton

Solution

	%		%
Nitrogen	78.03	Neon	0.0015
Oxygen	20.99	Helium	0.0005
Argon	0.94	Krypton	0.00011

Carbon dioxide 0.023 to 0.050%

Answer is (A) *OK*

PROBLEM 9-11

Oxygen is converted into ozone by

(A) great heat
(B) high pressure

(C) electric discharge
(D) a catalyst
(E) none of these processes

Solution

Ozone is formed by passing an electric discharge through oxygen.

Answer is (C)   *Ok*

PROBLEM 9-12

The phenomenon known as corrosion in metals is by nature

(A) chemical
(B) organic
(C) electrochemical
(D) inorganic
(E) electrical

Solution

The phenomenon is electrochemical. It is often called electrolytic corrosion.

Answer is (C)   *Ok*

PROBLEM 9-13

Radioactivity is a property of all elements with atomic numbers greater than

(A) 48
(B) 62
(C) 83
(D) 88
(E) 90   *Ok*

Solution

Radioactivity is the property of naturally occurring elements with atomic numbers greater than 83.

Answer is (C)

PROBLEM 9-14

A certain process has three factors which always occur: (1) a current flow; (2) substances that are formed at each electrode; and (3) matter that is transported through a solution or molten salt. This process is called

(A) evaporation
(B) decomposition
(C) hydrolysis
(D) condensation
(E) electrolysis

Solution

The process is called electrolysis.

Answer is (E)

PROBLEM 9-15

The process in which a solid changes directly to the gaseous state is called

(A) sublimation
(B) homogenization
(C) crystallization
(D) vaporization
(E) distillation

Solution

The process is sublimation. Solid CO_2, iodine, and naphthalene are examples of materials that sublime.

Answer is (A)

PROBLEM 9-16

Tetraethyl lead has been used as an antiknock compound in gasoline. The formula for tetraethyl lead is

(A) CH_4Pb
(B) C_2H_5Pb
(C) $(C_2H_5)_4Pb_4$

(D) $(C_2H_5)_4Pb$

(E) $(C_2H_4)_4Pb$

Solution

Ethyl is C_2H_5; hence tetraethyl lead is $(C_2H_5)_4Pb$, or, as it is often written, $Pb(C_2H_5)_4$.

Answer is (D) OK

PROBLEM 9-17

Carbon dioxide "snow" from a fire extinguisher puts out a fire because it

(A) lowers the kindling temperature of the material

(B) displaces the supply of oxygen

(C) raises the kindling temperature of the material

(D) cools the material below its kindling temperature

(E) acts as a catalyst to induce a chemical reaction

Solution

Carbon dioxide "snow" smothers a fire by keeping the oxygen away from the fire.

Answer is (B) OK

PROBLEM 9-18

Hardness in water supplies is primarily due to the solution in water of

(A) carbonates and sulfates of calcium and magnesium

(B) alum

(C) soda ash

(D) sodium sulfate

(E) sodium chloride

Solution

The compounds of calcium and magnesium cause hardness. These may include carbonates, bicarbonates, sulfates, and chlorides.

Answer is (A)

PROBLEM 9-19

Hydrogen-free carbon in the form of coke is burned with complete combustion, using 26.4% excess air. Assume the air contains 79 mole% nitrogen and 21 mole% oxygen.

Atomic weights: $C = 12$ $O = 16$ $N = 14$

If 100 moles of carbon are burned, the number of moles of CO_2 formed is nearest to

(A) 100
(B) 133
(C) 267
(D) 300
(E) 400

Solution

$$C + O_2 \rightarrow CO_2$$

This tells us that 100 moles of carbon, when burned with 100 moles of O_2, forms 100 moles of CO_2.

OK

Answer is (A)

PROBLEM 9-20

Helium gas and hydrogen gas are used in lighter-than-air craft. The lifting power of helium as compared to the same volume of hydrogen is nearest to

(A)   25%
(B)   50%
(C)   75%
(D)   90%
(E)   100%

Solution

In equal volumes of different gases, at the same temperature and pressure, the numbers of molecules present are equal. Thus density is proportional to molecular weight.

$$H_2 = 2 \times 1.008 = 2.016$$

$$He = 4.003$$

But the lifting power of gas is proportional to the difference between its density and the density of air. Since air has an average molecular weight of about 29, the lifting power of hydrogen is $29 - 2 = 27$ and of helium $29 - 4 = 25$.

   Thus the ratio of lifting power is

$$\frac{He}{H_2} = \frac{25}{27} = 93\%$$

Answer is (D)   *OK*

PROBLEM 9-21

Pure water has a hydrogen-ion concentration in moles per liter of approximately

(A) 1.0
(B) 0.1
(C) 0.001
(D) 0.00001
(E) 0.0000001

Solution

Pure water is neutral (pH = 7), though slightly lowered by dissolved CO_2.

$$pH = -\log [H^+] = 7$$

Since $-(-7) \log 10 = 7$, the hydrogen-ion concentration $[H^+] = 10^{-7}$

Answer is (E)   *OK*

PROBLEM 9-22

The common anesthetic chloroform, $CHCl_3$, is a colorless liquid boiling at 61.2°C.

Atomic weights: $C = 12$ $H = 1$ $Cl = 35.5$

The proportion by weight of chlorine it contains is nearest to

(A) 20%
(B) 35%
(C) 60%
(D) 75%
(E) 90%

Solution

Proportion of each element:

$$
\begin{array}{ll}
1 \text{ atom of carbon} & = 12 \\
1 \text{ atom of hydrogen} & = 1 \\
3 \text{ atoms of chlorine} = 3(35.5) & = 106.5 \\
\hline
\text{Molecular weight} & = 119.5
\end{array}
$$

Percent chlorine by weight $= \dfrac{3(35.5)}{119.5} \times 100 = 89\%$.

Answer is (E)

PROBLEM 9-23

The hydroxyl ion concentration of 1×10^{-11} moles/liter in water is equivalent to what pH number?

(A) 3
(B) 5
(C) 7
(D) 11
(E) 14

Solution

$$
pOH = -\log [OH] = -\log 10^{-11} = 11
$$

For water, $pH + pOH = 14$; therefore, $pH = 14 - 11 = 3$.

Answer is (A)

PROBLEM 9-24

The volume of 1 g-mole of oxygen at 0°C and 1 atmosphere of pressure is closest to

(A) 1.0 liter
(B) 16.0 liters
(C) 22.4 liters
(D) $0.1\,m^3$
(E) $1.0\,m^3$

Solution

The volume of 1 mole of any gas at standard conditions (0°C and 1 atm of pressure) is 22.4 liters.

Answer is (C)

PROBLEM 9-25

Which one of the following chemical equations is balanced? The valences of the various elements or radicals are as follows:

$$H^+ \quad Na^+ \quad Ag^+ \quad Ca^{2+} \quad Zn^{2+} \quad C^{4+}$$

$$OH^- \quad Cl^- \quad NO_3^- \quad O^{2-} \quad CO_3^{2-} \quad SO_4^{2-}$$

(A) $H_2 + O_2 \rightarrow H_2O$
(B) $NaCl + 2AgNO_3 \rightarrow NaNO_3 + AgCl$
(C) $CaO + H_2O \rightarrow Ca(OH)_3$
(D) $CaSO_4 + Na_2CO_3 \rightarrow CaCO_3 + Na_2SO_4$
(E) $ZnCO_3 \rightarrow ZnO + 2CO_2$

Solution

The correct equations are

(A) $2H_2 + O_2 \rightarrow 2H_2O$
(B) $NaCl + AgNO_3 \rightarrow NaNO_3 + AgCl$
(C) $CaO + H_2O \rightarrow Ca(OH)_2$
(D) correct as given
(E) $ZnCO_3 \rightarrow ZnO + CO_2$

Answer is (D)

PROBLEM 9-26

Which one of the following chemical equations is not balanced? The valences of the various elements or radicals are as follows:

$$Ag^+ \quad H^+ \quad Ca^{2+} \quad Ba^{2+} \quad Pb^{2+} \quad C^{4+}$$
$$C^- \quad Cl^- \quad NO_3^- \quad HCO_3^- \quad O^{2-} \quad SO_4^{2-} \quad CO_3^{2-}$$

(A) $2AgNO_3 + CaCl_2 \rightarrow 2AgCl + Ca(NO_3)_2$ OK
(B) $BaO + H_2SO_4 \rightarrow BaSO_4 + H_2O$ OK
(C) $CaCO_3 + HCl \rightarrow CaCl + H_2O + CO_2$ NOV
(D) $PbCl_2 + H_2SO_4 \rightarrow PbSO_4 + 2HCl$ OK
(E) $Ca(HCO_3)_2 \rightarrow CaCO_3 + H_2O + CO_2$ OK

Solution

Equation (C) should be

$$CaCO_3 + 2HCl \rightarrow CaCl_2 + H_2O + CO_2$$

Answer is (C)

PROBLEM 9-27

The action of nitric acid on silver gives silver nitrate, water, and nitric oxide gas. The balanced equation for this reaction is

(A) $Ag + HNO_3 \rightarrow AgNO_3 + NO + H_2O$
(B) $2Ag + H_2NO_3 \rightarrow AgNO_3 + NO + H_2O$
(C) $3Ag + 4HNO_3 \rightarrow 3AgNO_3 + NO + 2H_2O$
(D) $Ag + 4HNO_3 \rightarrow AgNO_3 + 4NO + 8H_2O$
(E) $Ag + 2H_2NO_3 \rightarrow AgNO_3 + NO + 2H_2O$

Solution

$$4H^+ + 3e^- + NO_3^- \rightarrow NO + 2H_2O$$
$$3Ag^0 - 3e^- \qquad\qquad \rightarrow 3Ag^+$$
$$\overline{}$$
$$3Ag + 4HNO_3 \qquad \rightarrow 3AgNO_3 + NO + 2H_2O$$

Answer is (C)

PROBLEM 9-28

The specific gravity of a substance is the weight of a given volume of the substance divided by the weight of an equal volume of water. If 2 kg of a liquid with a specific gravity of 0.800 are mixed with 1 kg of water, the specific gravity of the mixture is nearest to

(A) 0.833
(B) 0.850
(C) 0.857
(D) 0.867
(E) 1.000

Solution

The equivalent volume of liquid $= (2/0.8) + 1 = 3.5$. The weight of liquid $= 2 + 1 = 3$ kg.

$$\text{Specific gravity} = \frac{\text{Wt of liquid}}{\text{Wt of equiv. volume of water}}$$

$$= \frac{3}{3.5(1)} = 0.857$$

Answer is (C)

PROBLEM 9-29

How much water is required to set 100 kg of plaster of Paris? The formulas are $CaSO_4 \cdot \frac{1}{2}H_2O$ for plaster of Paris and $CaSO_4 \cdot 2H_2O$ for the set form of plaster of Paris (gypsum).

Atomic weights: $Ca = 40.1$ $S = 32.1$ $O = 16$ $H = 1$

The amount of water, in kg, is closest to

(A) 20
(B) 25
(C) 35
(D) 90
(E) 100

Solution

Molecular weights:

$$CaSO_4 \cdot \tfrac{1}{2}H_2O = 40.1 + 32.1 + 4(16) + \tfrac{1}{2}(2 + 16) = 145.2$$
$$H_2O = 2(1) + 16 = 18$$

The balanced reaction formula is

$$2(CaSO_4 \cdot \tfrac{1}{2}H_2O) + 3H_2O \rightarrow 2(CaSO_4 \cdot 2H_2O)$$

For one mole of $CaSO_4 \cdot \tfrac{1}{2}H_2O$, $1\tfrac{1}{2}$ moles of H_2O are needed to form $CaSO_4 \cdot 2H_2O$. Thus for 145.2 kg of plaster of Paris, $\tfrac{3}{2}(18) = 27$ kg of water are needed to form gypsum.

$$\frac{100}{145.2} = \frac{X}{27}$$

The amount of water X to set 100 kg of plaster of Paris is 18.6 kg.

Answer is (A)

PROBLEM 9-30

The number of moles of sulfuric acid required to dissolve 10 moles of aluminum according to the reaction

$$2Al + 3H_2SO_4 = Al_2(SO_4)_3 + 3H_2$$

is

(A) 3
(B) 6
(C) 10
(D) 15
(E) 30

Solution

$$\frac{10 \text{ moles Al}}{2Al} = \frac{X \text{ moles } H_2SO_4}{3H_2SO_4} \qquad X = 15 \text{ moles } H_2SO_4$$

Answer is (D)

PROBLEM 9-31

C_7H_{16} burns completely to form water and carbon dioxide. If 10 grams of it are burned, how many liters of carbon dioxide are formed at standard conditions?

$$\text{Atomic weights:}\quad C = 12 \qquad O = 16 \qquad H = 1$$
$$\text{Molecular weight:}\quad C_7H_{16} = 7(12) + 16(1) = 100$$

The amount of CO_2, in liters, is nearest to

(A) 15
(B) 20
(C) 22.4
(D) 45
(E) 90

Solution

By inspection, 1 g-mole of C_7H_{16} will yield, on complete combustion, 7 g-moles of CO_2, occupying $7 \times 22.4 = 156.8$ liters at standard conditions. Since 100 g (1 g-mole) of C_7H_{16} produce 156.8 liters of CO_2, 10 g will produce one-tenth that amount, or 15.7 liters of CO_2 at standard conditions.

Answer is (A)

PROBLEM 9-32

How many kilograms of limestone, 80% $CaCO_3$, would be required to make 1000 kg of CaO?

$$\text{Atomic weights:}\quad Ca = 40 \qquad C = 12 \qquad O = 16$$

The quantity of limestone, in kg, required is nearest to

(A) 1000
(B) 1400
(C) 1800
(D) 2200
(E) 2600

Solution

$$CaCO_3 \rightarrow CaO + CO_2$$

Thus 1 mole of $CaCO_3$ yields 1 mole of CaO.
   Molecular weights:

$$CaCO_3: \quad 40 + 12 + 3(16) = 100$$
$$CaO: \quad 40 + 16 = 56$$

So 100 kg of pure $CaCO_3$ yield 56 kg of CaO. For 1000 kg of CaO, the quantity of limestone required is

$$\frac{1000}{56} \times \frac{100}{0.8} = 2230 \text{ kg}$$

Answer is (D)

PROBLEM 9-33

If silver sells for $160 per kilogram, what is the maximum amount you could pay for a tank containing 1000 kg of silver nitrate solution that is 60% water by weight? The cost of recovering the silver from the silver nitrate solution is $1000.

Atomic weights: $Ag = 108$ $N = 14$ $O = 16$

The maximum amount that could be paid for the silver nitrate solution is closest to

  (A) $20,000
  (B) $30,000
  (C) $40,000
  (D) $47,000
  (E) $48,000

Solution

$$\text{Pure } AgNO_3 = 1000(0.40) = 400 \text{ kg}$$
$$\text{Molecular weight:} \quad AgNO_3 = 108 + 14 + 3(16) = 170$$
$$Zn + 2AgNO_3 \rightarrow 2Ag + Zn(NO_3)_2$$
$$170 \text{ kg} \rightarrow 108 \text{ kg}$$

$$\text{Weight of Ag in 400 kg of AgNO}_3 = \frac{400 \times 108}{170} = 254 \text{ kg}$$

Value of the silver nitrate solution $= 254 \text{ kg} \times \$160 - \$1000$

$$= \$39,600$$

Answer is (C)

PROBLEM 9-34

The number of liters of oxygen necessary to burn 10 liters of H_2S gas according to the reaction $2H_2S + 3O_2 \rightarrow 2H_2O + 2SO_2$ is

(A) 3
(B) 6
(C) 10
(D) 15
(E) 30

Solution

Using a properly balanced formula, we could easily calculate the molecular weights of the gases and of the resulting products. This is not necessary in this problem because we know that equal numbers of molecules of gases occupy equal volumes.

Here we have 2 molecules of hydrogen sulfide and 3 molecules of oxygen reacting. We see, therefore, that 2 volumes of H_2S will react with 3 volumes of O_2.

For 10 liters of H_2S, $\frac{3}{2} \times 10 = 15$ liters of oxygen will be required.

Answer is (D)

PROBLEM 9-35

A compound was found to contain 42.12 wt% carbon, 51.45 wt% oxygen, and 6.43 wt% hydrogen. Its molecular weight was determined to be approximately 340. What is the chemical formula of the compound?

Atomic weights: $C = 12$ $H = 1$ $O = 16$

(A) $C_4H_6O_5$
(B) C_6H_8

(C) $C_8H_{10}O_7$
(D) $C_{10}H_{20}O_8$
(E) $C_{12}H_{22}O_{11}$

Solution

One molecular weight of the compound contains these components:

$$
\begin{array}{lll}
\text{C} & 42.12\% \times 340 = 143.21/12 = 11.93 & \text{or} \quad 12 \text{ gram-atoms of C} \\
\text{O} & 51.45\% \times 340 = 174.93/16 = 10.93 & \text{or} \quad 11 \text{ gram-atoms of O} \\
\text{H} & \underline{6.43\% \times 340 = } 21.86/1 = 21.86 & \text{or} \quad 22 \text{ gram-atoms of H} \\
\phantom{\text{H}} \quad 100.00 & 340.00 &
\end{array}
$$

Thus the formula should contain the factors

$$12\text{C} \qquad 11\text{O} \qquad 22\text{H}$$

We might recognize this as the formula for plain sugar, which is correctly written as $C_{12}H_{22}O_{11}$.

$$\text{Answer is (E)}$$

PROBLEM 9-36

Calculate the molecular weight of a gas, 1.13 liters of which, collected over water at 24°C and 754 mm pressure, weighed 1.251 g when deprived of the aqueous vapor. The vapor pressure of water at 24°C is 22.2 mm. The gas constant $\bar{R} = 0.082$ liter-atm/g-mole-°K.
   The molecular weight is nearest to

(A) 18
(B) 22
(C) 25
(D) 28
(E) 31

Solution

To apply the equation $pV = (m/M)\bar{R}T$, we first must correct for the water vapor present:

$$P_{total} = P_{gas} + P_{water}$$

$$P_{gas} = P_{total} - P_{water} = 754 - 22.2 = 731.8 \text{ mmHg} = \frac{731.8}{760} \text{ atm}$$

The number of moles of the gas is

$$M = \frac{m\bar{R}T}{pV} = \frac{1.251 \text{ g} \times 0.082(24 + 273)}{(731.8/760) \times 1.13} = 28 \text{ g/mole}$$

Thus the molecular weight of the gas is 28.

Answer is (D)

PROBLEM 9-37

A dry mixture of clay and barium sulfate has a density of 3.45 g/cm^3. The respective densities of the constituents are 2.67 g/cm^3 and 4.10 g/cm^3.
The volume percent of clay in the mixture is nearest to

(A) 25%
(B) 35%
(C) 45%
(D) 55%
(E) 65%

Solution

The calculation will be based on 100 cm^3 of the mixture.

Let x = volume clay and $100 - x$ = volume of $BaSO_4$

Wt clay + wt $BaSO_4$ = total wt

$$2.67x + 4.10(100 - x) = 3.45(100)$$

$$2.67x + 410 - 4.10x = 345$$

$$-1.43x = -65$$

Therefore

$$x = 45.5 = \% \text{ vol. clay}$$

$$100 - x = 54.5 = \% \text{ vol. } BaSO_4$$

Although not required in this problem, one could also compute the weight percent of the two constituents in the mixture as follows:

$$\text{Weight} = \text{volume} \times \text{density}$$
$$\text{Wt clay} = (45.5)(2.67) = 121.5 \text{ g} = 35.2\% \text{ wt clay}$$
$$\text{Wt BaSO}_4 = (54.5)(4.10) = 223.5 \text{ g} = 64.8\% \text{ wt BaSO}_4$$
$$\text{Total wt} = 345.0 \text{ g} = 100.0\%$$

Answer is (C)

PROBLEM 9-38

Which of the following metals has the slowest rate of combining with oxygen to form metallic oxide?

(A) tin
(B) copper
(C) iron
(D) magnesium
(E) zinc

Solution

A ranking of these metals, based on order of activity in the electromotive series, would be

magnesium
zinc
iron
tin
copper

with the most active metal placed first.

Metals of the copper subgroup (Cu, Ag, and Au) are known for their inactivity as a result of their low oxidation potentials. These are the coinage metals. As the oxidation potential of Cu is low and nearest to that for the reduction potential of oxygen in the presence of water, it will thus have the slowest rate of reaction.

Answer is (B)

PROBLEM 9-39

Which of the following is the most chemically active (highest in electro-motive series)?

(A) cesium
(B) potassium
(C) sodium
(D) strontium
(E) magnesium

Solution

Electromotive Series

1. lithium	9. aluminum
2. potassium	10. manganese
3. cesium	11. zinc
4. barium	.
5. strontium	.
6. calcium	.
7. sodium	silver
8. magnesium	gold

Answer is (B)

PROBLEM 9-40

The concentration of H_2SO_4 is given in terms of moles per liter as 0.2 M H_2SO_4. The normal concentration of this solution would be designated as

(A) 0.05 N H_2SO_4
(B) 0.1 N H_2SO_4
(C) 0.2 N H_2SO_4
(D) 0.4 N H_2SO_4
(E) 0.5 N H_2SO_4

Solution

A molar solution contains a gram-molecular weight of the solute in 1 liter of solution. A normal solution contains a gram-equivalent weight of the solution in 1 liter of solution.

Since valence = atomic weight/equivalent wt and in H_2SO_4 the radical is bivalent, it follows that a 1 M solution of H_2SO_4 is also a 2 N solution. In this problem the 0.2 M solution could also be called a 0.4 N solution.

Answer is (D)

PROBLEM 9-41

A solution of 0.1 N hydrochloric acid is required. The number of liters of water that should be added to 1 liter of 2.0 N hydrochloric acid in order to obtain the desired solution is nearest

(A) 19
(B) 20
(C) 21
(D) 22
(E) 23

$$\text{Atomic weights:}\quad H = 1 \qquad O = 16 \qquad Cl = 35.5$$

Solution

A normal solution contains a gram-equivalent weight of the solute in 1 liter of solution.

$$x = \text{total quantity}$$

$$2.0(1 \text{ liter}) = 0.1(x \text{ liters})$$

$$x = \frac{2.0}{0.1} = 20 \text{ liters}$$

Therefore 1 liter of 2.0 N HCl yields 20 liters of 0.1 N HCl. So add $20 - 1 = 19$ liters of water.

$$\text{Answer is (A)}$$

PROBLEM 9-42

What volume of 2 N H_2SO_4 must be mixed with 6 N H_2SO_4 to yield 700 ml of a 3 N H_2SO_4 solution. Assume all volumes are additive. The volume, in ml, of 2 N H_2SO_4 is nearest to

(A) 175
(B) 250
(C) 350
(D) 425
(E) 525

Solution

Let x = ml of $2\,N\ H_2SO_4$ and y = ml of $6\,N\ H_2SO_4$. Now we can write two expressions:

$$\text{Total volume:} \qquad x + y = 700 \tag{1}$$
$$\text{Normality} \times \text{volume:} \quad 2x + 6y = 3(700) \tag{2}$$

Substitute Eq. (1) into Eq. (2)

$$2(700 - y) + 6y = 2100$$
$$1400 - 2y + 6y = 2100$$
$$y = \tfrac{700}{4} = 175 \text{ ml of } 6\,N\ H_2SO_4$$
$$x = 700 - 175 = 525 \text{ ml of } 2\,N\ H_2SO_4$$

Answer is (E)

PROBLEM 9-43

Commercial preparation of magnesium metal is by electrolysis of a molten mixture of 70 wt% magnesium chloride and 30 wt% sodium chloride.

Atomic weights: Mg = 24.32   Cl = 35.45   Na = 23

Five kg of this mixture will yield an amount of magnesium, in kg, closest to

(A) 0.90
(B) 1.10
(C) 1.30
(D) 2.00
(E) 3.50

Solution

Five kg of mixture contains $0.70 \times 5 = 3.5$ kg of $MgCl_2$. The weight of Mg in

$$3.5 \text{ kg of } MgCl_2 = \frac{24.32}{24.32 + 2(35.45)} \times 3.5 = 0.89 \text{ kg}$$

Answer is (A)

PROBLEM 9-44

It is desired to produce 100 g-moles of compound AD by reacting compounds A_2B and CD_2 at a constant temperature of 25°C.

$$A_2B + CD_2 \rightarrow BC + 2AD \pm Q \text{ kcal}$$

The heats of formation of these compounds from the elements A, B, C, and D at 25°C are as follows:

A_2B	100 kcal/g-mole
CD_2	200 kcal/g-mole
BC	300 kcal/g-mole
AD	150 kcal/g-mole

Assume that the reactants react completely to form the products. How much heat must be added or removed during the reaction to maintain the desired temperature of 25°C? Use − sign if heat is removed and + sign if heat is added.

The heat required, in kcal, is closest to

(A)   −7,500
(B) −15,000
(C)   0
(D) +7,500
(E) +15,000

Solution

50 g-moles A_2B + 50 g-moles $CD_2 \rightarrow$ 50 g-moles BC + 100 g-moles AD

50(100 kcal) + 50(200 kcal) $\rightarrow$ 50(300 kcal) + 100(150 kcal) + Q kcal

5,000 kcal + 10,000 kcal $\rightarrow$ 15,000 kcal + 15,000 kcal + Q kcal

For a heat balance

$$15,000 \text{ kcal} \rightarrow 30,000 \text{ kcal} - 15,000 \text{ kcal}$$

Therefore

$$Q = -15,000 \text{ kcal}$$

Answer is (B)

PROBLEM 9-45

Naturally occurring boron consists of two isotopes, $_5B^{10}$ and $_5B^{11}$. It has an average atomic weight of 10.81. The mole% $_5B^{10}$ in the naturally occurring mixture is nearest to

(A) 7%
(B) 14%
(C) 19%
(D) 23%
(E) 37%

Solution

Let x = mole fraction $_5B^{10}$. A mass balance yields

$$x(10) + (1 - x)(11) = 1(10.81)$$
$$10x + 11 - 11x = 10.81$$
$$x = 0.19 = 19\%$$

Answer is (C)

PROBLEM 9-46

If common salt, NaCl, is dissolved in water, which of the following statements is true about the behavior of the solution relative to the behavior of pure water?

(A) boiling point is lowered
(B) freezing point is decreased
(C) ionization of the water is decreased
(D) ionization of the water is increased
(E) none of the above is true

Solution

Adding solute (salt) to solvent (water) decreases the water freezing point and elevates its boiling point. Since there are no ions from NaCl common with those of water, ionization of the water is unaffected.

Answer is (B)

PROBLEM 9-47

The solubility product of MgF_2 at 18°C is 7.1×10^{-9}. Solubility of MgF_2 in grams per liter of water is nearest to

(A) 0.07
(B) 0.15
(C) 0.69
(D) 1.80
(E) 3.70

Atomic weights: $Mg = 24.3$ $F = 19.0$

Solution

The small amount of MgF_2 that dissolves in water produces Mg^{2+} and F^- ions, according to $MgF_2 \rightarrow Mg^{2+} + 2F^-$. The solubility product equals $[Mg^{2+}][F^-]^2$, where the brackets indicate the molar concentration of the ions.

Let $x = $ moles/liter Mg^{2+}.

$$7.1 \times 10^{-9} = x(2x)^2$$
$$4x^3 = 7.1 \times 10^{-9}$$
$$x = 1.21 \times 10^{-3}$$

The solution contains 1.21×10^{-3} moles MgF_2 per liter.

$$\text{Molecular weight } MgF_2 = 24.3 + 2(19.0) = 62.3 \text{ g/mole}$$
$$\text{Solubility} = 62.3 \times 1.21 \times 10^{-3} = 0.075 \text{ g/liter}$$

Answer is (A)

PROBLEM 9-48

If K is the equilibrium constant for decomposition of nitrogen tetroxide according to $N_2O_4 = 2NO_2$, which is nitrogen dioxide, the expression in partial pressures for the equilibrium constant is

(A) $K = \dfrac{p_{NO_2}}{p_{N_2O_4}}$

(B) $K = p_{NO_2} p_{N_2O_4}$

(C) $K = \dfrac{p_{N_2O_4}}{p_{NO_2}}$

(D) $K = \dfrac{(p_{NO_2})^2}{p_{N_2O_4}}$

(E) $K = \dfrac{p_{NO_2}}{(p_{N_2O_4})^2}$

Solution

For a reaction of the form $aA + bB = cC + dD$, an expression for the equilibrium constant, expressed in partial pressures of each component in the system, is

$$K = \frac{(p_C)^c (p_D)^d}{(p_A)^a (p_B)^b}$$

Answer is (D)

PROBLEM 9-49

If the ionization constant of acetic acid, CH_3COOH, is 1.76×10^{-5} at 25°C and acetic acid ionizes according to $CH_3COOH = CH_3COO^- + H^+$, the H^+ concentration in moles/liter produced by 1 M CH_3COOH in water is nearest to

(A) 1×10^{-7}
(B) 2×10^{-5}
(C) 4×10^{-3}
(D) 7×10^{-2}
(E) 1

Solution

$K = [CH_3COO^-][H^+]/[CH_3COOH]$, where brackets indicate concentrations in moles per liter.

Let x = moles/liter of CH_3COOH ionized, yielding x moles/liter of H^+ and CH_3COO^-.

$$1 - x = \text{moles/liter of acid remaining and not ionized}$$

$$K = \frac{(x)(x)}{(1 - x)} = 1.76 \times 10^{-5}$$

Simplify by considering the denominator term $(1 - x) \cong 1$, since ionization is small.

$$x = (1.76 \times 10^{-5})^{1/2} = 4.19 \times 10^{-3} = [H^+]$$

(An exact solution yields a comparable answer.)

Answer is (C)

PROBLEM 9-50

At 300°C the equilibrium constant is 45 for the water gas shift reaction: $CO + H_2O = CO_2 + H_2$. An equimolar mixture of CO and steam comes to equilibrium at 1 atm total pressure after being passed over a catalyst. The mole fraction of hydrogen in the resulting gas is

(A) 0.06
(B) 0.15
(C) 0.27
(D) 0.44
(E) 0.73

Solution

The equilibrium constant for the shift reaction is

$$K = \frac{(p_{H_2})(p_{CO_2})}{(p_{CO})(p_{H_2O})} = 45$$

The partial pressure p of each component of the resulting mixture is total pressure p_t times mole fraction. Starting with one mole each of CO and H_2O, let x = moles of H_2 produced at equilibrium. From the equation, total moles does not change from two. Mole fractions are

$$CO = \frac{1-x}{2} \qquad H_2O = \frac{1-x}{2} \qquad CO_2 = \frac{x}{2} \qquad H_2 = \frac{x}{2}$$

$$\frac{(p_t x/2)(p_t x/2)}{[p_t(1-x)/2][p_t(1-x)/2]} = 45 = \frac{x^2}{(1-x)^2}$$

Take the square root of the entire equation and solve to obtain

$$\sqrt{45} = \frac{x}{1-x} \qquad x = \frac{\sqrt{45}}{1+\sqrt{45}} = 0.870$$

Mole fraction H_2 is $x/2 = 0.435$.

Answer is (D)

10

ELECTRICITY

The study and use of electrical energy in an ever-increasing variety of ways is the goal of an entire engineering discipline, that of electrical engineering. And as this field continues its rapid growth, the importance of a fundamental knowledge of electric circuits and electromechanical energy conversion grows with it. The problems collected here primarily emphasize a few basic principles of direct current (dc) and alternating current (ac) circuits. Both single-phase and three-phase circuits and motors are mentioned. Some features of the operational amplifier are also described.

dc CIRCUITS

In dc circuitry a potential difference or voltage V across the terminals of some resistive device causes an electric current I, in amperes (A), to flow through it. According to Ohm's law,

$$V = IR \qquad (10\text{-}1)$$

where R is the resistance, in ohms (Ω), of the element to current flow, and the current is measured in amperes. Furthermore, power must be expended in sustaining this current flow in the amount

$$P = VI = I^2R = V^2/R \qquad (10\text{-}2)$$

The units of this equation are volt-amperes or watts (W). The driving force for such current flows is called an electromotive force (emf). Typical sources of emf are batteries, generators, and motors.

For resistors of uniform geometric shape, the resistance R is a function of material properties and the size of the element. For a uniform wire, for example,

$$R = \rho \frac{L}{A} \tag{10-3}$$

where ρ is the resistivity (a material property) and L and A are, respectively, the length and cross-sectional area of the wire.

Kirchhoff's two laws define the relations that exist when individual elements are combined to form an electric circuit:

1. The sum of all currents flowing into a junction must equal the sum of all currents flowing away from the junction.
2. The algebraic sum of the emfs and the voltage drops around a closed loop is equal to zero.

Rule 1 assures that charge is conserved at a point. In a large circuit or network containing n junctions, this rule must be applied $n - 1$ times; the rule is then automatically satisfied at the last junction. Rule 2 is in a way a generalized application of Ohm's law. In the use of these rules, care must be taken so that a chosen sign convention is used consistently.

In dc circuits, resistors and emfs may be connected in a seemingly bewildering variety of ways for different purposes. Most, but not all, resistor combinations are sequences of either the basic series or parallel arrangements of individual resistors. In these cases the single equivalent series resistance R_e is

$$R_e = \sum_{n=1}^{N} R_n \tag{10-4}$$

for a series of N individual resistors and

$$\frac{1}{R_e} = \sum_{n=1}^{N} \frac{1}{R_n} \tag{10-5}$$

for a parallel combination of N resistors. The Y-Δ transformation, an important exception, will be discussed later and also examined briefly in some problems.

Mesh analysis is normally applied to the solution of larger networks or networks that are not easily simplified by use of equivalent resistances. The technique concentrates on writing loop equations or node equations for a network; only loop equations are considered here. If the common intersection of two or more lines in a network is called a junction J, and a line connecting two junctions and containing an electrical element is called a branch B, then the number of independent loop equations N required to solve the network is $N = 1 + B - J$. First the loops and the N independent unknown loop currents are identified. Then Kirchhoff's voltage law (rule 2) is written for each loop, and the resulting set of equations is solved. As Example 2 will show, these loop equations could also be written directly by inspection.

Example 1

Solve for the five branch currents shown (Fig. 10-1) in this diagram of a dc network.

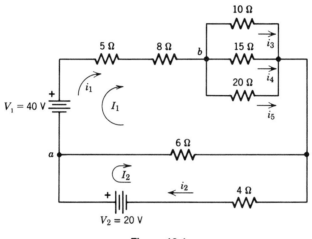

Figure 10-1

Solution

We begin by applying Kirchhoff's first rule. This network may be considered to have only two junctions if the parallel combination of resistors is replaced by one equivalent resistor R_e in series. Examining the junction marked a, the branch current i_6 through the 6-Ω resistor to the left is equal to the difference in loop currents I_1 and I_2

$$i_6 = I_1 - I_2$$

Next we replace the three resistors by the one equivalent series resistor R_e:

$$\frac{1}{R_e} = \frac{1}{10} + \frac{1}{15} + \frac{1}{20} = \frac{13}{60}$$

$$R_e = \tfrac{60}{13} \ \Omega$$

Now we can apply Kirchhoff's second rule to the two loops to obtain

$$40 - (5 + 8 + \tfrac{60}{13})I_1 - 6(I_1 - I_2) = 0 \qquad (1)$$

$$20 - 6(I_2 - I_1) - 4I_2 = 0 \qquad (2)$$

From Eq. (2), $I_2 = 0.1(20 + 6I_1)$. Substituting the relation into Eq. (1) gives $I_1 = 2.6$ A. Then we find $I_2 = 3.56$ A.

Returning to the parallel combination of resistors, the voltage drop across this combination is, by use of Ohm's law,

$$V_e = I_1 R_e = (2.6)(\tfrac{60}{13}) = 12 \text{ V}$$

This voltage drop occurs across each of these three resistors, whereas the use of Kirchhoff's first rule at point b shows that

$$I_1 = i_1 = i_3 + i_4 + i_5$$

Repeated use of Ohm's law gives

$$i_3 = \frac{V_e}{10} = 1.2 \text{ A}$$

$$i_4 = \frac{V_e}{15} = 0.8 \text{ A}$$

$$i_5 = \frac{V_e}{20} = 0.6 \text{ A}$$

Example 2

Find the three loop currents in the dc network shown in Fig. 10-2.

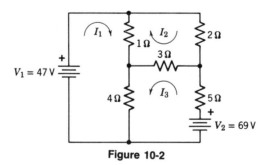

Figure 10-2

Solution

This network has $B = 6$ branches and $J = 4$ junctions; hence the number of required loop equations is $N = 1 + B - J = 3$. Tracing each loop in the direction of the indicated current flow I and applying Kirchhoff's voltage law yields

$$1(I_1 - I_2) + 4(I_1 + I_3) = 47 \tag{1}$$

$$1(I_2 - I_1) + 2I_2 + 3(I_2 + I_3) = 0 \tag{2}$$

$$5I_3 + 3(I_2 + I_3) + 4(I_1 + I_3) = 69 \tag{3}$$

which could be rearranged to the form

$$(1 + 4)I_1 - \quad\quad 1I_2 + \quad\quad 4I_3 = 47 \tag{1'}$$

$$-1I_1 + (1 + 2 + 3)I_2 + \quad\quad 3I_3 = 0 \tag{2'}$$

$$4I_1 + \quad\quad 3I_2 + (3 + 4 + 5)I_3 = 69 \tag{3'}$$

In Eq. (1') the coefficient of I_1 is the sum of the resistances in loop 1 and is called the self-resistance of the loop; the coefficients multiplying I_2 and I_3 are called mutual resistances and are the resistances common to loops 1 and 2, and 1 and 3, respectively. Equations (2') and (3') can be similarly interpreted. Modest practice will allow one to write these equations directly by inspection.

The method of substitution can be used to solve sets of equations, but here Cramer's method is used (see Chapter 2):

$$I_1 = \frac{D_1}{D} \quad\quad I_2 = \frac{D_2}{D} \quad\quad I_3 = \frac{D_3}{D}$$

where

$$D = \begin{vmatrix} 5 & -1 & 4 \\ -1 & 6 & 3 \\ 4 & 3 & 12 \end{vmatrix} = 183 \quad\quad D_1 = \begin{vmatrix} 47 & -1 & 4 \\ 0 & 6 & 3 \\ 69 & 3 & 12 \end{vmatrix} = 1098$$

$$D_2 = \begin{vmatrix} 5 & 47 & 4 \\ -1 & 0 & 3 \\ 4 & 69 & 12 \end{vmatrix} = -183 \quad\quad D_3 = \begin{vmatrix} 5 & -1 & 47 \\ -1 & 6 & 0 \\ 4 & 3 & 69 \end{vmatrix} = 732$$

This leads directly to

$$I_1 = 6\text{ A} \quad\quad I_2 = -1\text{ A} \quad\quad I_3 = 4\text{ A}$$

ac CIRCUITS

Ohm's law expresses only one of three possible relations between voltage and current, that resulting from the action of a resistive element. Two other relations exist and are characterized by two electrical elements: the inductance and the capacitance.

For an inductance

$$v = L \frac{di}{dt} \tag{10-6}$$

where L is the inductance in henries (H) when i is in amperes and v is in volts. For a sinusoidal alternating current the current lags $90°$ behind the voltage in a purely inductive circuit, as can be seen by assuming a sinusoidal form for i and computing v.

Capacitance, the third electrical element, directly relates the charge q to the voltage v. Thus the current-voltage relation is

$$i = \frac{dq}{dt} = C \frac{dv}{dt} \tag{10-7}$$

and C is measured in farads (F). In contrast to the inductive circuit, the current leads the voltage by $90°$ in a purely capacitive ac circuit. Each phase relation contrasts with the purely resistive element, which causes no phase shift. The handy mnemonic "ELI the ICEman" can help one recall that voltage E leads current I for an inductor, whereas E lags behind I for a capacitor.

In ac circuits we are usually interested in effective or rms values of the current or voltage rather than the maximum value of the quantity over one full cycle. For a sinusoidal current or voltage the effective values are

$$I = \frac{I_m}{\sqrt{2}} = 0.707 I_m \quad \text{and} \quad V = \frac{V_m}{\sqrt{2}} = 0.707 V_m$$

where the subscript m stands for the maximum value over a cycle. The effective values are unsubscripted since they are the commonly used quantities. In some instances the average value is desired over only a half cycle; for a sine wave over the first half cycle, for example, the results

$$I_h = \frac{2}{\pi} I_m = 0.637 I_m \quad \text{and} \quad V_h = \frac{2}{\pi} V_m = 0.637 V_m$$

are obtained.

Resistance, inductance, and capacitance are all generally present in ac circuits. When elements are connected in series, a voltage drop between two points is the sum of the individual *phasor* voltage drops; for a parallel combination of elements, the total current at a junction is the *phasor* sum of individual currents through the parallel elements. Phasor algebra is the same as vector algebra since phasors have a magnitude and orientation, but we must remember that the angle is a phase angle between two sinusoidally varying quantities and not a definite direction in space.

The relation between current and voltage is conveniently given by

$$V = ZI \qquad (10\text{-}8)$$

where these quantities are all phasor quantities, and Z is the impedance. If the inductive and capacitive reactances X_L and X_C are defined in terms of frequency f, inductance L, and capacitance C as

$$X_L = 2\pi fL \qquad X_C = -\frac{1}{2\pi fC} \qquad (10\text{-}9)$$

then the relations between impedance Z, reactance X, and resistance R, as shown in Fig. 10-3 for $X > 0$, are $Z^2 = R^2 + X^2$, $\tan\theta = X/R$ with the reactance $X = X_L + X_C$. Note, however, that X may be either positive or negative.

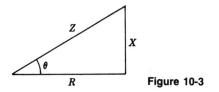

Figure 10-3

An alternative to the use of the impedance triangle is the volt-ampere method, which is based on ac power considerations. Most ac equipment is rated in volt-amperes (VA) even though this power quantity is dimensionally the same as watts. Because of the phase shift θ, however, the useful average power in an ac circuit is $I^2R = P = VI\cos\theta$, where effective values are used for current and voltage. Cos θ is called the power factor. The reactive power is $Q = I^2X = VI\sin\theta$; its units are volt-amperes reactive (VAR). These relations are summarized in Fig. 10-4, the volt-ampere triangle.

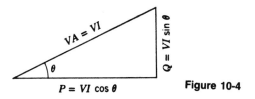

Figure 10-4

Equations (10-9) show that reactance is a function of frequency. By selecting this frequency so that $X = 0$, one can maximize the current in the system and at the same time maximize the power by operating with a power factor of unity. This is called the resonant frequency; at this frequency the response to a given voltage input will also be greatest.

Example 3

In a 440-V ac circuit there is a resistive and inductive load that requires 30 A at 0.8 power factor. Calculate

  (a) volt-amperes
  (b) average power
  (c) reactive volt-amperes
  (d) resistance R
  (e) impedance Z
  (f) reactance X_L

Solution

We apply the volt-ampere method here.

  (a) Volt-amperes $VI = (440)(30) = 13{,}200$ VA
  (b) Average power $P = VI \cos \theta = (440)(30)(0.8) = 10{,}560$ W
  (c) Reactive volt-amperes $Q = VI \sin \theta = (440)(30)(0.6) = 7920$ VAR
  (d) The resistance R is found from $P = I^2 R$, or

$$R = \frac{P}{I^2} = \frac{10{,}560}{(30)^2} = 11.73 \ \Omega$$

  (e) The impedance relation is $I^2 Z = VI$. The impedance magnitude is

$$Z = \frac{VI}{I^2} = \frac{V}{I} = \frac{440}{30} = 14.67 \ \Omega$$

with a phase angle $\theta = \cos^{-1} 0.8 = 37°$.
  (f) The expression $I^2 X_L$ is an alternative expression for reactive volt-amperes. Hence

$$X_L = \frac{Q}{I^2} = \frac{7920}{(30)^2} = 8.80 \ \Omega$$

Example 4

A coil with an inductance of 1 H and a resistance of 50 Ω is connected in series with a 0.01 μF capacitor and a variable frequency supply, as in Fig. 10-5.

  (a) At what frequency f_0 will the maximum voltage appear across the capacitor if the supply voltage is held constant as the frequency is varied?
  (b) Using the frequency f_0, what is the maximum supply voltage that may be used in this circuit if the capacitor voltage rating is 200 V?

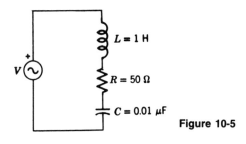

Figure 10-5

Solution

(a) The voltage is maximized at resonance when $X = 0$:

$$X = 2\pi f_0 L - \frac{1}{2\pi f_0 C} = 0$$

Solving,

$$2\pi f_0 = \frac{1}{(LC)^{1/2}} = \frac{1}{[1(10^{-8})]^{1/2}} = 10^4 \text{ rad/sec}$$

$$f_0 = \frac{10^4}{2\pi} = 1592 \text{ Hz}$$

(b) At the frequency f_0 the current $I_0 = V/R$. The maximum capacitor voltage $V_c = 200 = I_0 X_C$, where $X_C = -(2\pi f_0 C)^{-1}$. Combining these relations and solving for V yields

$$V = I_0 R = \frac{V_c}{X_c} R = 2\pi f_0 C V_c R$$

$$V = (10^4)(10^{-8})(200)(50) = 1.00 \text{ V}$$

In three-phase ac circuits the circuit elements are normally connected in either a Y or Δ (delta) configuration. We consider here only balanced loads. In solving problems it is sometimes convenient to switch from one configuration to the other; for balanced loads the impedance relation is then simply $Z_\Delta = 3Z_Y$. In the Y-arrangement the line current I_L is the same as the current in that leg of the load, called the phase current, but the line voltage V_L is $\sqrt{3}$ times the phase voltage owing to the 30° phase shift in the voltages. In the Δ-configuration there is no central node, so the line and phase voltages are equal, but the line current is $\sqrt{3}$ times the phase current with a corresponding 30° phase shift.

For a balanced load the three-phase power relation is

$$P_{\text{Total}} = \sqrt{3} V_L I_L \cos \phi \qquad (10\text{-}10)$$

in which ϕ is the load phase angle. The total power is also the sum of $I^2 R$ for the three phases, which should check the result given by Eq. (10-10).

Example 5

A three-phase source voltage of 440 V at 60 Hz is applied to a Y load. Each leg of the Y has an inductive load of $4.0 + j3.0 \, \Omega$ at 60 Hz. What is the line current and the total power dissipated in the load?

Solution

A three-phase source voltage is normally presumed to be the voltage V_L. Therefore the phase voltage is $V_{phase} = V_{line}/\sqrt{3} = 440/1.732 = 254$ V lagging the line voltage by 30°.

For a Y the line and phase currents are the same and have the value $V_{phase}/|Z_{phase}| = 254/|4.0 + j3.0| = 254/5.0 = 50.8$ A with a lagging phase angle with respect to the phase voltage of $\tan^{-1}(3/4) = 36.9°$. The power per phase is $P_{phase} = I^2R = (50.8)^2(4.0) = 10{,}320$ W, and the total power is $3P_{phase} = 31.0$ kW. This should match the result obtained from Eq. (10-10), which is

$$P_{Total} = \sqrt{3}(440)(50.8) \cos 36.9° = 31.0 \text{ kW}$$

Example 6

The source voltage in Example 5 is applied to the same load, but the load is in a delta configuration. What is the line current and the total power dissipated in the load?

Solution

Now the line and phase voltages are the same. Hence the phase current is $V_{phase}/|Z_{phase}| = 440/|4.0 + j3.0| = 440/5.0 = 88.0$ A lagging with a phase angle of 36.9°. The line current is $\sqrt{3}I_{phase} = \sqrt{3}(88.0) = 152.4$ A with a phase angle of $30° - 36.9° = 6.9°$ lagging. The power per phase is $P_{phase} = (I_{phase})^2R = (88.0)^2(4.0) = 31.0$ kW. The total power is $3P_{phase} = 93.0$ kW. If Eq. (10-10) were used, then

$$P_{Total} = \sqrt{3}(440)(152.4) \cos 36.9° = 93.0 \text{ kW}$$

THE OPERATIONAL AMPLIFIER

For simplicity the operational amplifier or "op amp" will be considered here as merely a black box without considering in detail the internal electronics. Many kinds of operational amplifiers exist, but the most common one is the voltage type, including the 741 op amp. The primary characteristics of this kind of device are listed here:

1. The output voltage signal is typically 10^5 to 10^6 times larger than the difference in input voltages. The ratio of output to input is the gain of the amplifier. To keep the output voltage within the usual limit of 10–15 V, the input voltage difference must be very small.

2. The input impedance is very large so that little input current is required.

3. The output impedance is very low, usually much less than 100 Ω.

4. The gain of the operational amplifier is nearly constant between 0 Hz (dc operation) and several hundred thousand kHz.

The symbol for the operational amplifier and a diagram for the idealized amplifier at relatively low frequency are shown in Fig. 10-6. For the 741 op amp the power connections to the amplifier are pin 7 (+15 V relative to ground) and pin 4 (−15 V relative to ground).

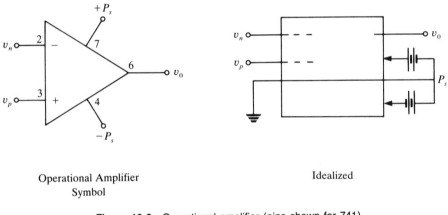

Operational Amplifier
Symbol

Idealized

Figure 10-6 Operational amplifier (pins shown for 741)

The inverted summing mode is a common, practical use for the op amp; the setup is shown in Fig. 10-7. In this mode of operation the positive input pin is grounded, and the negative input pin is connected to a nearby node. This node accepts input from n resistors, as shown, and a feedback resistor R_f is connected between the output and this same node. By simply summing currents at this node, the fundamental relation for the inverted summing mode is obtained as

$$v_0 = -(R_f/R_1)v_1 - (R_f/R_2)v_2 - \cdots - (R_f/R_n)v_n \qquad (10\text{-}11)$$

By using other circuit elements and configurations in combination with the operational amplifier, many different kinds of circuit behavior can be achieved.

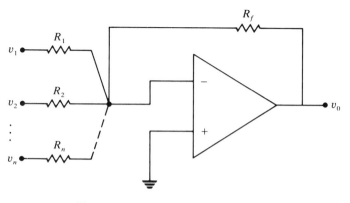

Figure 10-7 Inverted summing mode

Example 7

Determine the output voltage for the idealized operational amplifier shown in Fig. 10-8.

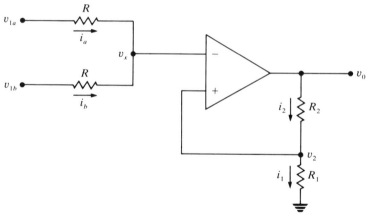

Figure 10-8

Solution

Any of several solution methods may be used in circuit analysis. In this problem we will use superposition. We begin by setting $v_{1b} = 0$ and determining the output voltage v_{0a}; we continue by setting $v_{1a} = 0$ and finding the corresponding output v_{0b}. Finally we superpose, i.e. add, the two solutions.

To begin, temporarily let $v_{1b} = 0$. First we analyze the input circuit to the left of the op amp. Writing Kirchhoff's voltage law around the line from a to x to b, $v_{1a} - i_a R = v_x$ and $v_x + i_b R = v_{1b} = 0$. Since the current from point x

into the op amp is essentially zero, we see that $i_a = -i_b$ at point x. From the second equation $i_b = -v_x/R$; then from the first equation $v_{1a} - (v_x/R)R = v_x$ and $v_x = v_{1a}/2$.

Next we turn our attention to the lower right portion of Fig. 10-8. Between the output and the ground $v_{0a} - i_2 R_2 - i_1 R_1 = 0$. The current to the positive input pin of the op amp is effectively zero, so $i_1 = i_2$. Since $v_2 = i_1 R_1$, we find $v_2 = [v_{0a}/(R_1 + R_2)]R_1 = v_{0a} R_1/(R_1 + R_2)$. One can usually assume the difference in input voltages to the op amp is nearly zero, which means that effectively $v_2 = v_x$. The input-output voltage relation, by substitution, is thus $v_{0a} = 0.5(1 + R_2/R_1)v_{1a}$.

Now we set $v_{1a} = 0$ and repeat the analysis to obtain $v_{0b} = 0.5(1 + R_2/R_1)v_{1b}$. By superposition the full output is

$$v_0 = v_{0a} + v_{0b} = 0.5(1 + R_2/R_1)(v_{1a} + v_{1b})$$

The circuit can be called a linear noninverting adder.

SIMPLE TRANSIENTS

Books have been written on this topic; we will look only at linear, first-order circuit transients, one described by a first-order differential equation. If a constant input is supplied as the initial condition at some point in a circuit, the output will undergo a transient and eventually also become constant. It is the determination of the transient behavior during this time interval that is our goal here.

Only the capacitor and the inductor can store energy in an electric circuit. For each of these two circuit elements a first-order differential relation exists between voltage and current, as Eqs. (10-6) and (10-7) show. If both are present in the same circuit, a second-order differential equation will be required to describe circuit behavior. These equations also make clear that neither the current in an inductor nor the voltage across a capacitor can change discontinuously (jump) in time.

In the typical problem one first describes the circuit or network with one or more equations, usually differential equations. As shown in Fig. 10-9, a simple one loop RC circuit could be supplied with a constant voltage V_B by a battery after the switch is closed at time $t = 0$. We assume the capacitor is initially uncharged so that $v_C(t = 0) = 0$, and we seek the behavior of v_C for time $t > 0$. We follow the convention that lower case letters represent quantities that are functions of time, as is usual in this subject. The loop voltage equation is

$$R_1 i + v_C + R_2 i = V_B$$

Using Eq. (10-7) to replace $i(t)$, we obtain the differential equation

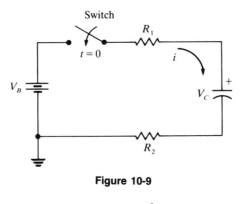

Figure 10-9

$$V_B = (R_1 + R_2)C \frac{dv_C}{dt} + v_C$$

The solution of this equation is

$$v_C = V_B(1 - e^{-t/T}) \quad \text{with} \quad T = (R_1 + R_2)C$$

The exponential decay is the expected behavior. If the capacitor is initially charged, then the initial condition for v_C at $t = 0$ should be changed to the given value.

Example 8

For the circuit in Fig. 10-9 assume the battery voltage is $V_B = 10$ V and that $R_1 = 1000\,\Omega$, $R_2 = 2000\,\Omega$, and $C = 3000\,\mu$F. Determine the voltage between the top of the capacitor and ground exactly 15 sec after the switch closes.

Solution

The time constant T for this circuit is

$$T = (R_1 + R_2)C = (1.0 + 2.0)(10^3)(3.0)(10^{-3}) = 9.0 \text{ sec}$$

After 15 sec the voltage across the capacitor is

$$v_C = 10(1 - e^{-15/9}) = 8.11 \text{ V}$$

The current is then $i = (V_B - v_C)/(R_1 + R_2) = (10.0 - 8.11)/(3000) = 0.63$ mA, and the voltage across R_2 is $iR_2 = (0.00063)(2000) = 1.26$ V. Thus the voltage between the top of the capacitor and ground is

$$v_C + iR_2 = 8.11 + 1.26 = 9.37 \text{ V}$$

ELECTRICAL MACHINERY

Both electric motors and generators utilize the same fundamental principles in their operation. For this reason the analysis of either type of machine is essentially the same. All these machines are basically usable as *either* a motor or a generator; it is only our optimization of the primary and secondary operating characteristics of a given machine so that it performs better in one role than in the other that has led us to separate the two.

Both ac and dc machines are widely used. By various means electric and magnetic fields, fields set up by the current flow through the machine, create a mechanical torque, or vice versa. However, there are sufficiently many different variations in the actual design of the major types of machinery that it is impractical to outline them here. One sample case is presented next; others may be found in the problems.

Example 9

A dc shunt motor running at no load draws 5.0-A armature current and 1.0-A field current from a 120-V line. Its speed is 1190 rpm. When loaded, the armature current increases to 50 A without adjustment of the field. Neglecting the effect of armature reaction, what will be the speed under this load condition? The armature resistance is 0.2 Ω.

Solution

One primary characteristic of this motor is the relative constancy of the armature voltage over a wide speed range.

Letting V_a = armature voltage, I_a = armature current, R_a = armature resistance, and V_t = terminal voltage, we can write

$$V_a = V_t - I_a R_a$$

At no load

$$V_a = 120 - (5.0)(0.2) = 119 \text{ V}$$

Under load

$$V_a = 120 - (50)(0.2) = 110 \text{ V}$$

For a particular machine the armature voltage V_a is proportional to the product of the field strength ϕ and the armature speed n in rpm, or

$$V_a = K\phi n$$

where K is a proportionality constant dependent on the machine. Here ϕ is also constant so that

$$\frac{V_{a_1}}{V_{a_2}} = \frac{n_1}{n_2}$$

$$n_2 = \frac{110}{119}(1190) = 1100 \text{ rpm}$$

PROBLEM 10-1

A 0.02 μF capacitor consists of two parallel plates of 100 in.^{2} each, with an air gap of 0.001 in. between them. If the air gap is increased to 0.0015 in., the resulting capacitance, in μF, will be

(A) 0.0133
(B) 0.0300
(C) 0.0090
(D) 0.0450
(E) 0.0100

Solution

Capacitance is proportional to A/d. Hence

$$\frac{C_2}{C_1} = \left(\frac{A_2}{d_2}\right)\left(\frac{d_1}{A_1}\right) \quad \text{and} \quad C_2 = C_1\left(\frac{d_1}{d_2}\right)$$

since A is constant.

$$C_2 = (0.02)\left(\frac{0.001}{0.0015}\right) = 0.0133 \ \mu F$$

Answer is (A)

PROBLEM 10-2

A capacitor is rated at 4000 F (farads) and is given a charge of 80 C (coulombs). The potential difference, in V, between the plates is

(A) 5.0
(B) 50.0

(C)   0.04
(D)   0.02
(E)   0.2

Solution

$$C = \frac{Q}{V}$$

where C = capacitance in farads, Q = quantity of electricity in coulombs, and V = difference in potential in volts. Thus

$$V = \frac{Q}{C} = \frac{80}{4000} = 0.02 \text{ V}$$

bk

Answer is (D)

PROBLEM 10-3

A copper rod 1 in. in diameter and 10 ft long is found to have a resistance of $10^{-4}\,\Omega$. If the rod is drawn into a wire with a uniform diameter of 0.05 in., the new resistance, in Ω, of the wire is nearest to

(A) 16.0
(B)   1.0
(C)   0.8
(D)   0.04
(E) 10^{-4}

Solution

Equation (10-3) applies here, and the volume of copper $\mathcal{V} = LA$ is constant in the problem. For a circular cross section the area $A = \pi d^2/4$, and Eq. (10-3) now gives

$$R = \rho\frac{L}{A} = \rho\frac{\mathcal{V}}{A^2} = \rho\mathcal{V}\left(\frac{4}{\pi}\right)^2\frac{1}{d^4} = Kd^{-4}$$

where K is a constant. If subscript 1 refers to the original wire and subscript 2 to the new wire, we have

$$R_1 d_1^4 = R_2 d_2^4 = K$$

and

$$R_2 = R_1\left(\frac{d_1}{d_2}\right)^4 = (10^{-4})\left(\frac{1}{0.05}\right)^4 = 16.0\,\Omega$$

Answer is (A)

PROBLEM 10-4

Cathode rays are

(A) high-energy X rays
(B) alpha particles
(C) protons
(D) electrons
(E) neutrons

Solution

Here we are dealing with a flow of electrons.

Answer is (D)

PROBLEM 10-5

A kilowatt-hour is a unit of

(A) momentum
(B) power
(C) acceleration
(D) energy
(E) impulse

Solution

A kilowatt-hour is a measure of work or energy. Energy = Power × Time. Power, on the other hand, is work per unit time; it might be measured in watts or kilowatts, not kilowatt-hours.

Answer is (D)

PROBLEM 10-6

A phonograph cartridge creates voltage by a process that is

  (A) electromechanical
  (B) electrochemical
  (C) thermoelectric
  (D) piezoelectric
  (E) photovoltaic

Solution

For completeness we cite an example of developing a voltage via each process. Both ac and dc generators use the electromechanical process of rotating a coil through the field of a magnet. Batteries use oxidation-reduction reactions in the cells in an electrochemical process. A thermocouple uses the thermoelectric effect caused by heating a junction of dissimilar metals and completing the circuit at a different temperature to create a dc voltage. The phonograph cartridge employs the piezoelectric effect, whereby some crystals develop voltage differences when they are deformed by applied forces. Photovoltaic cells are probably best known for converting sunlight into electricity in space vehicles.

$$\text{Answer is (D)}$$

PROBLEM 10-7

Permittivity is expressed by the Greek letter ε, and its defining equation is Coulomb's law

$$\varepsilon = \frac{QQ'}{4\pi FS^2}$$

where Q and $Q' = $ two point charges, $F = $ force between these charges, and $S = $ distance between these charges.
   Permittivity in the MKS system is

  (A) coulombs^2/newton-meter^2
  (B) coulombs^2/dyne-meter^2
  (C) statcoulombs^2/dyne-meters^2
  (D) coulombs^2/newton-centimeter^2
  (E) statcoulombs^2/gram-centimeter^2

Solution

Permittivity in the MKS system is coulombs^2/newton-meter^2

Answer is (A)

PROBLEM 10-8

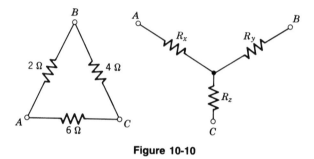

Figure 10-10

The Δ and Y circuits are intended to be equivalent. The resistance R_y, in Ω, is

(A) 0.50
(B) 0.67
(C) 1.00
(D) 1.67
(E) 2.67

Solution

First determine the line-to-line resistance for the Δ circuit:

$$AB: \quad \frac{1}{R_{AB}} = \frac{1}{2} + \frac{1}{6+4} = 0.6 \qquad R_{AB} = \frac{1}{0.6} = 1.67 \, \Omega$$

$$BC: \quad \frac{1}{R_{BC}} = \frac{1}{4} + \frac{1}{2+6} = 0.375 \qquad R_{BC} = \frac{1}{0.375} = 2.67 \, \Omega$$

$$CA: \quad \frac{1}{R_{CA}} = \frac{1}{6} + \frac{1}{4+2} = 0.333 \qquad R_{CA} = \frac{1}{0.333} = 3.0 \, \Omega$$

Similarly, determine the line-to-line resistance for the Y circuit:

$$AB: \quad R_{AB} = R_x + R_y$$

$$BC: \quad R_{BC} = R_y + R_z$$
$$CA: \quad R_{CA} = R_z + R_x$$

For the circuits to be equivalent, the line-to-line resistances must be equal. Therefore

$$R_{AB} = R_x + R_y \qquad = 1.67 \qquad (1)$$
$$R_{BC} = \qquad R_y + R_z = 2.67 \qquad (2)$$
$$R_{CA} = R_x \qquad + R_z = 3.0 \qquad (3)$$

Solving these three equations simultaneously,

$$
\begin{array}{ll}
(1) & R_x + R_y \qquad = \quad 1.67 \\
-(3) & -R_x \qquad - R_z = -3.0 \\
\hline
& R_y - R_z = -1.33 \\
(2) & R_y + R_z = \quad 2.67 \\
\hline
& 2R_y \qquad = \quad 1.34
\end{array}
$$

$$R_y = \frac{1.34}{2} = 0.67 \ \Omega$$

Completing the solution would yield

$$
\begin{array}{lll}
(2) & 0.67 + R_z = 2.67 & R_z = 2.00 \ \Omega \\
(1) & R_x + 0.67 \qquad = 1.67 & R_x = 1.00 \ \Omega
\end{array}
$$

$$\text{Answer is (B)}$$

PROBLEM 10-9

One has five capacitors that are each rated at 5 μF.

Question 1. When the five are connected in series, the equivalent capacitance, in μF, is nearest to

(A) 1
(B) 2
(C) 5
(D) 11
(E) 25

Solution

Capacitors in series add reciprocally:

$$\frac{1}{C_{eq}} = \frac{1}{C_1} + \frac{1}{C_2} + \frac{1}{C_3} + \frac{1}{C_4} + \frac{1}{C_5}$$

$$\frac{1}{C_{eq}} = 5\left(\frac{1}{5\ \mu F}\right) \qquad C_{eq} = 1\ \mu F$$

Answer is (A)

Question 2. When the five are connected in parallel, the equivalent capacitance, in μF, is nearest to

   (A) 1
   (B) 2
   (C) 5
   (D) 11
   (E) 25

Solution

Capacitors in parallel add directly:

$$C_{eq} = C_1 + C_2 + C_3 + C_4 + C_5 = 5(5\ \mu F) = 25\ \mu F$$

Answer is (E)

PROBLEM 10-10

One has n identical cells that each have emf V and internal resistance r. When they are connected to an external resistance R, they will produce a current I.

Question 1. If the cells are connected in series, the current I is

   (A) nV/R
   (B) V/r
   (C) $V/(R + r)$
   (D) $nV/(R + nr)$
   (E) $nV/(nR + r)$

Solution

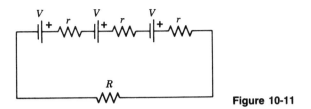

Figure 10-11

Figure 10-11 shows the series circuit for $n = 3$ cells. Each cell internal resistance r has been shown separately from the emf. Direct application of Kirchhoff's voltage law around the closed loop yields

$$nV - I(nr) - IR = 0$$

$$I = \frac{nV}{R + nr}$$

Answer is (D)

Question 2. If the cells are connected in parallel, the current I is

(A) V/nR
(B) V/R
(C) $V/(R + r)$
(D) $nV/(R + nr)$
(E) $nV/(nR + r)$

Solution

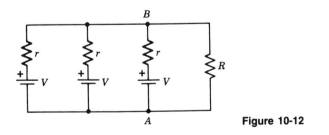

Figure 10-12

Figure 10-12 depicts the parallel circuit for $n = 3$ cells. The current through R is I, but the current from A to B can take any one of n identical paths through a cell; by Kirchhoff's current law the current through a cell is I/n. Application of the voltage law around a loop containing A, B and R gives

$$V - \left(\frac{I}{n}\right)r - IR = 0$$

$$I = \frac{V}{R + r/n} = \frac{nV}{nR + r}$$

Answer is (E)

PROBLEM 10-11

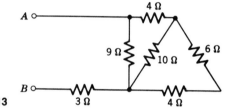

Figure 10-13

The total equivalent resistance, in Ω, between A and B is nearest to

(A) 3.2
(B) 6.5
(C) 7.5
(D) 10.2
(E) 36.0

Solution

The circuit is first redrawn in a more conventional form.

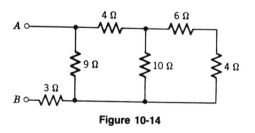

Figure 10-14

The 6-Ω and 4-Ω resistors in series may be added.

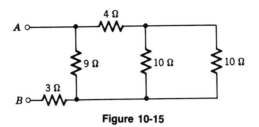

Figure 10-15

The equivalent resistance of the two 10-Ω resistors in parallel is

$$\frac{1}{R_e} = \frac{1}{10} + \frac{1}{10} \qquad R_e = 5\,\Omega$$

This 5-Ω resistor can now be added to the 4-Ω resistor in series to produce

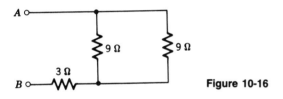

Figure 10-16

The equivalent resistance of the two parallel 9-Ω resistors is

$$\frac{1}{R_e} = \frac{1}{9} + \frac{1}{9} \qquad R_e = 4.5\,\Omega$$

When this resistor is added in series to the remaining 3-Ω resistor, we find the total equivalent resistance to be 7.5 Ω.

Answer is (C) ok

PROBLEM 10-12

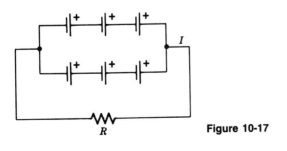

Figure 10-17

When Kirchhoff's laws are applied to three similar cells that are connected in series, and then they are connected in parallel with three other similar cells, as in the figure, the total current is

(A) $I = \dfrac{V}{r/3 + R}$

(B) $I = \dfrac{3V}{3r/2 + R}$

(C) $I = \dfrac{V}{3r/2 + R}$

(D) $I = \dfrac{3V}{3r + R}$

(E) $I = \dfrac{1}{6/r + R}$

V is the emf of each cell, r is the internal resistance of each cell, and R is the resistance of the external circuit.

Solution

Voltage and resistance are each directly additive in simple series. When this is done within each parallel limb, the equivalent circuit with the internal resistances shown explicitly is

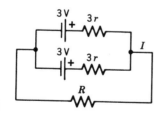

Figure 10-18

By symmetry the current in each parallel limb is $I/2$. By applying Kirchhoff's voltage law around the main circuit, one obtains

$$3V - \frac{I}{2}(3r) - IR = 0$$

$$I = \frac{3V}{3r/2 + R}$$

Answer is (B)

PROBLEM 10-13

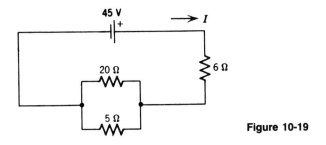

Figure 10-19

The current I, in A, in this circuit is nearest to

(A) 1.5
(B) 4.5
(C) 7.2
(D) 9.3
(E) 18.8

Solution

The equivalent resistance for the two parallel resistors is

$$\frac{1}{R_e} = \frac{1}{5} + \frac{1}{20} \qquad R_e = 4\,\Omega$$

The total resistance of the circuit is $4\,\Omega + 6\,\Omega = 10\,\Omega$. Applying Ohm's law gives

$$I = \frac{V}{R} = \frac{45}{10} = 4.5\text{ A}$$

Answer is (B)

PROBLEM 10-14

The resistance, in Ω, between A and B in Fig. 10-20 is nearest to

(A) 1.1
(B) 1.8
(C) 6.9
(D) 16.5
(E) 39.0

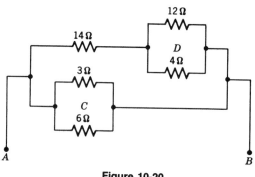

Figure 10-20

Solution

The two parallel pairs of resistors at C and D can each be replaced by an equivalent resistor:

$$\frac{1}{R_C} = \frac{1}{3} + \frac{1}{6} \quad \text{and} \quad \frac{1}{R_D} = \frac{1}{12} + \frac{1}{4}$$

$$R_C = 2\,\Omega \qquad R_D = 3\,\Omega$$

The total resistance in the upper branch is $14\,\Omega + 3\,\Omega = 17\,\Omega$. The circuit now reduces to

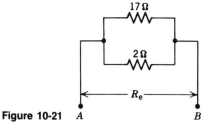

Figure 10-21 A B

$$\frac{1}{R_e} = \frac{1}{17} + \frac{1}{2} \qquad R_e = \frac{34}{19}\,\Omega = 1.79\,\Omega$$

Answer is (B)

PROBLEM 10-15

The circuit shown is composed of resistors that have equal resistances R. The resistance, in Ω, between points A and B is nearest to

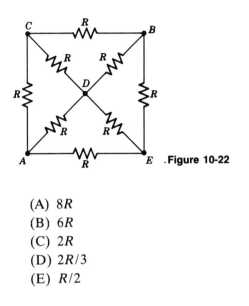

. **Figure 10-22**

(A) $8R$

(B) $6R$

(C) $2R$

(D) $2R/3$

(E) $R/2$

Solution

By examining Fig. 10-22 we find that the resistance encountered along paths ACB, ADB, and AEB between A and B are identical $(2R)$ and furthermore that the voltage at points C, D, and E must be the same. In other words, no current flows along the diagonal wire connecting C, D, and E, and therefore the resistors between C and D, and between D and E, have no effect on the circuit and can be removed. A simplified diagram can now be drawn.

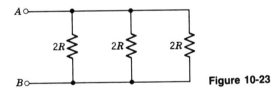

Figure 10-23

For parallel resistors the equation for equivalent resistance is

$$\frac{1}{R_e} = \frac{1}{R_1} + \frac{1}{R_2} + \frac{1}{R_3} \qquad \frac{1}{R_{AB}} = \frac{1}{2R} + \frac{1}{2R} + \frac{1}{2R} = \frac{3}{2R}$$

Therefore the resistance R_{AB} between A and B is

$$R_{AB} = \frac{2R}{3}\ \Omega$$

Answer is (D)

PROBLEM 10-16

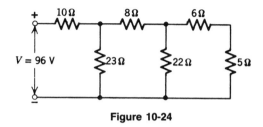

Figure 10-24

The current input, in A, to the circuit in Fig. 10-24 is nearest to

(A) 1.3
(B) 3.5
(C) 5.0
(D) 12.9
(E) 14.8

Solution

Add the resistors in series to obtain $6 + 5 = 11\ \Omega$.

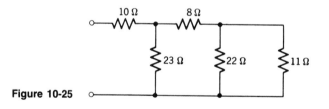

Figure 10-25

Now add the 11-Ω and 22-Ω resistors in parallel,

$$\frac{1}{R_e} = \frac{1}{11} + \frac{1}{22} \qquad R_e = 7\tfrac{1}{3}\ \Omega$$

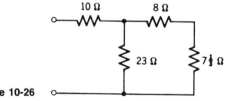

Figure 10-26

Again add the series resistors to obtain $7\frac{1}{3}\,\Omega + 8\,\Omega = 15\frac{1}{3}\,\Omega$, and find the equivalent resistance for the last pair of parallel resistors:

$$\frac{1}{R_e} = \frac{1}{15\frac{1}{3}} + \frac{1}{23} \qquad R_e = 9.2\,\Omega$$

Finally, we add the last resistors in series to find $9.2 + 10 = 19.2\,\Omega$ is the resistance of the entire combination. By Ohm's law

$$I = \frac{V}{R} = \frac{96}{19.2} = 5.0\,\text{A}$$

Answer is (C)

PROBLEM 10-17

The resistance, in Ω, of the circuit shown in the figure is

(A) 0.43
(B) 0.80
(C) 2.28
(D) 5.50
(E) 14.00

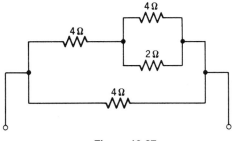

Figure 10-27

Solution

The 2-Ω and 4-Ω resistors in parallel can be replaced by R_e given by

$$\frac{1}{R_e} = \frac{1}{2} + \frac{1}{4} \qquad R_e = \frac{4}{3}\,\Omega$$

Adding this value to the 4-Ω resistor in series with it leads to

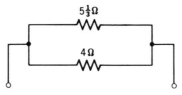

Figure 10-28

The remaining two resistors are in parallel and can be combined to yield

$$\frac{1}{R} = \frac{1}{4} + \frac{1}{5\frac{1}{3}} = 0.250 + 0.188 = 0.438$$

$$R = \frac{1}{0.438} = 2.28 \ \Omega$$

Answer is (C)

PROBLEM 10-18

The value of the resistor R, in Ω, is nearest to

(A) 1.9
(B) 4.0
(C) 5.0
(D) 8.0
(E) 9.8

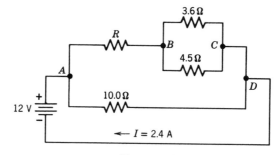

Figure 10-29

Solution

If we apply Ohm's law directly to the exterior branch of the circuit, we find the equivalent resistance R_{AD} to be given by

$$R_{AD} = \frac{V}{I} = \frac{12.0}{2.4} = 5.0 \ \Omega$$

Now we must find R_{AD} in terms of R. The resistance R_{BC} is found from

$$\frac{1}{R_{BC}} = \frac{1}{3.6} + \frac{1}{4.5} \qquad R_{BC} = 2.0\,\Omega$$

and in the same general manner

$$\frac{1}{R_{AD}} = \frac{1}{10.0} + \frac{1}{R + 2.0}$$

and also $1/R_{AD} = 1/5.0$. We find that

$$\frac{1}{5.0} - \frac{1}{10.0} = \frac{1}{10.0} = \frac{1}{10.0} = \frac{1}{R + 2.0}$$

and $R = 8.0\,\Omega$.

Answer is (D)

PROBLEM 10-19

A two-loop dc circuit is shown below.

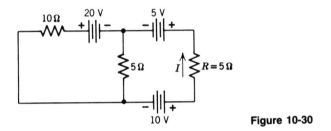

Figure 10-30

Question 1. The current I, in A, through the resistor R is closest to

(A) 1.8
(B) 1.4
(C) 1.0
(D) 0.4
(E) 0.2

Solution

This problem will be solved by using mesh equations. Let us assume a counterclockwise direction for the current.

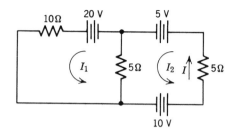

Figure 10-31

Around the first loop we obtain

$$(10 + 5)I_1 - 5I_2 = 20 \tag{1}$$

and for the second loop

$$-5I_1 + (5 + 5)I_2 = -5 + 10 \tag{2}$$

Add $3 \times$ Eq. (2) to Eq. (1) to obtain

$$25I_2 = 35 \qquad I_2 = I = \frac{35}{25} = 1.4 \text{ A}$$

Answer is (B)

Question 2. The power, in W, dissipated in resistor R is nearest to

(A) 25.0
(B) 16.2
(C) 9.8
(D) 5.0
(E) 0.8

Solution

The power dissipated is $P = I^2R = (1.4)^2(5) = 9.8$ W

Answer is (C)

PROBLEM 10-20

For the circuit shown, assume the inductance has negligible resistance.

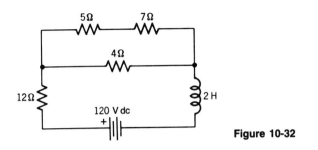

5Ω   7Ω

4Ω

12Ω   2H

120 V dc

Figure 10-32

The voltage drop, in V, across the 7-Ω resistor during steady-state operation is nearest to

(A) 12
(B) 14
(C) 26
(D) 30
(E) 31

Solution

Since this is a steady-state problem, the inductance plays no role, for it responds only to changes in the current. Rather than use the equivalent resistance approach, let us select loop currents and write two mesh equations.

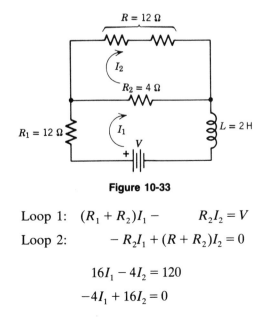

Figure 10-33

Loop 1: $(R_1 + R_2)I_1 - \quad R_2I_2 = V$

Loop 2: $- R_2I_1 + (R + R_2)I_2 = 0$

or

$$16I_1 - 4I_2 = 120$$
$$-4I_1 + 16I_2 = 0$$

From the second equation $I_1 = 4I_2$ so that the first equation gives $16(4I_2) - 4I_2 = 120$, $60I_2 = 120$ and $I_2 = 2A$. The voltage drop across the 7-Ω resistor is then

$$V_7 = I_2 R_7 = 2(7) = 14 \text{ V}$$

Answer is (B)

PROBLEM 10-21

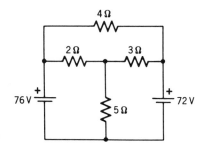

Figure 10-34

If it is known that the current through the 4-Ω resistor is 1 A, then the current, in A, through the 5-Ω resistor is closest to

(A) 1
(B) 3
(C) 6
(D) 9
(E) 12

Solution

Let us set up three loop equations for the loop currents I_1, I_2, and I_3:

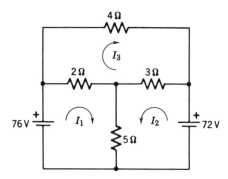

Figure 10-35

$$(2+5)I_1 + \qquad 5I_2 - \qquad\qquad 2I_3 = 76 \text{ V} \qquad\qquad (1)$$
$$5I_1 + (5+3)I_2 + \qquad\qquad 3I_3 = 72 \text{ V} \qquad\qquad (2)$$
$$-2I_1 + \qquad 3I_2 + (2+3+4)I_3 = \; 0 \text{ V} \qquad\qquad (3)$$

It is given that $I_3 = 1$ A; Eq. (3) requires that

$$I_1 = \frac{9+3I_2}{2}$$

Then from one of the other equations, say Eq. (1), we find

$$\tfrac{7}{2}(9+3I_2) + 5I_2 = 76+2 \quad \text{or} \quad I_2 = 3 \text{ A}$$

and

$$I_1 = \frac{9+3(3)}{2} = 9 \text{ A}$$

The current through the 5-Ω resistor is $I_1 + I_2 = 12$ A.

Answer is (E)

PROBLEM 10-22

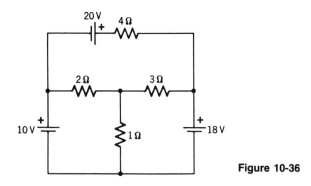

Figure 10-36

For the circuit shown, the current, in A, through the 1-Ω resistor is nearest to

(A) 2
(B) 4
(C) 5
(D) 6
(E) 10

Solution

We will write three loop equations. The loops are chosen so that only I_1 passes through the 1-Ω resistor; the third loop traverses the entire outer boundary.

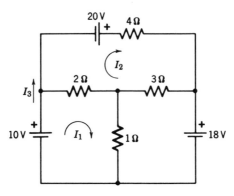

Figure 10-37

The loop equations are

$$(2 + 1)I_1 - \qquad\qquad 2I_2 \qquad = 10 \qquad\qquad (1)$$

$$-2I_1 + (2 + 4 + 3)I_2 + 4I_3 = 20 \qquad\qquad (2)$$

$$4I_2 + 4I_3 = 10 + 20 - 18 \qquad\qquad (3)$$

From Eq. (3), $4I_3 = 12 - 4I_2$. Substitution of this result into Eq. (2) yields

$$-2I_1 + 5I_2 = 8 \qquad\qquad (4)$$

Now multiply Eq. (1) by 5, multiply Eq. (4) by 2 and add the two results to find

$$11I_1 = 66$$

$$I_1 = 6 \text{ A}$$

Answer is (D)

PROBLEM 10-23

A 12-V car battery, with an internal resistance of 4 Ω, supplies a bank of 25 lights that are connected in parallel. Each light has an effective resistance of 500 Ω.

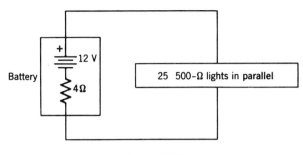

Figure 10-38

The power, in W, supplied to the bank of lights by the battery is nearest to

(A) 7.0
(B) 6.0
(C) 5.0
(D) 0.01
(E) 0.001

Solution

The resistance equivalent to the 25 lights in parallel is

$$\frac{1}{R_e} = 25\left(\frac{1}{500}\right) \qquad R_e = 20\ \Omega$$

The circuit current is then

$$I = \frac{V}{R} = \frac{12}{20 + 4} = 0.5\ \text{A}$$

and the power supplied *to the bank* (not the total power supplied) is

$$P = I^2 R_e = (0.5)^2(20) = 5.0\ \text{W}$$

Answer is (C)

PROBLEM 10-24

A "black box" has within it several batteries and resistors. An external resistor R and an ammeter are placed in series between the terminals A and B of the box as shown. When $R = 3.0\ \Omega$, the current flow indicated by the meter is 1.0 A. When $R = 9.0\ \Omega$, the indicated flow of current is 0.5 A.

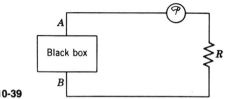

Figure 10-39

The resistance, in Ω, between terminals A and B is most closely

(A) 2.0
(B) 3.0
(C) 5.0
(D) 6.0
(E) 9.0

Solution

We can draw the circuit diagram.

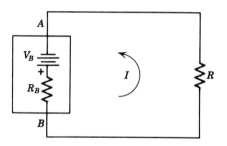

Figure 10-40

Using Kirchhoff's voltage law, we can write the loop equation for the circuit:

$$IR_B + IR = V_B$$

When $R = 3.0\,\Omega$ and $I = 1.0\,\text{A}$,

$$1.0\,R_B + 1.0(3.0) = V_B \tag{1}$$

When $R = 9.0\,\Omega$ and $I = 0.5\,\text{A}$,

$$0.5R_B + 0.5(9.0) = V_B \tag{2}$$

Subtracting Eq. (2) from Eq. (1) results in

$$0.5R_B - 1.5 = 0 \qquad R_B = 3.0\,\Omega$$

Answer is (B)

PROBLEM 10-25

A metal transport plane has a wingspread of 88 ft and flies horizontally with a speed of 150 miles/hr. At the elevation of the plane the vertical component of the earth's magnetic field is 0.65×10^4 tesla (T). The difference in potential, in V, that exists between the ends of the wings is most nearly

(A) 10^7
(B) 10^3
(C) 10^1
(D) 10^{-1}
(E) 10^{-3}

Solution

An emf is induced in a circuit whenever any of its conductors cuts magnetic flux. The equation is

$$e_i = BLv$$

where e_i = induced voltage (in units of 10^{-8} V)
B = flux density in teslas
L = conductor length in meters
v = velocity of conductor in m/sec

$$e_i = 0.65 \times 10^4 \left(\frac{88 \times 12 \times 2.54}{100} \right) \left(150 \times \frac{5280}{3600} \times \frac{12 \times 2.54}{100} \right)$$

$$e_i = 0.117 \text{ V}$$

Answer is (D)

PROBLEM 10-26

In a Y-connected circuit the line current equals

(A) the phase current

(B) $\frac{1}{\sqrt{3}}$ times the phase current

(C) $\sqrt{3}$ times the phase current

(D) $\frac{\sqrt{3}}{2}$ times the phase current

(E) $\frac{2}{\sqrt{3}}$ times the phase current

Solution

For the balanced three-phase Y connection, line current equals phase current.

Answer is (A)

PROBLEM 10-27

A bank of three transformers is to be connected to reduce the voltage from a three-phase, 12,000-V (line-to-line) distribution line to supply power for a small irrigation pump driven by a 440-V induction motor. A Y connection will be used for the primary and a Δ connection for the secondary.

Question 1. The primary voltage rating, in V, of each transformer is nearest to

    (A) 4,000
    (B) 7,000
    (C) 12,000
    (D) 21,000
    (E) 36,000

Solution

In a Y connection, $V_{\text{line}} = \sqrt{3}\, V_{\text{phase}}$. Hence

$$V_{\text{phase}} = \frac{12{,}000}{\sqrt{3}} = 6930\ \text{V} = \text{primary rating}$$

Answer is (B)

Question 2. The secondary voltage rating, in V, of each transformer is nearest to

    (A) 150
    (B) 250
    (C) 440
    (D) 760
    (E) 1320

Solution

In a Δ connection, $V_{\text{line}} = V_{\text{phase}} = 440\ \text{V} = \text{secondary rating}$.

Answer is (C)

PROBLEM 10-28

A circuit consisting of $R = 3\,\Omega$ and $X_L = 4\,\Omega$ in series is connected to a 100-V, 60-Hz source as shown.

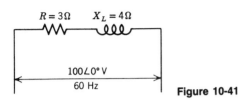

$R = 3\Omega$ $X_L = 4\Omega$

$100\angle 0°$ V
60 Hz

Figure 10-41

Question 1. The average power, in W, is nearest to

(A)   1200
(B) −1200
(C)   1600
(D) −1600
(E)   2000

Solution

$$\text{Impedance } \mathbf{Z} = R + jX_L = 3 + j4$$

In polar form

$$\mathbf{Z} = (3^2 + 4^2)^{1/2}\,\tan^{-1}\left(\tfrac{4}{3}\right) = 5\angle 53.1°\,\Omega$$

$$\text{Current } \mathbf{I} = \frac{\mathbf{V}}{\mathbf{Z}} = \frac{100\angle 0°}{5\angle 53.1°} = 20\angle -53.1°$$

The average power is

$$\mathbf{P} = \mathbf{V} \times \mathbf{I} = 100\angle 0° \times 20\angle -53.1° = 100 \times 20\cos(-53.1°)$$

$$\mathbf{P} = 2000 \times 0.6 = 1200 \text{ W}$$

Answer is (A)

Question 2. The reactive power, in VAR, is nearest to

(A)   1200
(B) −1200
(C)   1600
(D) −1600
(E)   2000

Solution

The reactive power or reactive volt-amperes is

$$Q = 1200 \tan(-53.1°) = 1200 \times (-\tfrac{4}{3}) = -1600 \text{ VAR}$$

Answer is (D)

Question 3. The total volt-amperes (VA) is nearest to

(A) 1150
(B) 1200
(C) 1600
(D) 2000
(E) 3500

Solution

The apparent or total volt-amperes is

$$VI = 100 \times 20 = 2000 \text{ VA}$$

Answer is (D)

The relations between these results are conveniently summarized by the volt-ampere triangle for this problem.

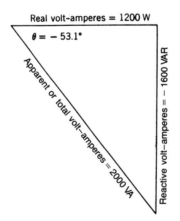

Figure 10-42

PROBLEM 10-29

Consider the circuit shown in Fig. 10-43.

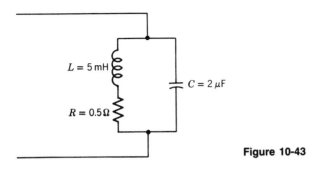

Figure 10-43

Question 1. The resonant frequency f_0, in Hz, is nearest to

(A) 50,000
(B) 10,000
(C) 8,000
(D) 4,000
(E) 1,590

Solution

The resonant frequency occurs when $X_L = X_C$, assuming $X_L \gg R$.

$$2\pi f_0 L = \frac{1}{2\pi f_0 C} \qquad 2\pi f_0 \times 0.005 = \frac{1}{2\pi f_0 \times 2 \times 10^{-6}}$$

$$f_0 = \left(\frac{10^6}{8\pi^2 \times 0.005} \right)^{1/2} = (2.525 \times 10^6)^{1/2} = 1{,}590 \text{ Hz}$$

Answer is (E)

Question 2. The figure of merit Q of the circuit at f_0 is nearest to

(A)   1
(B)   10
(C)   25
(D) 100
(E) 200

Solution

$$X_C = X_L = 2\pi \times 1590 \times 0.005 = 50 \ \Omega$$

The figure of merit is therefore

$$Q = \frac{X}{R} = \frac{50}{0.5} = 100$$

Answer is (D)

Question 3. The wavelength, in m, at the resonant frequency is nearest to

(A) 2.0×10^7
(B) 2.0×10^5
(C) 7.5×10^4
(D) 4.0×10^4
(E) 3.0×10^3

Solution

Using the value of c = velocity of light $= 300 \times 10^6$ m/sec, the wavelength is

$$\lambda = \frac{c}{f_0} = \frac{300 \times 10^6}{1590} = 1.89 \times 10^5 \text{ m}$$

Answer is (B)

PROBLEM 10-30

A choke coil having an inductance of 150 mH is connected across a 60-Hz source.

Question 1. To have the same reactance as the choke coil, the value of the capacitance C, in μF, that must be connected across a 50-Hz source is most nearly

(A) 5.6
(B) 17.6
(C) 39.2
(D) 56.3
(E) 354

Solution

The inductive reactance is $X_L = 2\pi f L = 2\pi \times 60 \times 0.150 = 56.5 \, \Omega$. The capacitive reactance is

$$X_C = \frac{1}{2\pi fC} = \frac{1}{2\pi \times 50C}$$

For equal reactance,

$$\frac{1}{2\pi \times 50C} = 56.5\ \Omega$$

$$C = \frac{1}{2\pi \times 50 \times 56.5} = \frac{1}{17{,}750} = 56.3 \times 10^{-6}\ F = 56.3\ \mu F$$

Answer is (D)

Question 2. Assuming the capacitor has negligible resistance, the series resonant frequency, in Hz, of L and C (part 1) is nearest to

(A) 50.0
(B) 54.8
(C) 55.0
(D) 56.3
(E) 60.0

Solution

The series resonant frequency can be found by equating X_L and X_C.

$$2\pi fL = \frac{1}{2\pi fC} \qquad 4\pi^2 f^2 = \frac{1}{LC} \qquad f = \frac{1}{2\pi(LC)^{1/2}}$$

$$f = \frac{1}{2\pi(0.15 \times 56.3 \times 10^{-6})^{1/2}} = \frac{10^3}{2\pi(8.45)^{1/2}} = 54.8\ \text{Hz}$$

Answer is (B)

PROBLEM 10-31

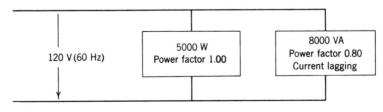

Figure 10-44

A circuit is shown in Fig. 10-44.

Question 1. The power, in W, is most nearly

    (A) 4,800
    (B) 5,000
    (C) 8,000
    (D) 11,400
    (E) 13,000

Solution

For each load, power $P = VI \cos \theta$

$$P = 5000(1.00) + 8000(0.80) = 11{,}400 \text{ W}$$

$$\text{Answer is (D)}$$

Question 2. The reactive power, in VAR, is most nearly

    (A) 4,800
    (B) 6,400
    (C) 8,000
    (D) 8,400
    (E) 11,400

Solution

For each load, reactive power $Q = VI \sin \theta$

$$Q = 5000(0.00) + 8000(0.60) = 4800 \text{ VAR}$$

$$\text{Answer is (A)}$$

Question 3. The total volt-amperes (VA) input is most nearly

    (A) 8,000
    (B) 9,800
    (C) 11,400
    (D) 12,400
    (E) 13,000

Solution

As can be seen from Fig. 10-45, the total volt-amperes is equal to

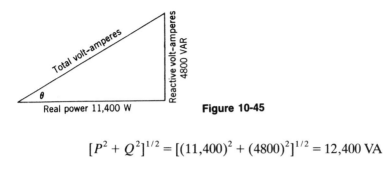

Figure 10-45

$$[P^2 + Q^2]^{1/2} = [(11,400)^2 + (4800)^2]^{1/2} = 12,400 \text{ VA}$$

Answer is (D)

Question 4. The power factor for the entire circuit is most nearly

(A) 0.80
(B) 0.88
(C) 0.92
(D) 0.95
(E) 1.00

Solution

$$\text{Power factor} = \cos \theta = \frac{11,400}{12,400} = 0.92 \text{ lagging}$$

Answer is (C)

PROBLEM 10-32

A rectifier circuit connected to a 60-Hz ac supply is shown below. Each secondary winding of the transformer has a voltage of 120 V rms. Selenium disks are used as the rectifying element; their voltage drop is negligible. The average voltage, in V, across the load is most nearly

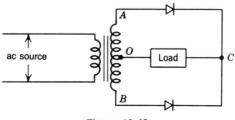

Figure 10-46

(A) 170
(B) 120
(C) 108
(D) 85
(E) 60

Solution

From the problem statement $V_{AO} = V_{OB} = 120$ V rms. The instantaneous voltage is therefore

$$V = 120\sqrt{2}\sin 2\pi(60)t \qquad 0 < t < \tfrac{1}{120}$$

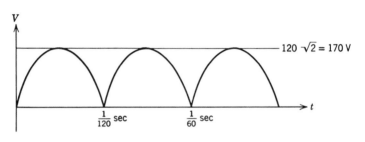

Figure 10-47

For a half cycle the average or effective voltage is

$$V_h = \frac{2}{\pi} V_m = \frac{2}{\pi}(120\sqrt{2}) = 108 \text{ V}$$

Answer is (C)

PROBLEM 10-33

In a series ac circuit with a lagging power factor, to increase the power factor you should

(A) increase the current
(B) increase the voltage
(C) increase the frequency
(D) add inductance to the circuit
(E) add capacitance to the circuit

Solution

A lagging power factor indicates that the current is behind the voltage by a phase angle θ. This is caused by the inductive load in the circuit, which causes the inductive reactance X_L to be larger than the capacitive reactance X_C. The power factor, $\cos \theta$, can be increased by adding capacitance to the circuit in such a way that X_C increases and the reactance X thereby decreases, as shown in Fig. 10-48.

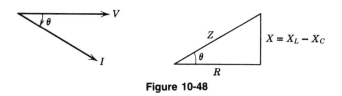

Figure 10-48

Answer is (E)

PROBLEM 10-34

Given the circuit shown in Fig. 10-49:

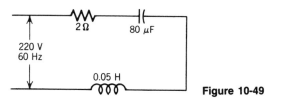

Figure 10-49

Question 1. The impedance is nearest to

    (A) $2.0\,\Omega\angle 0°$
    (B) $14.4\,\Omega\angle -82°$
    (C) $16.3\,\Omega\angle 2°$
    (D) $19.0\,\Omega\angle 84°$
    (E) $33.3\,\Omega\angle -87°$

Solution

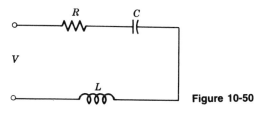

Figure 10-50

The impedance Z is

$$Z = R + jX_L - jX_C$$

where R = resistance (ohms)

$X_L = 2\pi fL$ (ohms) (inductive reactance)

$X_C = \dfrac{1}{2\pi fC}$ (ohms) (capacitive reactance)

$$Z = 2 + j(2\pi 60)(0.05) - j\left[\frac{1}{(2\pi 60)(80 \times 10^{-6})}\right]$$

$$= 2 + j(18.85) - j(33.2)$$

$$\cong 2 - j14.3 \ \Omega \text{ (in rectangular form)}$$

$$\cong 14.4 \ \Omega \angle - 82.0°$$

Answer is (B)

Question 2. The line current, in A, is closest to

(A) 110
(B)   15.2
(C)   13.5
(D)   11.6
(E)   6.6

Solution

$$\mathbf{I} = \frac{V\angle 0°}{\mathbf{Z}} = \frac{220\angle 0°}{14.4\angle - 82.0°} = 15.22\angle + 82.0°$$

Hence $|I| \cong 15.2$ A.

Answer is (B)

Question 3. The power factor is closest to

(A) 0.0
(B) 0.052
(C) 0.105
(D) 0.138
(E) 1.0

Solution

Power factor = $\cos 82.0° = 0.138$ (leading).

Answer is (D)

PROBLEM 10-35

Given the circuit and voltage measurements shown in Fig. 10-51:

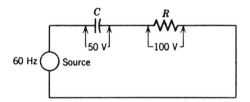

Figure 10-51

Question 1. The voltage, in V, of the 60-Hz source is nearest to

(A) 50
(B) 100
(C) 112
(D) 150
(E) 212

Solution

Here $IR = 100$ V and $I(-jX_C) = 50$ V. Therefore

$$\mathbf{V} = IR - jIX_C = 100 - j50 \text{ V (rectangular form)}$$
$$\mathbf{V} = 112\angle -26.6°$$

$|\mathbf{V}| = 112$ V

Answer is (C)

Question 2. The power factor of the circuit is nearest to

(A) 0.45
(B) 0.67
(C) 0.75
(D) 0.89
(E) 1.00

Solution

Power factor $= \cos 26.6° = 0.89$ (leading).

Answer is (D)

PROBLEM 10-36

For the difference amplifier shown in Fig. 10-52, the output voltage v_0, in V, is closest to

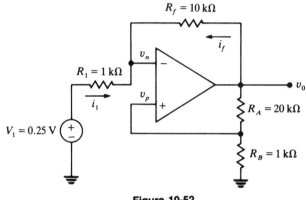

Figure 10-52

(A) -2.5
(B) 2.5
(C) -12.5
(D) -5.25
(E) 5.25

Solution

Between the source voltage V_1 and the negative pin of the op amp, $V_1 - i_1 R_1 = v_n$. Here we can assume $v_n = v_p$ since they differ by only a few millivolts and also that $i_1 + i_f = 0$ since the input current to the op amp is very small. The feedback loop current is $i_f = (v_0 - v_n)/R_f$. Between the output v_0 and the ground we have $v_0 = i(R_A + R_B)$ and also $v_p = iR_B$, which leads to $v_p = v_0 R_B/(R_A + R_B) = Kv_0$ with $K = 1/21$ for the given data. Now we substitute these results into the first equation to obtain

$$V_1 + i_f R_1 = v_p$$

$$V_1 + (v_0 - Kv_0)\frac{R_1}{R_f} = Kv_0$$

or

$$\left[K + (K-1)\,\frac{R_1}{R_f} \right] v_0 = V_1$$

$$\left[\frac{1}{21} - \frac{20}{21}\,\frac{1}{10} \right] v_0 = 0.25$$

$$v_0 = -5.25 \text{ V}$$

Answer is (D)

PROBLEM 10-37

For the operational amplifier and circuit configuration shown in Fig. 10-53,

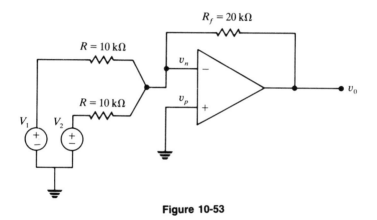

Figure 10-53

in which the two source voltages are each one volt, the output voltage v_0, in V, is closest to

(A) −4.0
(B)   4.0
(C) −1.0
(D)   1.0
(E)   0.5

Solution

Since $v_n = v_p = 0$ and $V_1 = V_2 = 1.0$ V,

$$v_0 = -\frac{R_f}{R}\,(V_1 + V_2) = -2.0(1.0 + 1.0) = -4.0 \text{ V}$$

Answer is (A)

PROBLEM 10-38

It is desired, by discharging a capacitance, to send an electric impulse through a 1000-Ω resistor so that the initial current will be 1.00 A. At the end of 0.010 sec (10 msec), the voltage across the resistance is to be 368 V. The correct value of the required capacitance, in μF, is most nearly

   (A) 1.0
   (B) 3.7
   (C) 6.3
   (D) 10.0
   (E) 37.0

Solution

The time constant T is the time in seconds needed for a capacitor to lose 63% of its voltage, that is, for the voltage to drop to 37% of its initial value. In this case $T = 0.01$ sec. For a capacitor the time constant is

$$T = RC \quad \text{or} \quad C = \frac{T}{R} = \frac{10 \times 10^{-3}}{10^3} = 10 \times 10^{-6} \, \text{F} = 10 \, \mu\text{F}$$

Answer is (D)

PROBLEM 10-39

A 50-μF capacitor is initially charged to 10 V; then a switch is closed at time $t = 0$ and the capacitor discharges through a 100-Ω resistor. The magnitude of the current, in mA, through the resistor after 5 msec is nearest to

   (A) 37
   (B) 63
   (C) 100
   (D) 200
   (E) 500

Solution

The discharge current through a resistor connected in series with a capacitor is given by $i = I_0 e^{-t/T}$, in which T is the RC time constant and I_0 is the initial current when the switch is closed. Here $I_0 = V_0/R = 10/100 = 0.1$ A and $T = RC = (100)(50 \times 10^{-6}) = 5$ msec. Hence $i = 0.1 e^{-1.0} = 0.037$ A $= 37$ mA.

Answer is (A)

PROBLEM 10-40

A radio circuit oscillates at a frequency $f = 1/[2\pi(LC)^{1/2}]$. A given circuit has an inductance L of 0.1 H, but its capacitance is not exact, being given as 0.1 μF with a 10% tolerance.

Question 1. The resonant frequency, in Hz, is most nearly

(A)   1.6
(B)   160
(C)   1,600
(D) 10,000
(E) 62,800

Solution

$$f = \frac{1}{2\pi(LC)^{1/2}} = \frac{1}{2\pi(0.1 \times 0.1 \times 10^{-6})^{1/2}} = \frac{1}{2\pi(1 \times 10^{-8})^{1/2}} = \frac{10^4}{2\pi}$$

$$f = 1592 \text{ Hz}$$

Answer is (C)

Question 2. The variation in the resonant frequency, in Hz, caused by the capacitor tolerance is most nearly

(A)   0.2
(B)   10
(C)   20
(D)   80
(E) 160

Solution

$$f = \frac{1}{2\pi(L)^{1/2}} C^{-1/2}$$

Hence

$$\frac{df}{dC} = \frac{1}{2\pi L^{1/2}} \left(-\tfrac{1}{2}C^{-3/2}\right)$$

$$= \frac{-1}{2C} \frac{1}{2\pi(LC)^{1/2}} = -\frac{1}{2C} f$$

Using finite increments,

$$\left|\frac{\Delta f}{\Delta C}\right| = \frac{1}{2C} f \quad \text{or} \quad |\Delta f| = \frac{f}{2}\frac{\Delta C}{C}$$

The variation in frequency is therefore

$$\Delta f = \left(\frac{1592}{2}\right)(0.1) = 80 \text{ Hz}$$

Answer is (D)

PROBLEM 10-41

In a series circuit containing resistance, capacitance, and inductance, to increase the resonant frequency it is necessary to

(A) increase the resistance
(B) decrease the inductance
(C) increase the capacitance
(D) decrease the resistance
(E) increase the voltage

Solution

At resonance, $2\pi fL = \dfrac{1}{2\pi fC}$

$$2\pi f = \frac{1}{(LC)^{1/2}} \quad \text{and} \quad f = \frac{1}{2\pi(LC)^{1/2}}$$

Therefore to increase f, L or C must be decreased.

Answer is (B)

PROBLEM 10-42

For a dc shunt motor operating from a constant potential supply, one of the following will happen:

(A) Increasing the shunt field resistance will cause the motor speed to increase.
(B) Increasing the shunt field resistance will cause the motor speed to decrease.
(C) Increasing the shunt field resistance will cause the shunt field current to increase.

(D) Decreasing the shunt field resistance will cause the shunt field current to decrease.

(E) Increasing the shunt field resistance will cause the armature current to decrease.

Solution

The practical way to control the speed of a dc shunt motor is to place a rheostat in series with the shunt field. Increasing the resistance will decrease the field current I_f, which is proportional to the flux ϕ_f. This can be seen in the equation for motor speed,

$$\text{Speed} = \frac{V - I_a R_a}{K\phi_f}$$

Answer is (A)

PROBLEM 10-43

A four-pole synchronous motor operating from a 50-Hz supply will have a synchronous speed, in rpm, of

(A) 3600
(B) 3000
(C) 1800
(D) 1500
(E) 1200

Solution

Frequency $f = (\text{no. of poles}/2) \times (\text{rpm}/60)$

$$50 = \frac{4}{2} \times \frac{\text{rpm}}{60}$$

$$\text{rpm} = 1500$$

Answer is (D)

PROBLEM 10-44

Two wattmeters are properly connected in a three-phase circuit to read the power supplied to a 220-V, 15-hp, Δ-connected, three-phase induction motor running at no load. Voltmeters and ammeters are also properly connected in the circuit to supply lines A, B, and C. The readings are

Voltmeter (V)	Ammeter (A)	Wattmeter (W)
$V_{AB} = 221$	$I_A = 4.04$	810
$V_{BC} = 221$	$I_C = 4.04$	-200

Question 1. The power input to the motor, in W, is nearest to

(A) 200
(B) 610
(C) 810
(D) 893
(E) 1010

Solution

The power in a three-phase circuit can be determined by the use of two wattmeters. The total power is the algebraic sum of the two meter readings.

$$P_{\text{total}} = W_1 + W_2 = 810 + (-200) = 610 \text{ W}$$

Answer is (B)

Question 2. The power factor of the motor, with current lagging, is nearest to

(A) 0.394
(B) 0.523
(C) 0.577
(D) 0.683
(E) 0.907

Solution

Current lags for the induction motor. The relation for total power is $P_t = (3)^{1/2} VI \cos \theta$. Hence

$$\cos \theta = \frac{P_t}{(3)^{1/2} VI} = \frac{610}{(3)^{1/2} \times 221 \times 4.04} = 0.394$$

Power factor $= 0.394$ current lagging

Answer is (A)

PROBLEM 10-45

The armature resistance of a 50-hp, 550-V, dc shunt wound motor is 0.35 Ω. The full-load armature current of this motor is 76 A.

Question 1. So that the initial starting current is 150% of the full-load value, the resistance of the starting coil, in Ω, should be closest to

(A) 7.59
(B) 7.24
(C) 6.89
(D) 4.82
(E) 4.47

Solution

The starting current is to be limited to 1.50(76) = 114 A. The total resistance is then

$$R = \frac{V}{I_s} = \frac{550}{114} = 4.82 \, \Omega$$

and the resistance of the starter is $4.82 - 0.35 = 4.47 \, \Omega$.

Answer is (E)

Question 2. If the field current under full load is 3 A, the overall efficiency of the motor is nearest to

(A) 81%
(B) 84%
(C) 86%
(D) 89%
(E) 91%

Solution

Line current = armature current + field current = 76 + 3 = 79 A.

$$\text{Efficiency} = \frac{\text{output}}{\text{input}} = \frac{50 \times 746}{79 \times 550} = 0.859 = 85.9\%$$

Answer is (C)

PROBLEM 10-46

The rpm of an ac electric motor

    (A) varies directly as the number of poles
    (B) varies inversely as the number of poles
    (C) is independent of the number of poles
    (D) is independent of the frequency
    (E) is directly proportional to the square of the frequency

Solution

The speed varies directly as the frequency and inversely as the number of poles.

Answer is (B)

11

STRUCTURE OF MATTER AND MATERIALS SCIENCE

The structure of matter is concerned with the description of nuclei, atoms and molecules; in doing so, it overlaps portions of the subjects of chemistry (see Chapter 9) and physics. This chapter also presents an introduction to materials science. The dominant theme is the explanation of material properties by understanding the microstructural characteristics of the materials. Altogether these subjects are a major and growing field, albeit a relatively new field for the EIT examination. Although our review of these fields cannot be truly comprehensive, it will sample the major ideas and principles of the subject under four headings: the structure of matter, the principles that determine the structure of materials, the properties of materials themselves, and the processing of materials.

Some information that could also be placed in this chapter will be found in the chapters on mechanics of materials (Chapter 6), thermodynamics (Chapter 8) and electrical properties (Chapter 10).

STRUCTURE OF MATTER

The realization in the 19th century that matter is made of atoms contributed much to the knowledge of the chemistry of substances. Several fundamental problems in the structure of matter remained unexplained by physics a century ago.

One problem was the difficulty of explaining black body radiation; Planck found that the variation of black body radiation intensity with wavelength at a given temperature was explainable if the electromagnetic radiation energy E existed only in bundles, or quanta, depending on its frequency ν according to the equation

$$E = h\nu \qquad (11\text{-}1)$$

where Planck's constant is $h = 0.6626 \times 10^{-33}$ J-sec. Soon thereafter Einstein explained a feature of the photoelectric effect by expressing the kinetic energy E_{kin} of the electron emitted from a metal on which light of frequency ν has fallen as $E_{\text{kin}} = h\nu - W$, where W is the work function. About the same time Bohr developed a quantitative explanation of the spectra of the hydrogen atom; he stated that the angular momentum of an electron is quantized and that the transition of an electron between two energy levels E_1 and E_2 is directly related to its spectral frequency via the relation $E_1 - E_2 = h\nu$. When Michelson and Morley found the velocity of light to be independent of the direction of motion, it became possible for the theory of relativity to connect energy and mass by the equation

$$E = mc^2 \qquad (11\text{-}2)$$

The combination of these ideas allows one to express the momentum p of a photon as $p = E/c = h\nu/c$, in which $c = 3 \times 10^8$ m/sec is the velocity of light. The actual existence of the energy and momentum carried by photons in units of Planck's constant was shown by Raman and Compton in their respective studies of the scattering of light and x-rays. These ideas were combined to produce a new concept known as the stimulated emission of radiation. By this concept, an electron could be induced by an incident photon to move from a higher energy level to a lower energy level, an idea that proved indispensable in the later discovery of lasers.

The combination of classical physics and the quantum hypothesis to explain these experiments produced tremendous confusion while a unified theory was lacking. The basis for such a theory was established when de Broglie proposed that all particles must exhibit wave behavior and gave the relation for their wavelength as $\mathbf{p} = h\mathbf{k}/2\pi$ in which the momentum vector $\mathbf{p}$ of the particle is related to the wave vector $\mathbf{k}$. Schroedinger searched for a differential equation that represented these waves. He then proceeded systematically to present the solution for the hydrogen atom and substantiated all of Bohr's work and added more predictions. Furthermore, his equation provided the basis for solutions to many problems in physics, and the new field of wave mechanics appeared in physics. Important to one's understanding of atoms is the introduction of quantum numbers and how the electron configuration of atoms could be understood via the Pauli exclusion principle, which stated that no two electrons could occupy the same energy level with the same set of quantum numbers. The mystery of the periodic arrangement of atoms was resolved by the successful inclusion of electron spin in prescribing one of the quantum numbers.

Heisenberg, Born and Jordan approached problems in quantum mechanics in terms of frequency, for which they were compelled to track two energy levels. Their use of matrices in this work produced an important concept which is today referred to as Heisenberg's uncertainty principle. According to this principle, it is not possible to have simultaneous knowledge of the momentum and position of a particle to within an accuracy $\Delta p\,\Delta x \geqslant h/2\pi$. A complete theory of quantum mechanics is today couched in the language of matrices.

There are three essential features in quantum mechanics that an engineer can use. Information about any system can be written in a mathematical expression called a wave function. This wave function, when multiplied by its complex conjugate, gives one the probability of the position of the particle at any given time and position. Most experiments in a laboratory correspond in mathematics to what are called operators. Performing an experiment corresponds to operating on the wave function; the kind of operator one uses is analogous to the type of experiment one does. The results of these experiments are numbers. The outcome of operating on the wave function also provides numbers which are called eigenvalues. The average properties of a system may always be obtained by integration over the point values of these quantities.

Example 1

The wave function of a particle is given by $\Psi = \exp\left[-i(\omega t - 2\pi x/\lambda)\right]$. What is the momentum of the wave?

Solution

The operator corresponding to the measurement of momentum is $(-ih/2\pi)\,d/dx$. Carrying out the indicated differentiation, one obtains

$$-i\,\frac{h}{2\pi}\,\frac{d\Psi}{dx} = -i\,\frac{h}{2\pi}\left(\frac{2\pi i}{\lambda}\right)\exp\left[-i\left(\omega t - \frac{2\pi x}{\lambda}\right)\right] = \frac{h}{\lambda}\,\Psi$$

In this case the eigenvalue is the momentum and is h/λ.

Exploration of the structure of matter has proceeded unrelentingly to discover many subatomic particles. A systematic classification produces a list of at least twelve fundamental constituents of matter. These are called either leptons or quarks. Leptons are characterized by independent existence and include the electron neutrino, electron, muon nutrino, muon, tau neutrino and tau. The quarks are always thought to be part of larger particles such as protons and neutrons. The six quarks have the names up, down, charm, strange, top/truth, and bottom/beauty. The various particles interact with each other through at least four different types of forces. The carrier that conveys this force is also given a name. The force of gravity is said to be carried by the graviton. Electromagnetic forces are carried by the photon. Weak forces are carried by weak sector bosons. Strong forces are carried by gluons. Of all these forces, the electromagnetic forces are most important owing to the range over which they act and the strength of their interaction. The other forces are either not strong or extend their influence only over very short distances.

PRINCIPLES THAT DETERMINE THE STRUCTURE OF MATERIALS

Determination of the chemical composition of a solid is the first step in understanding its structure. Most materials can first be categorized as a metal, a ceramic or a polymer; some will be composite materials with intermediate properties, and a small number will be semiconductors. In each group the arrangement of atoms will be governed by interatomic force versus distance relations; the net bonding force will be determined by the equilibrium between a coulombic attractive force and a repulsive force. The kind of bonding between atoms is mostly determined by the behavior of the electrons in the atoms. The bonds may be ionic which involves electron transfer, or covalent which involves electron sharing, or metallic which involves a sharing of a large number of nonlocal electrons in an electron cloud. The differences in these bonds can also be described in terms of a coordination number, which is the number of nearest-neighbor atoms which a single atom can have for a given material structure; ionic and metallic bonding have high coordination numbers, but covalent bonding has a low coordination number. Polymers and semiconductors have covalent bonding, whereas ceramics have ionic bonding and metals have metallic bonding. If bonding occurs without electron sharing or transfer, it is a secondary bonding called a van der Waals bond.

Most materials have a crystalline structure; there exist seven different unit cell shapes, and the possible arrangements of atoms in these cells are described by the 14 Bravais lattices. Crystal structures are defined by lattice positions, directions and planes. As examples, Fig. 11-1 shows the simple cubic, body-centered cubic (bcc) and face-centered cubic (fcc) lattices. Lattice positions are described notationally as fractions of the (x, y, z)

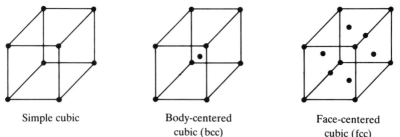

Simple cubic   Body-centered   Face-centered
                      cubic (bcc)   cubic (fcc)

Figure 11-1

coordinate positions in a unit cell. The eight corner positions in the simple cubic cell are $(0\,0\,0)$, $(1\,0\,0)$, $(0\,1\,0)$, $(0\,0\,1)$, $(1\,1\,0)$, $(1\,0\,1)$, $(0\,1\,1)$ and $(1\,1\,1)$. The additional lattice point in the body-centered cubic cell is at $(\frac{1}{2}\,\frac{1}{2}\,\frac{1}{2})$. Lattice directions follow a somewhat similar notation with the directions given as the set of smallest integers which are the intercepts with the coordinate axes; they are enclosed in square brackets. Negative directions are indicated by a bar over the numeral. Examples are $[1\,0\,0]$, $[1\,1\,\bar{1}]$, $[\bar{1}\,0\,1]$, and $[0\,0\,1]$.

Example 2

What is the packing density of an element that has crystallized in a face-centered cubic structure?

Solution

Let the volume of a face-centered cubic unit cell be a^3 and the radius of each model atom be r. As Fig. 11-2 shows schematically for one face, this crystal

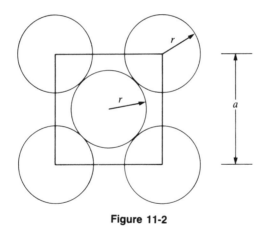

Figure 11-2

has one atom at each of the eight cell corners and one more atom at the middle of each of the six cell faces. The corner and face-centered atoms will touch along the face of the diagonal, so $4r = \sqrt{2}a$. The volume of each atom is $4\pi r^3/3$. A unit cell contains one eighth of each corner atom, plus one half of each midface atom. Thus the cell contains $8(1/8) + 6(1/2) = 4$ atoms, and the packing density is the ratio of the volume occupied by atoms to total cell volume:

$$\frac{4(4\pi r^3/3)}{a^3} = \frac{16\pi r^3/3}{(4r/\sqrt{2})^3} = 0.74$$

Approximately a quarter of the cell volume is not occupied by atoms.

X-ray diffraction is used to identify crystal structures. A set of closely spaced crystal layers acting as a diffraction grating causes light scattering in well-defined directions given by the Bragg equation

$$n\lambda = 2d \sin \theta \qquad (11\text{-}3)$$

In this equation the scattering angle θ is also called the Bragg angle whereas 2θ is called the diffraction angle which is actually measured, d is the spacing between layers, and λ is the wavelength of the x-rays.

Nearly all real materials will not have all the structural uniformity that was just described for crystals. Often a small amount of one compound will dilute the purity of the majority compound in the form of a solid solution; the result will be a changed set of material properties. In many instances the effect is a desirable change in properties; such is the case with metal alloys, including the various kinds of steel made from the parent iron. The solutions are of two kinds: when both kinds of atoms are similar, a substitutional solution forms, but when the kinds of atoms are dissimilar, an interstitial solution forms. The phase diagram can be useful in describing the micro-structure that is present for a certain material composition and temperature.

The other major form of imperfection in crystals is the existence of point, line, surface or volume defects in the lattice. It may be the absence of an atom at a lattice site, or it may be the dislocation, presence or absence of a part of a plane of atoms in the crystal. The description of edge, screw, and mixed dislocations is aided by the use of the Burgers vector **b**. This vector is the displacement needed to close a loop around the defect. For edge and screw dislocations the Burgers vector is perpendicular and parallel to the dislocation line, respectively. The defects in materials move in response to applied forces. Thermal energy from the environment, mechanical forces acting on the material and chemical forces from interactions with other atoms can provide the driving force for defects to move. The change is occasionally slow in order to avoid harm, but more frequently it is rapid and can cause material reliability problems. The behavior of defects can be used to explain several material properties, including ductile and brittle behavior.

PROPERTIES OF MATERIALS

An engineer selects a material primarily for its ability to carry the desig-nated load and transmit forces without excessive deflection, without de-formation and to remain free from fracture within the lifetime of the product. Many properties, including the material's density, elastic modulus, strength, toughness, fracture toughness, thermal conductivity, thermal ex-pansion, damping capacity and specific heat are of interest to most en-gineers. The longevity of an engineering product depends on its mode of use and on its physical and chemical environment; each of these factors may affect the properties of the material.

Some major mechanical properties of materials, including stress, strain and the elastic (Young's) modulus, are reviewed in Chapter 6, but a few additional ones are mentioned here. The ultimate strength or tensile strength is the highest stress a material can withstand before complete failure. Related to this is ductility, which is the percent elongation at failure, and toughness is the total area under the stress-strain curve (cf. Fig. 6-7). Rockwell and Brinell hardness numbers are empirical values that are related to indentations of specified geometry in specimens. An analog of toughness is impact energy, most often measured by the Charpy test. The value of fracture toughness is specified by $K_{IC} = Y\sigma_f\sqrt{\pi a}$, in which Y is a scale factor near unity, σ_f is the applied stress at failure and a is the depth of a surface crack or half the length of an internal crack. After a large number of loading cycles (10^8 cycles is usual) the fatigue strength is commonly found to be between 1/2 and 1/4 the tensile strength. Finally, creep tests record the plastic deformation that occurs under steady load with time at an elevated temperature.

Some materials can provide special features that permit unique devices to be built. Conducting, semiconducting, insulating or superconducting materials provide the medium in which information can be generated, stored, modified, transmitted and received. The unique properties of these materials almost permit the design of devices to start at the atomic level. The magnetic behavior of materials contributes substantially to the generation of energy and also saves humankind from much manual labor with its contribution to motors. In addition, magnetic properties have been a major factor in providing entertainment and storing massive amounts of information. Finally, optical properties of materials have begun to create their own technological revolution.

PROCESSING OF MATERIALS

Materials must be available before they can be used in engineering; they are accessible in nature in the earth in shapes, forms, and combinations that very few can use. The principles of the extraction, refining, melting and consolidation of materials are important here only insofar as they influence the properties of materials. Most engineering materials are cast at some time, and the solidification process has an important effect on its subsequent behavior. In particular, annealing and other heat treatments will affect the development of material phases, precipitation and the growth of grain size within a phase, with definite effects on material hardening. Even materials that were never subjected to a melting process, as in powder technology, have features from the manufacturing process that are vital to their properties.

Materials must be processed to achieve a predetermined size, shape, and properties. Many procedures can remove, add and/or modify material pieces. The principles used in altering a material can be explained in terms of the basic defect structure of the material, although this is made difficult when several processes and variables interact simultaneously to influence the defect structure. Often progress is made merely by identifying the important parameters that influence the defect structure.

PROBLEM 11-1

Any product is made from materials with properties that are often the result of compromises. Therefore, yield losses in repetitive manufacturing are caused by

(A) variability of reproducing the geometry of the product
(B) a lack of correlation between properties and their relation to product performance
(C) the variation of material properties with time
(D) the sensitivity of properties to material structure
(E) a lack of reliable data about the properties of materials

Solution

The fundamental reason for yield loss is the structure sensitivity of material properties.

Answer is (D)   0K

PROBLEM 11-2

An engineer who is concerned with the strength of a material may assume that the strength has a specific value. This assumption is generally invalid for ceramics because

(A) heat treatment strengthens the material
(B) the cohesion between particles changes with treatment
(C) it is rare that samples can be produced with closely similar properties
(D) heat treatment weakens the material
(E) composition variations cannot be controlled

Solution

Ceramics, owing to the porosity that is invariably present, are difficult to produce with reproducible properties.

Answer is (C)   0K

PROBLEM 11-3

Engineers have spent considerable effort in refining their methods so they may be more objective and accurate in designing complex structures. The fact that failures of engineering structures and products still occur implies

(A) a lack of knowledge of the appropriate material properties
(B) a failure to calculate stresses and strains accurately for a complex structure
(C) a use of the wrong materials
(D) a use of the structures and products beyond their specified ranges of stresses
(E) a lack of understanding at the atomic level that can have a large impact on the strength of large structures.

Solution

The failure to understand that atomic events can have a large influence on the strength of materials has led to a number of disasters.

Answer is (E)

PROBLEM 11-4

An electron volt is

(A) the voltage required to accelerate an electron to a velocity of 1 m/sec
(B) the voltage required to give an electron 1 J of energy
(C) the kinetic energy acquired by an electron in crossing a potential difference of 1 V
(D) the potential energy of an electron in a static electric field of 1 V/m
(E) the product of mass times the acceleration of an electron in 1 V

Solution

The electron volt is a common energy unit used in discussing the behavior of electrons in atoms; 6.24×10^{18} eV $= 1$ J or 1 eV $= 1.603 \times 10^{-19}$ J.

Answer is (C)

PROBLEM 11-5

The wavelength, in meters, of a hydrogen atom moving with a velocity of 10^3 m/sec is nearest to

(A) 1.5×10^{-9}
(B) 4.0×10^{-10}
(C) 7.3×10^{-7}
(D) 2.5×10^{-8}
(E) 5.3×10^{-13}

Solution

According to de Broglie, the wavelength of an atom is the ratio of Planck's constant to the momentum of the atom. The mass of a hydrogen atom is 1.66×10^{-27} kg. Hence

$$\lambda = \frac{h}{mv} = \frac{0.6626 \times 10^{-33}}{(1.66 \times 10^{-27}) \times (10^3)} = 3.99 \times 10^{-10} \text{ m}$$

Answer is (B)

PROBLEM 11-6

When light falls on a metal during photoemission, the number of electrons and the energies of the emitted electrons depend, respectively, on the light's

(A) intensity only
(B) wavelength and intensity
(C) wavelength and velocity
(D) intensity and wavelength
(E) momentum and frequency

Solution

Considering light as photons, Einstein was able to explain the photo-electric effect by relating the number of emitted electrons to the intensity of light and their energy to the wavelength of the light.

Answer is (D)

PROBLEM 11-7

The number of electrons in a ferric ion is 23. Therefore the outermost orbit of this ion has

(A) three electrons in the 3p shell

(B) five electrons in the 3d shell

(C) two electrons in the 4s shell and one in the 4p shell

(D) three electrons in the 4p shell

(E) seven electrons in the 4d shell

Solution

The orbital configuration of the ferric ion is

$$(1s)^2(2s)^2(2p)^6(3s)^2(3p)^6(3d)^5$$

Answer is (B)

PROBLEM 11-8

In the stimulated emission of radiation, for every photon that causes this transition there must be

(A) another photon emitted at the same frequency as the incident photon

(B) a photon of higher frequency absorbed than the one that is emitted

(C) a reversal of the direction of motion of the photon that is incident on the atom

(D) two photons absorbed of the same frequency as the emitted photon

(E) two photons emitted in opposite directions

Solution

The stimulated emission process ejects one photon of the same energy and frequency as the incident photon.

Answer is (A)

PROBLEM 11-9

The wave function of a particle is given by $\Psi = A \sin(\omega t - 2\pi x/\lambda)$. At a given instant, the probability of finding the particle is greatest where x equals

(A) 0

(B) $\lambda/4$

(C) $\pi/2$

(D) π

(E) any value

Solution

The probability of finding a particle is determined from the product of the wave function and its complex conjugate; the resulting function is then evaluated at the time and position of interest. Here the product $\Psi \times \Psi$ is the sine squared function $A^2 \sin^2 (\omega t - 2\pi x/\lambda)$, which has its maximum value when $2\pi x/\lambda = \pi/2$ or $x = \lambda/4$.

Answer is (B)

PROBLEM 11-10

A transition metal atom is characterized by an electron configuration in which there are

(A) always two electrons in the outermost s orbital

(B) six electrons filling the outermost p shell

(C) an insufficient number of electrons to fill the d shell

(D) more than enough electrons to fill the d shell, but it is not full

(E) no spaces remaining in the d shell

Solution

Regardless of the electron configuration in the s and p shells, the main feature that characterizes transition elements is an insufficient number of electrons to fill the d shell.

Answer is (C)

PROBLEM 11-11

The wave vector of an electron describes

(A) the rate of change of phase of the wave with distance

(B) the direction of motion of the electron

(C) the direction in which waves carry energy

(D) the rate of change of phase with time

(E) the magnitude and direction of the momentum of the electron

Solution

The change of phase with distance is given by the wave number. The change of phase with time is given by the angular frequency. The direction and magnitude of the momentum are contained in the wave vector.

Answer is (E)

PROBLEM 11-12

According to the Pauli exclusion principle, the number of energy levels that can be accommodated by a principal quantum number $n = 3$ in a hydrogenic atom is

(A) 1
(B) 2
(C) 9
(D) 18
(E) 32

Solution

The hydrogen atom is characterized by energy levels which are defined by the quantum numbers n, l, m_l and m_s. For $n = 3$, l can assume values of 0, 1, and 2. For each l, m_1 can take integer values between -1 and $+1$. For each set of three quantum numbers n, l and m_l, there can be two energy levels, one for each value of m_s associated with the spin of the electron. Thus there are $2n^2$ energy levels for a given n, one for each distinct set of four quantum numbers.

Answer is (D)

PROBLEM 11-13

In classical physics a particle confined to a potential well cannot go outside of this well. However, in quantum mechanics the phenomenon of tunnelling is observed. In this phenomenon it is possible to observe whether the particle is tunnelling

(A) by observing the wave function in the forbidden region
(B) by measuring the energy of the particle in the forbidden region according to classical physics

(C) only indirectly, since Heisenberg's principle prohibits any meaningful experiment from succeeding

(D) by calculating the probability of finding a particle

(E) by insisting that the tunnelling can be directly observed when an experiment with sufficient accuracy becomes available

Solution

According to the uncertainty principle, there is no possibility of constructing an experimental apparatus that will directly measure the tunnelling phenomenon.

Answer is (C)

PROBLEM 11-14

All material particles exhibit wave behavior, according to de Broglie. Therefore a beam of electrons falling on a crystalline solid should

(A) show diffraction phenomena

(B) eject x-rays from the sample

(C) bounce back, giving its energy to the atoms in the solid

(D) emit light by exciting all the outer atoms in the solid

(E) melt the solid by heating

Solution

The electron will cause all these phenomena, but the only one relevant to its wave behavior is diffraction.

Answer is (A)

PROBLEM 11-15

The thermal energy of an atom at a room temperature of $300°K$ is nearest to

(A) 220 cal

(B) 1 J

(C) 1/25 eV

(D) 1000 kcal

(E) 300 erg

Solution

The thermal energy of an atom is $3k_B T/2$, where $k_B = 1.381 \times 10^{-23}$ J/°K is the Boltzmann constant and T is the Kelvin temperature. Thus the thermal energy is

$$\frac{3}{2} k_B T = \frac{3}{2} (1.381 \times 10^{-23} \text{ J/°K})(300°\text{K}) = 6.21 \times 10^{-21} \text{ J}$$

Since 1.60×10^{-19} J $= 1$ eV, the thermal energy in electron volts is 0.0388 or approximately 1/25 eV.

Answer is (C)

PROBLEM 11-16

Iron and its alloys form many engineering materials, including steels, cast irons and stainless steels. The chemical composition on which the classification of these three material groups, respectively, is based is

(A) $C < 1.3\%$, $C > 2.1\%$, $Cr > 11\%$
(B) $C > 1.3\%$, $C < 2.1\%$, $Cr < 11\%$
(C) $C < 1.3\%$, $C < 4.3\%$, $Ni > 8\%$
(D) $C < 1.3\%$, $C > 4.3\%$, $Ni < 8\%$
(E) $C > 1.3\%$, $Si < 2.1\%$, $Cr < 11\%$

Solution

The classification is based on the binary iron-carbon phase diagram in which the maximum solubility of carbon in iron is 2.1%. Alloys containing more carbon melt with a eutectic reaction at lower temperatures. Stainless steels are characterized by more than 11% Cr.

Answer is (A)

PROBLEM 11-17

Aluminum alloys are classified as heat treatable or nonheat treatable, depending on whether they respond to precipitation hardening. The heat treatable alloys contain

(A) elements that increase in solubility with decreasing temperature
(B) elements that do not include copper, magnesium, zinc or lithium

(C) concentrations of elements that are below the solid solubility limits at room temperature and moderately high temperatures
(D) elements that decrease in solubility with decreasing temperature
(E) none of these conditions

Solution

The heat treatable alloys contain elements that decrease in solubility with a decrease in temperature, and in concentrations that exceed the equilibrium solid solubilities at room and moderately high temperatures.

Answer is (D)

PROBLEM 11-18

Oxygen, nitrogen, hydrogen, sulphur, and phosphorus are generally kept to low concentrations in steels to avoid deleterious effects. Elements such as aluminum are added

(A) to promote the reaction of carbon to produce carbon monoxide
(B) to promote grain growth through the formation of aluminum nitride particles
(C) to keep the steel liquid at low temperatures
(D) to prevent the segregation of impurities
(E) to control the oxygen level in liquid steel

Solution

The primary function of aluminum is to control the oxygen content in liquid steel.

Answer is (E)

PROBLEM 11-19

Nitrogen is a liquid below $-198°C$. When liquid nitrogen is brought to room temperature,

(A) the covalent bonds between nitrogen atoms are destroyed
(B) thermal energy destroys the dipolar attraction between nitrogen molecules

(C) the van der Waals bonding between nitrogen atoms is progressively destroyed

(D) the covalent bonding is replaced by ionic bonding

(E) the thermal energy converts the liquid to atoms which then combine in the gas phase to form molecules

Solution

The van der Waals bonding is weak and is easily destroyed between nitrogen molecules, which is the reason for the low boiling point of nitrogen.

Answer is (C)

PROBLEM 11-20

Silicon is bonded to other silicon atoms by covalent bonds. If the silicon atom is replaced by a phosphorus atom, then silicon will

(A) become a p-type semiconductor

(B) become metallic

(C) become an n-type semiconductor

(D) change its crystal structure

(E) produce interstitial donor atoms and become an n-type semiconductor

Solution

Phosphorus with five electrons, when it replaces silicon with four outer electrons, gives its excess electron to the conduction band and makes silicon an n-type semiconductor.

Answer is (C)

PROBLEM 11-21

A whisker of a material strained to over 3% still shows elastic behavior, even though stress is not proportional to strain. The failure of Hooke's law to explain this behavior arises from

(A) the nonlinear interatomic force versus distance relation at high strains

(B) the permanent stretching of bonds

(C) the complicated effects of neighboring atoms at high strains

(D) the difficulty of applying stress uniformly

(E) the enormous heat that is generated at high strains

Solution

The relation between force and interatomic distance is nonlinear beyond about 0.05% strain. Normal materials deform plastically owing to the presence of dislocations at very low strains, which is avoided in whiskers of materials.

Answer is (A)

PROBLEM 11-22

In ionic solids the ratio of the radius of the smaller cation to the radius of the larger anion plays a key role in determining the number of neighbors and their spatial arrangement. For a sixfold coordination this ratio should equal or exceed

(A) 0.15

(B) 0.22

(C) 0.41

(D) 0.73

(E) 1.00

Solution

Figure 11-3a schematically shows one (dark) cation coordinated with six

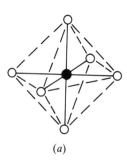

(a)

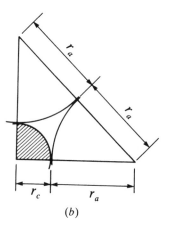

(b)

Figure 11-3

larger (light) anions. A one-quarter cross section of Fig. 11-3a is shown in Fig. 11-3b. Letting r_a and r_c represent the radius of the anion and cation, respectively, the Pythagorian theorem gives

$$2(r_a + r_c)^2 = (2r_a)^2$$

Dividing by r_a^2,

$$\left(1 + \frac{r_c}{r_a}\right)^2 = 2$$

$$\frac{r_c}{r_a} = \sqrt{2} - 1 = 0.41$$

Answer is (C)

PROBLEM 11-23

If 100 g of ethylene is converted into polyethylene, the amount of heat given off in kJ (the bond energies for the double and single bonds are C=C 680 kJ/mole and C–C 370 kJ/mole) is nearest to

(A) 1100
(B)   214
(C) −1100
(D) −214
(E) −168

Solution

The double bond in ethylene (C_2H_4) is eliminated and two single carbon-carbon bonds are formed in polyethylene. Hence the energy released is

$$([680 - 2(370)]\ kJ/mole)(100/28\ moles) = -214\ kJ$$

In this calculation one multiplies the energy per mole (gram-molecular weight) by the number of gram-molecular weights of ethylene, which is 100 g divided by the molecular weight which is $2 \times 12 + 4 \times 1 = 28$.

Answer is (D)

PROBLEM 11-24

The bonding between atoms may be displayed by sketching the force between atoms as a function of the distance of separation. At the equilibrium separation, the total force between atoms is

(A) maximum
(B) minimum
(C) negative
(D) positive
(E) zero

Solution

The attractive and repulsive forces balance at the equilibrium distance, so the total or net force is zero.

Answer is (E)

PROBLEM 11-25

The indices of direction that represent the line of intersection between planes normal to the indices $[1\,1\,1]$ and $[1\,\bar{1}\,1]$ are

(A) $[1\,1\,0]$
(B) $[1\,0\,\bar{1}]$
(C) $[\bar{1}\,\bar{1}\,\bar{1}]$
(D) $[\bar{1}\,\bar{1}\,0]$
(E) $[1\,0\,\bar{1}]$

Solution

Figure 11-4 shows the two given normals and the associated planes. From the diagram the intersection line AB is clearly seen to be $[1\,0\,\bar{1}]$.

Answer is (E)

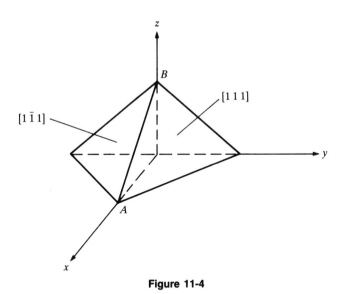

Figure 11-4

PROBLEM 11-26

Copper x-rays of wave length 1.542 Å (angstroms) fall on copper powder. Copper crystals are face centered cubic with a lattice constant of 3.62 Å. The angle, in degrees, at which diffraction peaks from the (1 1 1) plane will be observed, is nearest to

(A) 21.6
(B) 43.2
(C) 68.3
(D) 34.2
(E) 10.4

Solution

Equation (11-3) relates the x-ray wavelength to the crystal spacing and scattering angle as

$$n\lambda = 2d \sin \theta$$

Here $n = 1$ and we obtain

$$1.542 = 2 \frac{3.62}{\sqrt{1^2 + 1^2 + 1^2}} \sin \theta$$

and $\theta = 21.6°$. The diffraction peak is observed at $2\theta = 43.2°$.

Answer is (B)

PROBLEM 11-27

Normal metals and alloys are crystalline when cooled slowly from the liquid state. Useful properties such as strength and corrosion resistance may be obtained by cooling the molten material at rates in excess of 10^4 °K/sec. The property changes are primarily caused by

(A) the lack of change in composition during cooling
(B) the amorphous nature of the alloys
(C) the presence of metastable phases
(D) the extreme thinness of the product
(E) the fine grain size

Solution

The amorphous nature of the resulting structure determines most of the properties.

Answer is (B)

PROBLEM 11-28

Iron in its face centered cubic (fcc) form has less empty space than body centered cubic (bcc) iron. In spite of this fact, carbon is more soluble in fcc iron than in bcc iron. This occurs because

(A) atoms which are smaller than 0.41 times the diameter of iron atoms can be inserted in fcc iron
(B) atoms which are smaller than 0.29 times the diameter of iron atoms can be inserted in bcc iron
(C) a carbon atom which is 0.6 times the size of the iron atom can only be inserted at high temperatures
(D) the strain in the bcc lattice is excessive
(E) the bonding in the bcc form is different than in the fcc form

Solution

The tetrahedral and octahedral sites are the only sites in which an interstitial atom can go. For the geometry of these two structures, the carbon atom can fit much more easily in the fcc form than in the bcc form.

Answer is (A)

PROBLEM 11-29

Substitutional solid solutions of A and B when (a) A and B are indifferent to one another, (b) A atoms have a preference to be with their own kind, and (c) A atoms prefer B atoms, are respectively called

  (A) random, clustered, and ordered
  (B) a mixture, random, and ordered
  (C) random, ordered, and clustered
  (D) two phases, compound, and ordered
  (E) random, a phase mixture, and a compound

Solution

Solid solutions are random when there is no preference for one type of neighbor, are clustered with like-atom preference, and show a tendency towards ordering if unlike atoms attract.

Answer is (A)

PROBLEM 11-30

A dislocation in a crystal is defined by specifying its sense and Burgers vector. One of the most important mechanisms for the removal of dislocations produced by plastic deformation is the

  (A) annihilation of dislocations of opposite sense and opposite Burgers vector
  (B) annihilation of dislocations of opposite sense but same Burgers vector
  (C) formation of stacking faults by dislocations
  (D) combination of dislocations and vacancies
  (E) disappearance of dislocations at grain boundaries

Solution

Answer is (B)

PROBLEM 11-31

The Burgers vector and the sense of the dislocation line define a plane known as the slip plane in which the dislocation can glide and eventually produce plastic deformation. Cross slip is possible

(A) if screw dislocations glide
(B) when edge dislocations glide
(C) when a dislocation bound by a vacancy can glide
(D) when a dislocation meets a grain boundary
(E) when dislocations of a mixed nature glide

Solution

The screw dislocation and its sense are parallel so there is no unique glide plane. They can therefore glide on any plane.

Answer is (A)

PROBLEM 11-32

Grain boundaries exist in a single-phase material and have special properties. The importance of the boundaries arises from

(A) the low energy of the atoms at the boundary in comparison with the atoms within the grain
(B) the tendency of oversized atoms to concentrate preferentially at the boundary
(C) the slowness with which atoms move along the boundary
(D) the tendency of atoms to fit well at the boundary owing to the relaxation of the crystal structure rules
(E) the existence of a glass-like amorphous region

Solution

Atoms at the boundary have high energy and are extremely mobile. Atoms tend to segregate at the boundaries, and the fit at the boundary is far from perfect. No evidence is known for the existence of a glassy phase at the boundary.

Answer is (B)

PROBLEM 11-33

The boundary between two phases is called an interphase boundary. This boundary is called

(A) coherent if there is continuity in composition but not in structure
(B) incoherent if there are voids in the boundary
(C) semicoherent if there are line defects in the boundary
(D) coherent if there is no change in orientation and composition
(E) epitaxial if the structure is a single crystal

Solution

Coherent boundaries may change composition but not structure. The boundary can maintain coherency by slightly straining the lattice. Excessive strain results in the formation of line defects at the boundary which makes it semicoherent. When there is total incompatibility in the structure and composition, the boundary is called incoherent.

Answer is (C)

PROBLEM 11-34

The smaller the grain size in a material,

(A) the lower is the hardness of the material
(B) the higher is its elastic modulus
(C) the higher is the temperature at which the material melts
(D) the slower is its creep rate at high temperatures
(E) the higher is the yield strength of the material

Solution

Smaller grains result in an increased yield strength but can also result in excessive creep at high temperatures.

Answer is (E)

PROBLEM 11-35

The shapes of grains are affected by the manner in which the material is processed

  (A) During rolling, grains are flattened and elongated in a direction perpendicular to the rolling direction.
  (B) During casting, grains are preferentially elongated in the direction of high heat loss.
  (C) During thin film deposition, grains grow parallel to the substrate.
  (D) During heating, grains do not change shape.
  (E) During the application of a magnetic field, grains rotate to the direction of the field.

Solution

Each process is unique and produces different shapes of grains. The solidification always produces grains elongated in the direction of high heat flow.

Answer is (B)

PROBLEM 11-36

Pure iron changes its crystal structure from bcc to fcc at 914°C. In mild steel on heating

  (A) no such change in structure occurs
  (B) the same structural change at the same temperature occurs
  (C) the same structural change occurs but at different temperatures depending on carbon concentration
  (D) a martensitic structure forms
  (E) a bainitic structure forms

Solution

Polymorphism is the change in crystal structure as a function of temperature in a pure elemental solid. Alloying can influence the temperature of transition.

Answer is (C)

PROBLEM 11-37

A binary alloy of A and B, in which there is complete mutual solid solubility in the solid and liquid states, is cooled from the liquid state. The melting point of A is higher than that of B:

(A) solidification occurs at the melting point of A
(B) the solid that forms from the liquid is pure A
(C) the solid that forms has the composition of the alloy
(D) during solidification the solid-phase composition is independent of the liquid-phase composition
(E) solidification occurs over a range of temperatures

Solution

Solidification begins, but does not occur completely, at the melting point of A. The solid that forms will contain both A and B. While the liquid and solid solutions coexist, the composition of the two phases is not identical to the overall system composition; this composition can be found by a mass balance calculation. This calculation also shows the compositions of the two phases are interdependent.

Answer is (E)

PROBLEM 11-38

The speed of a structural transformation that requires a diffusion of atoms, such as solidification, has a maximum in the rate at which the transformation occurs; it is at some critical temperature T_c below the equilibrium temperature T_e. This maximum occurs because

(A) a larger value of $T_e - T_c$ indicates a larger driving force for change
(B) a larger value of $T_e - T_c$ implies a lower mobility of atoms
(C) the critical size is not reached unless the temperature is considerably below the equilibrium temperature
(D) the driving force that affects both the change and the speed atoms move produces the fastest transformation rate
(E) there is a critical velocity at which transformation occurs

Solution

The speed of transformation is a product of two factors: the energy to transport atoms across the barrier from one phase to another and the driving force that promotes that change. These two opposing effects are balanced at the critical temperature.

Answer is (D)

PROBLEM 11-39

A liquid of a pure substance is contained in a crucible; when cooled, it solidifies

(A) at the melting point
(B) a few degrees below the melting point by homogeneous nucleation
(C) considerably below the melting point by heterogeneous nucleation
(D) very close to the melting point by heterogeneous nucleation
(E) at a temperature which is tens of degrees below the melting point owing to the cooled surface of the liquid

Solution

Nucleation is normally heterogeneous with a small amount of undercooling.

Answer is (D)

PROBLEM 11-40

Martensitic transformations are characterized by

(A) a lack of change in composition
(B) the absence of any crystallographic relationship between the martensite and the parent phase
(C) a lack of change in shape
(D) their occurrence only in steels
(E) their slow dependence on time at the transformation temperature

Solution

Martensitic or diffusionless transformations are not confined to steels and occur with definite crystallographic relationships and a characteristic shape change. There is no change in composition, however.

Answer is (A)

PROBLEM 11-41

The phases present in the microstructure of white cast iron are

   (A) cementite and ferrite
   (B) graphite and ferrite
   (C) graphite, cementite and ferrite
   (D) graphite and pearlite
   (E) graphite and martensite

Solution

The large amount of cementite in white cast iron makes it brittle.

Answer is (A)

PROBLEM 11-42

Semiconductors are characterized by an energy gap

   (A) greater than 4 eV
   (B) less than 4 eV
   (C) about the same as thermal energy
   (D) that is direct
   (E) that is indirect

Solution

Semiconductors have energy gaps less than 4 eV. Both direct and indirect gaps can occur in semiconductors.

Answer is (B)

PROBLEM 11-43

The time for an atom in a metal to travel 1 mm at the melting point is approximately ($D = 10^{-9} \, cm^2/sec$)

   (A) 0.4 sec
   (B) 4 sec
   (C) 4 min

(D) 4 months

(E) 3 years and 3 months

Solution

By using the diffusion relation $x^2 = Dt$, one finds

$$(0.1 \, \text{cm})^2 = (10^{-9} \, \text{cm}^2/\text{sec}) \times t$$

which yields $t = 10^7$ sec or approximately 4 months.

Answer is (D)

PROBLEM 11-44

Engineering materials have a wide range of strengths (0.1 to 10^4 MPa). The intrinsic resistance of the material structure to plastic shear is

(A) very high in metals in which the bonds between atoms are nondirectional

(B) very low in ceramics since the covalent bonds appear strong in some directions

(C) very high in glasses owing to the disordered nature of its structure

(D) very weak in polymers since it requires only the breaking of van der Waals bonds

(E) very strong in ionic solids owing to the nondirectional nature of the bonds

Solution

The unit step in plastic shear requires the breaking of bonds, which is difficult for covalent bonds but less so for nondirectional ionic and metallic bonds. The van der Waals bonding is very weak. In glasses the unit step in shear requires the breaking of many bonds simultaneously.

Answer is (C)

PROBLEM 11-45

The ideal strength of a solid is reached when the bonds between atoms are stretched by more than 10%. Strengths on the order of 10% of Young's modulus are not achievable in

(A) glasses

(B) a single crystal of silicon

(C) metals and alloys

(D) whiskers of metals

(E) polymer fibers

Solution

Metals and alloys invariably have dislocations and deform plastically at very low strength levels. The other materials are extremely strong when they are prepared free of dislocations.

Answer is (C)

PROBLEM 11-46

The density of solids varies from about 0.1 to 30 Mg/m³. The factor that is primarily responsible for variations in the density of metals is the

(A) atomic weight

(B) efficiency with which atoms are packed

(C) size of the atom

(D) number of electrons in an atom

(E) special character of the metallic bond

Solution

The density of a solid is determined by the atomic weight, the size of the atom, and the way they are packed. But the size and packing density do not vary significantly between atoms.

Answer is (A)

PROBLEM 11-47

A system that is constrained and subjected to a temperature change undergoes thermal stresses which can cause cracking and bending. The ability of the material to withstand thermal stresses is measured by its

(A) thermal expansion coefficient α

(B) normalized yield strength σ_y/E

(C) Biot modulus, which is the ratio of a typical dimension to its thermal conductivity

(D) ratio of the normalized yield strength to the thermal expansion coefficient

(E) ratio of fracture strength to the thermal expansion coefficient

Solution

The thermal shock resistance is measured most commonly by $\sigma_y/\alpha E$.

Answer is (D)

PROBLEM 11-48

The thermal conductivity describes the rate at which heat can be transmitted through a solid. Lowest conductivities are observed in

(A) metals, in which electrons carry heat
(B) ceramics, in which vibrating atoms carry heat
(C) glasses, in which the free path of the conducting species is small
(D) polymers, in which the velocity of phonons is very low
(E) porous materials, in which heat is transferred by radiation through cell walls

Solution

Heat can be transmitted through solids in two ways, by the elastic vibrations of atoms and by the movement of free electrons. Electron movement is much more efficient in transferring heat. In nonmetals the thermal conductivity k can be described in terms of an elastic thermal wave through a so-called phonon gas by $k = c_p v \lambda /3$, in which c_p is the heat capacity of the phonons, v is the speed of elastic waves in the solid, and λ is the phonon mean free path. For porous materials, such as foams, the factors in this equation are smallest.

Answer is (E)

PROBLEM 11-49

An important material property that influences the selection of materials that are subject to vibration is the specific damping capacity. The highest damping occurs in materials where strain lags behind stress owing to

(A) the heating and cooling associated with the vibrating stresses

(B) the oscillatory movement of dislocations

(C) the jumping of atoms from one site to another

(D) a strain-induced martensitic transformation

(E) the rubbing of cracked surfaces against one another

Solution

All of these mechanisms produce damping, but the martensitic transformation (see Problem 11-40) produces the largest damping.

Answer is (D)

PROBLEM 11-50

Materials with a very high yield stress are useful only if they remain plastic and do not fail by rapidly fracturing. The resistance to the propagation of a crack is measured by the

(A) fracture toughness

(B) stress intensity factor

(C) surface energy of a crack

(D) blunting of a crack at its tip by plastic deformation

(E) size of the crack

Solution

The appropriate measure is fracture toughness; values range between 0.01 and 100 MPa m$^{1/2}$. Conventional design uses materials with a fracture toughness above 20 MPa m$^{1/2}$.

Answer is (A)

PROBLEM 11-51

Materials selected for springs should never deform plastically while in use. Therefore the materials used for springs have high values of

(A) Young's modulus E

(B) yield strength σ_y

(C) σ_y / E

(D) σ_y^2/E

(E) specific damping capacity

Solution

The best material will store the most elastic energy before it yields. Energy is the product of stress and strain; in turn, strain is stress divided by Young's modulus. Thus energy can be written as $\sigma_y \times (\sigma_y/E)$.

Answer is (D)

PROBLEM 11-52

The high intrinsic strength of ceramics implies that only a very small crack length can be created in the material. The fracture toughness of a ceramic is $2\,\text{MPa m}^{1/2}$. If the tensile strength is 200 MPa, the longest stable crack, in mm, is nearest to

(A) 2.0

(B) 6.4

(C) 60.0

(D) 0.02

(E) 0.064

Solution

The longest microcrack of length $2a$ is related to the tensile strength σ and fracture toughness K_{IC} by

$$\sigma = \frac{K_{IC}}{\sqrt{\pi a}}$$

Hence the crack length is

$$2a = \frac{2}{\pi}\left(\frac{K_{IC}}{\sigma}\right)^2 = \frac{2}{\pi}\left(\frac{2}{200}\right)^2 = 0.064\,\text{mm}$$

Answer is (E)

PROBLEM 11-53

Ceramics are generally brittle, since they almost always contain cracks. The cause of the cracks is

(A) thermal stress caused by rapid cooling

(B) due to pores from ceramics sintered to full density

(C) not corrosion since ceramics such as glass are not attacked by water

(D) not due to dust particles because ceramics are strong

(E) not the anisotropy of plastic deformation that restricts the number of planes on which slip occurs

Solution

Full-density ceramics do not have pores. Thermal stress is the principal cause of crack formation. Water, dust particles, and the anisotropy of plastic deformation are all contributing factors to the formation of cracks.

Answer is (A)

PROBLEM 11-54

The electrical resistivity of copper increases by 0.4% for a temperature increase of 1°C near 0°C. At what centigrade temperature will the resistance of a copper conductor be twice that at 0°C:

(A) -15

(B)   15

(C)   40

(D) -250

(E)   250

Solution

The resistance of a metal increases linearly with an increase in temperature. If R_0 and T_0 are the initial resistance and temperature and T_1 is the new temperature, then $[1 + 0.004(T_1 - T_0)]R_0 = 2R_0$, and $T_1 = 250°C$.

Answer is (E)

PROBLEM 11-55

An intrinsic semiconductor has an energy gap of 0.72 eV. If the conductivity at 300°K is σ, then the semiconductor has conductivity 2σ at a centigrade temperature nearest to

(A) 400
(B) 300
(C) 40
(D) 6.5
(E) 2.5

Solution

The conductivity of a semiconductor increases exponentially with absolute temperature T according to

$$\sigma = \sigma_0 \exp\left(-\frac{E_g}{2k_B T}\right)$$

in which E_g is the gap energy and $k_B = 1.381 \times 10^{-23}$ J/°K $= 8.62 \times 10^{-5}$ eV/°K is Boltzmann's constant:

$$\sigma_1 = \sigma_0 \exp\left(-\frac{E_g}{2k_B T_1}\right)$$

$$\sigma_2 = \sigma_0 \exp\left(-\frac{E_g}{2k_B T_2}\right)$$

$$\frac{\sigma_1}{\sigma_2} = \frac{1}{2} = \exp\left[\left(\frac{E_g}{2k_B}\right)\left(-\frac{1}{T_1} + \frac{1}{T_2}\right)\right]$$

$$-\ln 2 = -0.693 = \left(\frac{0.72}{2(8.62 \times 10^{-5})}\right)\left(-\frac{1}{300} + \frac{1}{T_2}\right)$$

$$T_2 = 316°K = 43°C$$

Answer is (C)

PROBLEM 11-56

A metal rod of length 10 cm and diameter 2 cm is pulled in tension by a force of 10 N. The length of the rod changes to 11 cm. The true stress, in 10^4 Pa, is closest to

(A) 3.5
(B) 24
(C) 21
(D) 100
(E) 300

Solution

During plastic deformation the material volume is conserved. If A_0 is the initial area and L_0 is the initial length, then $A_0 L_0 = AL$, where A and L are the final area and length, respectively. Therefore the true stress is

$$\sigma = \frac{F}{A} = \frac{FL}{A_0 L_0}$$

where F is the tensile force. Thus

$$\sigma = \frac{10(0.11)}{(\pi/4)(0.02)^2(0.10)} = 3.5 \times 10^4 \, \text{Pa}$$

Answer is (A)

PROBLEM 11-57

An aluminum rod of initial length 10 cm is stretched to a length of 15 cm. Later the same rod is compressed to 12 cm. The true strain is nearest

(A) 0.75
(B) 0.63
(C) 0.25
(D) 0.20
(E) 0.18

Solution

The true strain e is equal to $\ln(L/L_0)$, where L_0 is the initial length and L is the final length. True strains add; thus

$$\ln(15/10) - \ln(12/15) = \ln 1.2 = 0.182$$

Answer is (E)

PROBLEM 11-58

The defect-tolerant approach to the prediction of fatigue life requires a knowledge of

(A) the fatigue limit
(B) crack growth rate versus stress intensity

(C) the fracture toughness of the material

(D) the size of the crack

(E) the plastic behavior of the material

Solution

In this approach the rate of crack growth depends on the range of stress intensity that the crack experiences in each load cycle.

Answer is (B)

PROBLEM 11-59

Unexpected corrosion failures are more likely to occur by a localized corrosion attack than by uniform corrosion. The combination of stress and corrosion can be particularly devastating. Corrosion failure is uniform only in the case of

(A) stress corrosion cracking

(B) corrosion fatigue

(C) intergranular attack

(D) corrosion pitting

(E) galvanic corrosion

Solution

All these forms of corrosion, except galvanic corrosion, are localized attacks.

Answer is (E)

PROBLEM 11-60

The yield strength of a metal or alloy may be raised by increasing the resistance of internal dislocations to movement. The most effective way of increasing this strength is by

(A) alloying the metal

(B) producing precipitates in the alloy that are coherent with the matrix

(C) producing dispersoids in the alloy that are incoherent with the matrix

(D) increasing the number of dislocations in the material

(E) reducing the grain size of the material

Solution

The production of precipitates with coherent interfaces is very effective as a dislocation barrier.

Answer is (B)

PROBLEM 11-61

Age-hardening alloys are frequently given a double aging treatment: a low temperature age followed by a high temperature age. This procedure

  (A) eliminates the precipitate-free zone adjacent to the high angle boundary
  (B) increases the number of precipitates and minimizes the formation of a precipitate-free zone
  (C) forms incoherent precipitates in the shortest possible time
  (D) prevents the surface from showing excessive ripples
  (E) avoids the formation of Guinier-Preston zones

Solution

At low temperature the formation of a precipitate-free zone is inhibited and the number of precipitates increases.

Answer is (B)

PROBLEM 11-62

An age-hardenable alloy, which is soaked at high temperature to take the alloying element into solid solution, is rapidly cooled to obtain a solid solution which is supersaturated with both solute elements and vacancies. To obtain good mechanical properties, it is necessary to reheat the alloy

  (A) for a short period of time above the solvus
  (B) for at least an hour below the solvus
  (C) at a temperature which favors the formation of cluster regions that are coherent with the matrix
  (D) to produce the equilibrium precipitate in an hour
  (E) to allow the formation of an intermediate metastable precipitate

Solution

The decomposition of a supersaturated solid solution proceeds by the clustering together of solute atoms to form coherent precipitates. Longer aging periods produce one or more transition structures before the stable phase is formed. The strength is highest when the precipitates are coherent.

Answer is (C)

PROBLEM 11-63

Polyethylene is a thermoplastic. It softens when heated and does not have a unique melting point. This occurs because

- (A) it has a range of molecular weights and packing geometries
- (B) polyethylene is made by adding together monomer units to form long chains
- (C) the bonding between molecules is weak
- (D) each polymer molecule is heavily branched
- (E) the polymers are never pure

Solution

The range of molecular weights and geometric arrangements in polyethylene aids the molecules in moving more easily.

Answer is (A)

PROBLEM 11-64

Thermosetting polymers cannot be hot worked but decompose on heating. This limitation arises because the

- (A) resin and hardener react and harden
- (B) polymers are amorphous
- (C) polymers have extensive cross linking
- (D) secondary bonds between molecules melt
- (E) polymers are crystalline

Solution

Such polymers have a network type of molecular structure with much cross linking between molecules.

Answer is (C)

PROBLEM 11-65

Elastomers have a rubbery behavior. They are

(A) elastic because they melt when stretched
(B) flexible molecules with weak interatomic forces
(C) mixed with elastic additives
(D) composed of molecules which form and break bonds easily
(E) long chain molecules with occasional cross links

Solution

The initial stretching of elastomers occurs with the breaking of weak secondary bonds; the molecules themselves elongate as the covalent primary bonds stretch. Molecules in long chains with limited cross links behave in this way.

Answer is (E)

PROBLEM 11-66

The time-temperature transformation diagram of steel is useful in determining

(A) the rate of tempering of martensite
(B) the temperature and time at which undercooled austenite decomposes to pearlite or bainite
(C) the temperature at which bainite forms from martensite
(D) the temperature of formation of pearlite from martensite
(E) the temperature at which bainite becomes martensite

Solution

Pearlite and bainite are not formed from martensite but only from under-cooled austenite.

Answer is (B)

PROBLEM 11-67

The purpose of adding alloying elements to steel is to

(A) increase the hardness of steel
(B) improve carbon solubility
(C) provide corrosion protection for the steel
(D) adjust the temperature and compositions over which austenite is stable
(E) prevent the formation of martensite

Solution

The primary purpose of alloying is to increase hardenability. Corrosion protection is an additional benefit. All of this is possible because one can control the temperatures at which austenite is stable.

Answer is (D)

PROBLEM 11-68

The critical cooling rate is the rate above which steel will transform completely to martensite. The critical cooling rate

(A) increases with an increase in carbon content
(B) is higher for a smaller austenite grain size
(C) increases when alloying elements that stabilize austenite are added
(D) increases when alloying elements that stabilize ferrite are added
(E) depends on the section size of the steel

Solution

It is easier to form martensite if the critical cooling rate is low. A large austenite grain size, alloying elements that stabilize ferrite, and a higher carbon content all shift the time-temperature transformation curve to the right.

Answer is (B)

PROBLEM 11-69

Alloys which have densities below $4.5 \, Mg/m^3$ are usually based on aluminum, titanium, or magnesium as the major element. The strength in these alloys is derived primarily from

(A) solid solution strengthening
(B) work hardening
(C) grain boundary strengthening
(D) precipitation hardening
(E) stacking fault strengthening

Solution

Precipitation hardening is the primary mechanism in enhancing the strength of these alloys.

Answer is (D)

PROBLEM 11-70

The tempering of steel improves toughness with only a moderate decrease in hardness. During the tempering of steel

(A) the martensite structure relaxes to relieve its stresses
(B) the supersaturated solution of iron with carbon promotes the formation of numerous fine cementite precipitates
(C) the martensite structure changes to ferrite
(D) the cracks formed during martensite formation are healed
(E) dislocations are eliminated from the structure

Solution

During tempering, thermal energy provides the driving force for the supersaturated solid solution to adjust its carbon content and form numerous precipitates of cementite.

Answer is (B)

12

ENGINEERING ECONOMICS*

Engineering economics is the title given to a group of techniques for the systematic examination of alternative courses of action. Being money based, the analysis gives guidance for economically efficient decision making. This is the one topic on the EIT examination that is excluded from many undergraduate engineering programs. Thus a rather detailed discussion of engineering economics is presented here.

CASH FLOW

In examining alternative ways of solving a problem we recognize the need to resolve the various consequences (both favorable and unfavorable) of each alternative into some common unit. One convenient unit, and the one typically used in economic analysis, is money. Those factors which cannot readily be reduced to money are called intangible, or irreducible, factors. Intangible or irreducible factors are not included in any monetary analysis but are considered in conjunction with such an analysis when making the final decision on proposed courses of action.

A cash flow table shows the "money consequences" of a situation and its timing. For example, a simple problem might be to list the year-by-year consequences of purchasing and owning a used car:

---

* The text in this chapter is a modified version of *Engineering Economics Review* by Donald G. Newnan. It is reproduced here by permission of Engineering Press, Inc., the copyright owner.

	Year	Cash Flow	
Beginning of first Year	0	−$4500	Car purchased "now" for $4500 cash; the minus sign indicates a disbursement
End of Year	1	−350 ⎤	
End of Year	2	−350 ⎬	Maintenance costs are $350 per year
End of Year	3	−350 ⎦	
End of Year	4	⎰ −350 ⎱ +2000	The car is sold at the end of the 4th year for $2000; the plus sign represents a receipt of money

This same cash flow may be represented graphically (Fig. 12.1):

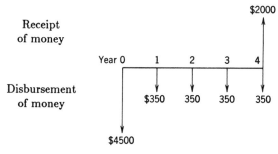

Figure 12-1 Graphical representation of a cash flow

The upward arrow represents a receipt of money, and the downward arrows represent disbursements.

Example 1

In January 1995 a firm purchases a used typewriter for $500. Repairs cost nothing in 1995 or 1996. Repairs are $85 in 1997, $130 in 1998, and $140 in 1999. The machine is sold in 1999 for $300. Set up the cash flow table.

Solution

Unless otherwise stated in problems, the customary assumption is a beginning-of-year purchase, followed by end-of-year receipts or disbursements, and an end-of-year resale or salvage value. Thus the typewriter repairs and the typewriter sale are assumed to occur at the end of the year. Letting a minus sign represent a disbursement of money and a plus sign a receipt of money, we can set up a cash flow table:

Year	Cash Flow
Beginning of 1995	−$500
End of 1995	0
End of 1996	0
End of 1997	−85
End of 1998	−130
End of 1999	+160

At the end of 1999 the cash flow table shows +160 which is the net sum of −140 and +300. If we define Year 0 as the beginning of 1995, the cash flow table becomes

Year	Cash Flow
0	−$500
1	0
2	0
3	−85
4	−130
5	+160

From this cash flow table, the definitions of Year 0 and Year 1 become clear. Year 0 is defined as the *beginning* of Year 1. Year 1 is the *end* of Year 1. Year 2 is the *end* of Year 2, and so forth.

TIME VALUE OF MONEY AND EQUIVALENCE

When the money consequences of an alternative occur in a short period of time, say less than one year, we might simply add the various sums of money and obtain the net result. But we cannot treat money in this way over longer periods of time because money has a different value now than the same amount has at some future time. As an example, consider this question: Which would you prefer, $100 today or the assurance of receiving $100 a year from now? Clearly you would prefer $100 today. If you had the money today, rather than a year from now, you could use it for the year. And if you had no use for it, you could lend it to someone who would pay interest for the privilege of using your money for the year.

We see that money at different points in time ($100 today or $100 one year hence) may superficially be equal in the sense that they both are $100, but $100 a year hence is *not* an acceptable substitute for $100 today. When we have acceptable substitutes, we say they are *equivalent* to each other. Thus at 8% annual interest, $108 a year hence is equivalent to $100 today.

Example 2

At a 10% per year interest rate, $500 now is equivalent to how much three years hence?

Solution

Five hundred dollars now will increase by 10% in each of the three years.

$$
\begin{aligned}
\text{Now } &= & \$500 \\
\text{End of 1st year } &= 500 + 0.10(500) = & 550 \\
\text{End of 2nd year } &= 550 + 0.10(550) = & 605 \\
\text{End of 3rd year } &= 605 + 0.10(605) = & 665.50
\end{aligned}
$$

Thus $500 now is *equivalent to* $665.50 at the end of three years.

Equivalence is an essential element in engineering economic analysis. Suppose we wish to select the better of two alternatives. First we must compute their cash flows. An example would be

Year	A	B
0	−$2000	−$2800
1	+800	+1100
2	+800	+1100
3	+800	+1100

The larger investment in alternative B results in larger subsequent benefits, but we have no direct way of knowing if alternative B is better than alternative A. Therefore we do not know which alternative should be selected. To make a decision we must resolve the alternatives into equivalent sums so they may be compared accurately and a decision can be made.

COMPOUND INTEREST FORMULAS

To aid equivalence computations, a series of compound interest factors are derived here, and their use is illustrated by subsequent examples. The following symbols are used:

i = interest rate per interest period; in equations the interest rate is stated as a decimal (that is, 8% interest is 0.08)

n = number of interest periods

P = a present sum of money

F = a future sum of money; the future sum F is an amount, n interest periods from the present, that is equivalent to P with interest rate i

A = an end-of-period cash receipt or disbursement in a uniform series continuing for n periods, the entire series being equivalent to P or F at interest rate i

G = uniform period-by-period increase in cash flows; the arithmetic gradient

The following *functional notation* is in common use in engineering economics calculations:

	To Find	Given	Functional Notation
Single Payment			
Compound amount factor	F	P	$(F/P, i, n)$
Present worth factor	P	F	$(P/F, i, n)$
Uniform Payment Series			
Sinking fund factor	A	F	$(A/F, i, n)$
Capital recovery factor	A	P	$(A/P, i, n)$
Compound amount factor	F	A	$(F/A, i, n)$
Present worth factor	P	A	$(P/A, i, n)$
Arithmetic Gradient			
Gradient uniform series	A	G	$(A/G, i, n)$
Gradient present worth	P	G	$(P/G, i, n)$

From the table one can see that functional notation is a shorthand which follows the scheme (To Find/Given, i, n). Thus, if we wished to find the future sum F, given a uniform series of receipts A, the proper compound interest factor to use would be $(F/A, i, n)$.

Single Payment Formulas

Suppose a present sum of money P is invested for one year at interest rate i. At the end of the year we receive back our initial investment P together with interest equal to Pi for a total amount $P + Pi$. Factoring P, the sum at the end of one year is $P(1 + i)$. If we let our investment remain for subsequent years, the progression is as follows:

	Amount at Beginning of Period	+	Interest for the Period	=	Amount at End of the Period
1st year	P		$+ \ Pi$		$= P(1 + i)$
2nd year	$P(1 + i)$		$+ \ Pi(1 + i)$		$= P(1 + i)^2$
nth year	$P(1 + i)^{n-1}$		$+ \ Pi(1 + i)^{n-1}$		$= P(1 + i)^n$

The present sum P increases in n periods to $P(1 + i)^n$. This gives a relationship between a present sum P and its equivalent future sum F.

$$\text{Future sum} = (\text{present sum})(1 + i)^n$$
$$F = P(1 + i)^n$$

This is the *single payment compound amount factor*. In functional notation it is

$$F = P(F/P, i, n)$$

Solving for P, the equation is rewritten as

$$P = F(1 + i)^{-n}$$

This is the *single payment present worth factor*. It is written

$$P = F(P/F, i, n)$$

Example 3

At a 10% per year interest rate, $500 now is equivalent to how much three years hence?

Solution

This problem was solved in Example 2. Now it will be solved with a single payment formula. Here $P = \$500$, $n = 3$ years, $i = 10\%$, and F is unknown.

$$F = P(1 + i)^n = 500(1 + 0.10)^3 = \$665.50$$

This problem may also be solved by using the compound interest tables in Appendix C.

$$F = P(F/P, i, n) = 500(F/P, 10\%, 3)$$

From the 10% compound interest table, read $(F/P, 10\%, 3) = 1.331$.

$$F = 500(F/P, 10\%, 3) = 500(1.331) = \$665.50$$

Example 4

To raise money for a new business, a man asks you to loan him some money. He offers to pay you $3000 at the end of four years. How much should you give him now if you want 12% interest per year on your money?

Solution

Here P is unknown, $n = 4$ years, $i = 12\%$, and $F = \$3000$.

$$P = F(1 + i)^{-n} = 3000(1 + 0.12)^{-4} = \$1906.55$$

Alternate computation using compound interest tables:

$$P = F(P/F, i, n) = 3000(P/F, 12\%, 4)$$
$$= 3000(0.6355) = \$1906.50$$

The solution using the compound interest table is slightly different from the exact solution using a hand calculator. In engineering economics the compound interest tables are always considered to be sufficiently accurate.

Uniform Payment Series Formulas

Consider the situation shown in Fig. 12-2.

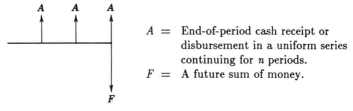

A = End-of-period cash receipt or disbursement in a uniform series continuing for n periods.

F = A future sum of money.

Figure 12-2

Using the single payment compound amount factor, we can write an equation for F in terms of A

$$F = A + A(1 + i) + A(1 + i)^2 \tag{1}$$

In our situation, with $n = 3$, Eq. (1) may be written in a more general form

$$F = A + A(1 + i) + A(1 + i)^{n-1} \tag{2}$$

Multiply Eq. (2) by $(1 + i)$

$$(1 + i)F = A(1 + i) + A(1 + i)^{n-1} + A(1 + i)^n \tag{3}$$
$$F = A + A(1 + i) + A(1 + i)^{n-1} \tag{2}$$

$(3) - (2)$: $\qquad iF = -A + A(1 + i)^n$

$$F = A\left[\frac{(1 + i)^n - 1}{i}\right] \qquad \text{Uniform series} \\ \text{Compound amount factor}$$

Solving this equation for A,

$$A = F\left[\frac{i}{(1+i)^n - 1}\right] \quad \begin{array}{l} \text{Uniform series} \\ \text{Sinking fund factor} \end{array}$$

Since $F = P(1+i)^n$, we can substitute this expression for F in the equation and obtain

$$A = P\left[\frac{i(1+i)^n}{(1+i)^n - 1}\right] \quad \begin{array}{l} \text{Uniform series} \\ \text{Capital recovery factor} \end{array}$$

Solving the equation for P,

$$P = A\left[\frac{(1+i)^n - 1}{i(1+i)^n}\right] \quad \begin{array}{l} \text{Uniform series} \\ \text{Present worth factor} \end{array}$$

In functional notation the uniform series factors are

Compound amount	$(F/A, i, n)$
Sinking fund	$(A/F, i, n)$
Capital recovery	$(A/P, i, n)$
Present worth	$(P/A, i, n)$

Example 5

If $100 is deposited at the end of each year in a savings account that pays 6% interest per year, how much will be in the account at the end of five years?

Solution

$$A = \$100 \qquad F = ? \qquad n = 5 \text{ years} \qquad i = 6\%$$
$$F = A(F/A, i, n) = 100(F/A, 6\%, 5) = 100(5.637) = \$563.70$$

Example 6

A woman wishes to make a uniform deposit every three months to her savings account so she will have $10,000 in the account at the end of 10 years. If the account earns 6% annual interest, compounded quarterly, how much should she deposit each three months?

Solution

$$F = \$10,000 \qquad A = ? \qquad n = 40 \text{ quarterly deposits}$$
$$i = 1.5\% \text{ per quarter year}$$
$$A = F(A/F, i, n) = 10,000(A/F, 1.5\%, 40) = 10,000(0.0184) = \$184$$

Example 7

An individual is considering the purchase of a used automobile. The total price is $6200 with $1240 as a downpayment and the balance paid in 48 equal monthly payments with interest at 1% per month. The payments are due at the end of each month. Compute the monthly payment.

Solution

The amount to be repaid by the 48 monthly payments is the cost of the automobile *minus* the $1240 downpayment.

$$P = \$4960 \qquad A = ? \qquad n = 48 \text{ monthly payments} \qquad i = 1\% \text{ per month}$$

$$A = P(A/P, i, n) = 4960(A/P, 1\%, 48) = 4960(0.0263) = \$130.45$$

Example 8

A couple sold their home. In addition to cash, they took a mortgage on the house. The mortgage will be paid off by monthly payments of $232.50 for 10 years. The couple decide to sell the mortgage to a local bank. The bank will buy the mortgage, but it requires a 1% per month interest rate on their investment. How much will the bank pay for the mortgage?

Solution

$$A = \$232.50 \qquad n = 120 \text{ months} \qquad i = 1\% \text{ per month} \qquad P = ?$$

$$P = A(P/A, i, n) = 232.50(P/A, 1\%, 120)$$

$$= 232.50(69.701) = \$16,205.48$$

Arithmetic Gradient Formulas

At times one will encounter a situation where the cash flow series is not a constant amount A. Instead it is an increasing cash flow series in the form shown in Fig. 12-3.

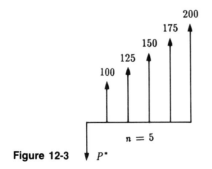

Figure 12-3

This cash flow may be resolved into two components (Fig. 12-4):

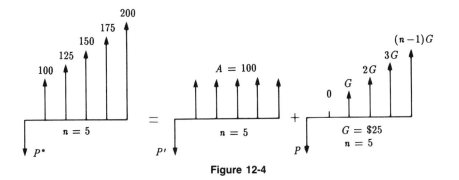

Figure 12-4

One can compute the value of P^* as equal to P' plus P. We already have an equation for P':

$$P' = A(P/A, i, n)$$

The value for P in the right-hand diagram is

$$P = G\left(\frac{(1+i)^n - in - 1}{i^2(1+i)^n}\right)$$

This is the arithmetic gradient present worth formula. In functional notation the relationship is $P = G(P/G, i, n)$.

Example 9

The maintenance on a machine is expected to be $155 at the end of the first year and increasing $35 each year for the following seven years. What present sum of money must be set aside now to pay the maintenance for the eight-year period? Assume 6% interest.

Solution

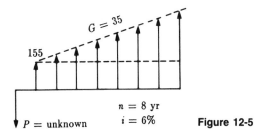

Figure 12-5

$$P = 155(P/A, 6\%, 8) + 35(P/G, 6\%, 8)$$
$$= 155(6.210) + 35(19.841) = \$1657$$

If an equivalent uniform series A is desired instead of the present sum P, the problem becomes

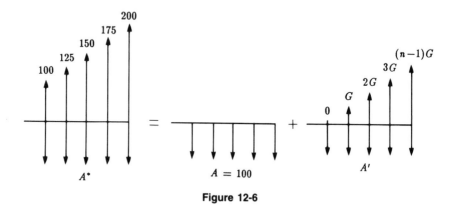

Figure 12-6

The relationship between A' and G in the right-hand diagram is

$$A' = G\left(\frac{(1+i)^n - in - 1}{i(1+i)^n - i}\right)$$

In functional notation, the arithmetic gradient (to) uniform series factor is

$$A = G(A/G, i, n)$$

Note carefully the diagrams for the two arithmetic gradient series factors. In both cases the first term in the arithmetic gradient series is zero and the last term is $(n-1)G$. But n is used in both the equations and the functional notation. The derivations (not shown here) are done on this basis, as are the arithmetic gradient series compound interest tables.

Example 10

For the situation in Example 9 we wish to know the uniform annual maintenance cost. Compute an equivalent A for the maintenance costs.

Solution

The equivalent uniform annual maintenance cost is

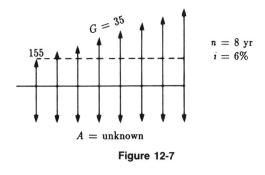

Figure 12-7

$$A = 155 + 35(A/G, 6\%, 8)$$
$$= 155 + 35(3.195) = \$266.83$$

NOMINAL AND EFFECTIVE INTEREST

Nominal interest is the annual interest rate without the effect of compounding. *Effective interest* is the annual interest rate with the effect of compounding during the year.

Frequently an interest rate is described as an annual rate, even though the interest period may be something other than one year. A bank may pay 1.5% interest on the amount in a savings account every three months. The *nominal* interest rate in this situation is 6% ($4 \times 1.5\% = 6\%$). But if you deposited $1000 in such an account, would you have 106%(1000) = $1060 in the account at the end of one year? The answer is no; you would have more. The amount in the account would increase as follows:

		Amount in Account
At beginning of year		= $1000.00
End of 3 months:	1000.00 + 1.5%(1000.00) =	1015.00
End of 6 months:	1015.00 + 1.5%(1015.00) =	1030.23
End of 9 months:	1030.23 + 1.5%(1030.23) =	1045.68
End of 1 year:	1045.68 + 1.5%(1045.68) =	1061.37

The actual interest rate on the $1000 would be the interest, $61.37, divided by the original $1000, or 6.137%. This is the *effective interest rate*.

Effective interest rate per year $= (1 + i)^m - 1$ in which $i =$ interest rate per interest period, and $m =$ number of compoundings per year.

For continuous compounding, effective interest rate per year $= e^r - 1$ in which r is the nominal interest rate per year.

Example 11

A bank charges 1.5% per month on the unpaid balance for purchases made on its credit card. What nominal interest rate is it charging? What effective interest rate?

Solution

The nominal interest rate is simply the annual interest ignoring compounding, or $12(1.5\%) = 18\%$.

Effective interest rate $= (1 + 0.015)^{12} - 1 = 0.1956 = 19.56\%$.

Example 12

Given a nominal interest rate of 18% per year, what is the effective interest rate per year, based on continuous compounding?

Solution

$$\text{Effective interest rate per year} = e^r - 1$$
$$= e^{0.18} - 1 = 1.1972 - 1$$
$$= 0.1972 = 19.72\%$$

SOLVING ECONOMIC ANALYSIS PROBLEMS

The techniques presented so far illustrate the conversion of single amounts of money, and a uniform or gradient series of money, into some equivalent sum at another point in time. These compound interest computations are an essential part of engineering economics problems. A typical problem is the existence of several alternatives and a desire to select the best alternative. The customary method of solution is to express each alternative in some common form and then choose the best alternative, taking both the monetary and intangible factors into account.

Criteria

Economic analysis problems inevitably fall into one of three categories.

1. *Fixed input*: The amount of money or other input resources is fixed.
   Example: A project engineer has a budget of $450,000 to overhaul a building materials plant.
2. *Fixed output*: There is a fixed task or other output to be accomplished.

Example: A contractor has been awarded a fixed price contract to build a building.

3. *Neither input nor output fixed*: This is the general situation where neither the amount of money or other inputs nor the amount of benefits or other outputs is fixed.

Example: A consulting engineering firm has more work available than it can handle. It is considering paying the staff for working evenings to increase the amount of design work it can perform.

PRESENT WORTH

Present worth analysis converts all the monetary consequences of an alternative into an equivalent present sum. For the three categories the criteria become

Category	Present Worth Criterion
Fixed input	Maximize the present worth of benefits or other outputs
Fixed output	Minimize the present worth of costs or other inputs
Neither input nor output fixed	Maximize [present worth of benefits minus present worth of costs] or Maximize net present worth

Present worth analysis is most frequently used to determine the present value of future money receipts and disbursements. One might want to know, for example, the present worth of an income producing property like an oil well. The computation should provide an estimate of the price at which the property could be bought or sold.

An important restriction in the use of present worth calculations is the need to use a common analysis period when comparing alternatives. It would be incorrect, for example, to compare the present worth (PW) of cost of Pump *A*, expected to last 6 years, with the PW of cost of Pump *B*, expected to last 12 years.

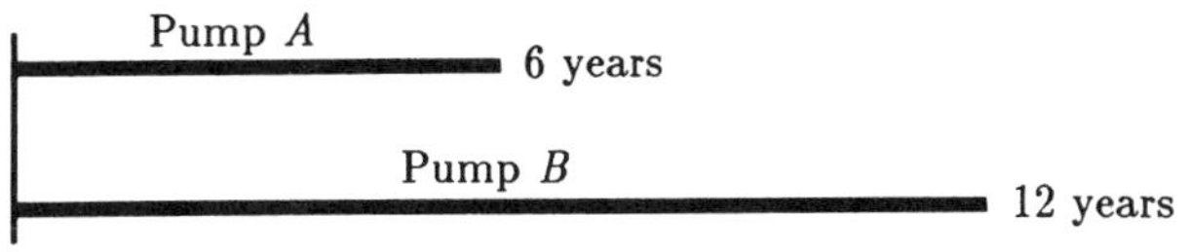

Figure 12-8 Improper present worth comparison

In situations like this, the solution is either to use some other analysis technique (generally the annual cost method is suitable) or to restructure the problem so there is a common analysis period. In the example in Fig. 12-8, a customary assumption would be that a pump is needed for 12 years and that Pump A will be replaced by an identical Pump A at the end of 6 years. This gives a 12-year common analysis period (Fig. 12-9). This approach is easy to

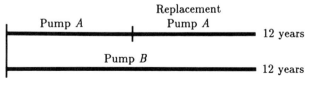

Figure 12-9 Correct present worth comparison

use when the different lives of the alternatives have a practical least common multiple life. When this is not true (for example, life of A equals 7 years and the life of B equals 11 years), some assumptions must be made to select a suitable common analysis period, or the present worth method should not be used.

Example 13

Machine X has an initial cost of $10,000, annual maintenance of $500 per year, and no salvage value at the end of its four-year useful life. Machine Y costs $20,000. In the first year there is no maintenance cost. In the second year maintenance is $100, and it increases $100 per year in subsequent years. The machine has an anticipated $5000 salvage value at the end of its 12-year useful life.

  If interest is 8%, which machine should be selected?

Solution

The analysis period is not stated in the problem. Therefore we select the least common multiple of the lives, or 12 years, as the analysis period.

  Present worth of cost of 12 years of machine X

$$= 10,000 + 10,000(P/F, 8\%, 4) + 10,000(P/F, 8\%, 8)$$
$$\quad + 500(P/A, 8\%, 12)$$
$$= 10,000 + 10,000(0.7350) + 10,000(0.5403) + 500(7.536)$$
$$= \$26,521$$

Present worth of cost of 12 years of machine Y

$$= 20{,}000 + 100(P/G, 8\%, 12) - 5000(P/F, 8\%, 12)$$

$$= 20{,}000 + 100(34.634) - 5000(0.3971)$$

$$= \$21{,}478$$

Choose machine Y with its smaller PW of cost.

Example 14

Two alternatives have the following cash flows:

Year	A	B
0	−$2000	−$2800
1	+800	+1100
2	+800	+1100
3	+800	+1100

At a 5% interest rate, which alternative should be selected?

Solution

Since neither input nor output is fixed, the net present worth of each of the alternatives will be computed.

$$\text{Net present worth (NPW)} = \text{PW of benefits} - \text{PW of cost}$$

$$NPW_A = 800(P/A, 5\%, 3) - 2000$$

$$= 800(2.723) - 2000$$

$$= +178.40$$

$$NPW_B = 1100(P/A, 5\%, 3) - 2800$$

$$= 1100(2.723) - 2800$$

$$= +195.30$$

To maximize NPW, choose B.

CAPITALIZED COST

In the special situation where the analysis period is infinite ($n = \infty$), a present worth analysis is called *capitalized cost*. Some public projects use such an analysis period. Other examples would be permanent endowments and cemetery perpetual care.

When $n = \infty$, a present sum P will accrue interest of Pi for every future interest period. For the principal sum P to continue undiminished (an essential requirement for n equal to infinity), the end-of-period sum A that can be disbursed is Pi.

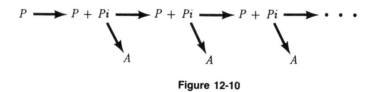

Figure 12-10

Consequently the fundamental relationship is $A = Pi$. This relation is used in some way in every problem with an infinite analysis period.

Example 15

In his will a man wishes to establish a perpetual trust to contribute to the maintenance of a small local park. If the contribution will be $7500 per year and the trust account can earn 5% interest, how much money must be set aside in the trust?

Solution

When $n = \infty$, $P = A/i$. The capitalized cost is

$$P = \frac{A}{i} = \frac{\$7500}{0.05} = \$150,000$$

Example 16

What initial investment will be required now to provide funding for a special project? The project begins at the end of year 12, and it is repeated indefinitely every 12 years. The special project cost is $2,200,000 and the interest is 10%.

Solution

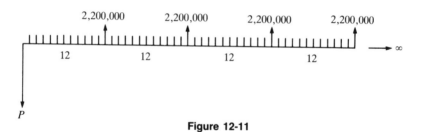

Figure 12-11

The $2,200,000 at the end of each 12 years must be converted to an equivalent annual disbursement A:

$$A = F(A/F, 10\%, 12) = 2,200,000(0.0497) = \$109,340$$

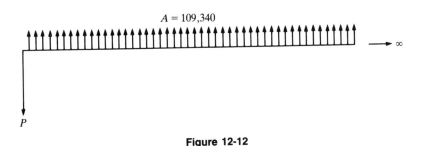

A = 109,340

∞

P

Figure 12-12

Now the initial investment for this continuing series of disbursements, that is, the capitalized cost, may be computed as

$$\text{Initial investment } P = \frac{A}{i} = \frac{109,340}{0.10} = \$1,093,400$$

FUTURE WORTH

In present worth analysis the comparison is made in terms of the equivalent *present* costs and benefits. But the analysis need not be made at the present time; it could be made at any point in time: past, present, or future. Although the numerical calculations may look different, the decision is unaffected by the choice of a reference time. When an analysis is based on some future point in time, it is called *future worth* analysis. In this situation the three criteria are to maximize the future worth of benefits, minimize the future worth of costs, or to maximize the net future worth.

Example 17

Two alternatives have the following cash flows:

Year	A	B
0	−$2000	−$2800
1	+800	+1100
2	+800	+1100
3	+800	+1100

At a 5% interest rate, which alternative should be selected?

Solution

In Example 14 this problem was solved by present worth analysis at Year 0. Here it will be solved by future worth analysis at the end of Year 3.

$$\text{Net future worth (NFW)} = \text{FW of benefits} - \text{FW of cost}$$

$$\text{NFW}_A = 800(F/A, 5\%, 3) - 2000(F/P, 5\%, 3)$$

$$= 800(3.152) - 2000(1.158)$$

$$= +205.60$$

$$\text{NFW}_B = 1100(F/A, 5\%, 3) - 2800(F/P, 5\%, 3)$$

$$= 1100(3.152) - 2800(1.158)$$

$$= +224.80$$

To maximize NFW, choose B.

ANNUAL COST

The annual cost method is more accurately described as the method of equivalent uniform annual cost (EUAC) or, where the computation is of benefits, the method of equivalent uniform annual benefits (EUAB).

For each of the three problem categories the annual cost criteria are to maximize the equivalent uniform annual benefits (EUAB), minimize the equivalent uniform annual cost (EUAC), or to maximize (EUAB − EUAC). Annual cost calculations can be made for a large range of problems, since a common analysis period may not be required. It is important, however, to understand the circumstances that justify comparing alternatives with different service lives.

Frequently an analysis is to provide for a more or less continuing requirement. One might need to pump water from a well, for example, as a continuing requirement. Whether the pump has a useful service life of 6 years or 12 years, one would select the one whose annual cost is a minimum. And this would still be the case if useful lives of the pumps were the more troublesome 7 and 11 years, respectively. Thus, if one can assume a continuing need for an item, an annual cost comparison among alternatives of differing service lives is valid.

The underlying assumption in these situations is that the shorter-lived alternative will be replaced at the end of its useful life with an identical item with identical costs. Therefore the EUAC of the initial alternative is equal to the EUAC for the continuing series of replacements. Engineer-In-Training examination problems are often readily solved by the annual cost method. And the underlying "continuing requirement" is often present, so that an annual cost comparison of unequal-lived alternatives is an appropriate method of analysis.

On the other hand, if there is a specific requirement in some situation to pump water for 10 years, then each pump must be evaluated to see what costs will be incurred during the analysis period and what salvage value, if any, may be recovered at the end of the analysis period. The annual cost comparison must consider the actual circumstances of the situation.

Example 18

Consider the following alternatives:

	A	B
First cost	$5000	$10,000
Annual maintenance	500	200
End-of-useful-life salvage value	600	1000
Useful life	5 years	15 years

Based on an 8% interest rate, which alternative should be selected?

Solution

Assuming both alternatives perform the same task and there is a continuing requirement, the goal is to minimize EUAC.

$$EUAC_A = 5000(A/P, 8\%, 5) + 500 - 600(A/F, 8\%, 5)$$
$$= 5000(0.2505) + 500 - 600(0.1705) = \$1650$$
$$EUAC_B = 10,000(A/P, 8\%, 15) + 200 - 1000(A/F, 8\%, 15)$$
$$= 10,000(0.1168) + 200 - 1000(0.0368) = \$1331$$

To minimize EUAC, select B.

RATE OF RETURN

Present worth and annual cost calculations start with the assumption of an interest rate or a minimum attractive rate of return. Sometimes we want to know the prospective rate of return on an investment rather than simply to determine whether it meets some fixed standard of attractiveness. In the rate of return method the typical situation is where there is a cash flow representing both costs and benefits. The rate of return may be defined as the interest rate where

$$\text{PW of cost} = \text{PW of benefits,}$$

$$\text{EUAC} = \text{EUAB, or}$$

$$\text{PW of cost} - \text{PW of benefits} = 0$$

These calculations frequently require a trial and error solution.

Example 19

A man invested \$10,000 in a project. At the end of three years he received \$9000, and at the end of six years he received \$6000. What rate of return did he receive on the project?

Solution

$$\text{PW of cost} = \text{PW of benefits}$$

$$10,000 = 9000(P/F, i, 3) + 6000(P/F, i, 6)$$

Try $i = 12\%$.

$$10,000 = 9000(0.7118) + 6000(0.5066)$$

$$= 6406 + 3040 = 9446$$

The PW of benefits is lower than the PW of cost, indicating that the interest rate i is too high.

Try $i = 10\%$.

$$10,000 = 9000(0.7513) + 6000(0.5645)$$

$$= 6762 + 3387 = 10,149$$

Using linear interpolation between 10% and 12%,

$$\text{Rate of return } i = 10\% + (2\%)\left(\frac{10,149 - 10,000}{10,149 - 9446}\right) = 10.42\%$$

RATE OF RETURN ANALYSIS

In this analysis one computes the incremental rate of return on the cash flow representing the difference between two alternatives. Since we want to look at increments of *investment*, the cash flow for the difference between the alternatives is computed by taking the higher initial-cost alternative *minus* the lower initial-cost alternative. If the incremental rate of return is greater than or equal to the predetermined minimum attractive rate of return (MARR), then we will choose the higher-cost alternative; otherwise we will choose the lower-cost alternative.

Example 20

Two alternatives have the following cash flows:

Year	A	B
0	−$2000	−$2800
1	+800	+1100
2	+800	+1100
3	+800	+1100

If 5% is considered the minimum attractive rate of return (MARR), which alternative should be selected?

Solution

These two alternatives were previously examined in Examples 14 and 17 by present worth and future worth analysis. This time the alternatives will be examined with rate of return analysis. This problem statement specifies a 5% minimum attractive rate of return (MARR), while Examples 14 and 17 referred to a 5% interest rate. These are really two different ways of saying the same thing: the minimum acceptable time value of money is 5%.

First we tabulate the cash flow that represents the increment of investment between the alternatives. This is done by taking the higher initial-cost alternative minus the lower initial-cost alternative:

Year	A	B	B − A
0	−$2000	−$2800	−$800
1	+800	+1100	+300
2	+800	+1100	+300
3	+800	+1100	+300

Then we compute the rate of return on the increment of investment represented by the difference between the alternatives:

$$\text{PW of cost} = \text{PW of benefits}$$
$$800 = 300(P/A, i, 3)$$
$$(P/A, i, 3) = \frac{800}{300} = 2.67$$
$$i \approx 6.1\%$$

Since the incremental rate of return exceeds the 5% MARR, the increment of investment is desirable. We would select the higher-cost alternative, *B*.

Before leaving this example problem, we should note something that relates to the rates of return of the two alternatives.

Rate of Return	
A	9.7%
B	8.7%

The correct answer to this problem has been shown to be *B*, and this is true even though *A* has a higher rate of return. The higher-cost alternative may be viewed as the lower-cost alternative, plus the increment of investment between them. Looked at this way, the higher-cost alternative *B* is equal to the desirable lower-cost alternative *A* plus the desirable differences between the alternatives.

The important conclusion is that computing the rate of return for each alternative does *not* provide the basis for choosing between alternatives. Instead, incremental analysis is required.

BREAKEVEN ANALYSIS

In business, "breakeven" is the point where income just covers the associated costs. In engineering economics, the breakeven point is more precisely defined as the point where the two alternatives are equivalent.

Example 21

A city is considering a new $50,000 snowplow. The new machine will operate at a saving of $600 per day in comparison with the present equipment. Assume the minimum attractive rate of return (interest rate) is 12%, and the machine's life is 10 years with zero resale value at that time. How many days per year must the machine be used to make the investment economical?

Solution

This breakeven problem may be readily solved by annual cost computations. We will set the equivalent uniform annual cost of the snowplow equal to its annual benefit and solve for the required annual utilization.

Let X = breakeven point = days of operation per year.

$$\text{EUAC} = \text{EUAB}$$
$$50,000(A/P, 12\%, 10) = 600X$$
$$X = \frac{50,000(0.1770)}{600} = 14.7 \text{ days per year}$$

DEPRECIATION AND VALUATION

Depreciation of capital equipment is an important component of many after-tax economic analyses. For this reason one must understand the fundamentals of depreciation accounting.

Depreciation is defined, in its accounting sense, as the systematic allocation of the cost of a capital asset over its useful life. *Book value* is the original cost of an asset, minus the accumulated depreciation of the asset. In computing a schedule of depreciation charges, three items are considered:

1. cost of the property P
2. useful life in years n
3. salvage value of the property at the end of its useful life S

For *straight-line depreciation*,

$$\text{Depreciation charge in any year} = \frac{P - S}{n}$$

For *sum-of-years digits depreciation*,

$$\text{Depreciation charge in any year} = \frac{\begin{array}{c}\text{Remaining useful life}\\ \text{at beginning of year}\end{array}}{\begin{array}{c}\text{Sum-of-years digits}\\ \text{for total useful life}\end{array}} (P - S)$$

$$\text{where sum-of-years digits} = 1 + 2 + 3 + \cdots + n = \frac{n}{2}(n + 1).$$

For *double declining balance depreciation*,

$$\text{Depreciation charge in any year} = \frac{2}{n}(P - \text{Depreciation charges to date})$$

Example 22

A piece of machinery costs $5000 and has an anticipated $1000 salvage value at the end of its five-year depreciable life. Compute the depreciation schedule for the machinery by

(a) straight-line depreciation,
(b) sum-of-years-digits depreciation,
(c) double declining balance depreciation.

Solution

Straight-line depreciation:

$$\frac{P-S}{n} = \frac{5000-1000}{5} = \$800 \text{ per year}$$

Sum-of-years-digits depreciation:

$$\text{Sum-of-years digits} = \frac{n}{2}(n+1) = \frac{5}{2}(6) = 15$$

$$\text{1st year depreciation} = \frac{5}{15}(5000-1000) = \$1333$$

$$\text{2nd year depreciation} = \frac{4}{15}(5000-1000) = 1067$$

$$\text{3rd year depreciation} = \frac{3}{15}(5000-1000) = 800$$

$$\text{4th year depreciation} = \frac{2}{15}(5000-1000) = 533$$

$$\text{5th year depreciation} = \frac{1}{15}(5000-1000) = \underline{267}$$

$$\$4000$$

Double declining balance depreciation:

$$\text{1st year depreciation} = \frac{2}{5}(5000-0) = \$2000$$

$$\text{2nd year depreciation} = \frac{2}{5}(5000-2000) = 1200$$

$$\text{3rd year depreciation} = \frac{2}{5}(5000-3200) = 720$$

$$\text{4th year depreciation} = \frac{2}{5}(5000-3920) = 432$$

$$\text{5th year depreciation} = \frac{2}{5}(5000-4352) = \underline{259}$$

$$\$4611$$

Since the problem specifies a $1000 salvage value, the total depreciation may not exceed $4000. The double declining balance depreciation must be stopped in the 4th year when it totals $4000.

The depreciation schedules computed by the three methods are as follows:

Year	Straight Line	Sum-of-Years Digits	Double Declining Balance
1	$800	$1333	$2000
2	800	1067	1200
3	800	800	720
4	800	533	80
5	800	267	0
	$4000	$4000	$4000

Example 23

For sum-of-years-digits depreciation, what is the book value of the machinery in Example 22 at the end of three years?

Solution

The valuation of the machinery, that is, its book value, is the original cost minus its accumulated depreciation. In this case

$$\text{Book value} = 5000 - (1333 + 1067 + 800) = \$1800$$

ANNUAL COST COMPUTED BY STRAIGHT-LINE DEPRECIATION PLUS AVERAGE INTEREST

At times an average annual cost is computed rather than an equivalent uniform annual cost (EUAC). The principal advantage of average annual cost is that it may be computed without compound interest tables.

In this approximate method the decline in value of the asset, over its useful life, is distributed equally to each year of asset life. This amounts to straight-line depreciation, $(P - S)/n$. In addition, there must be a charge to reflect the investment in the unrecovered cost of the asset. This is an interest charge on the money still invested in the asset. In the first year the interest is $(P - S)i + Si$, and in the last year of the useful life it is $(P - S)i/n + Si$. The average interest charge is half the sum of the first and last year's interest charges, or

$$\frac{(P - S)i + Si + \left(\dfrac{P - S}{n}\right)i + Si}{2} = (P - S)\left(\frac{i}{2}\right)\left(\frac{n + 1}{n}\right) + Si$$

Thus the straight-line depreciation plus average interest method is based on two equations:

$$\text{Straight-line depreciation} = \frac{P - S}{n}$$

$$\text{Average interest} = (P - S)\left(\frac{i}{2}\right)\left(\frac{n+1}{n}\right) + Si$$

This method should *not* be used in the engineering fundamentals examination unless the problem asks for a computation of annual cost by straight-line depreciation plus average interest.

ADDITIONAL ENGINEERING ECONOMICS INFORMATION

For an expanded discussion of engineering economics, a textbook written by one of the authors should be helpful: Newnan, *Engineering Economic Analysis* (Engineering Press, Inc., P.O. Box 1, San Jose, CA 95103-0001).

PROBLEM 12-1

How long will it take for $10,000 invested in a bank savings account to double in value? If the bank pays 6% interest, compounded semi-annually, the time, in years, for the $10,000 to double is nearest to

(A) 8
(B) 12
(C) 14
(D) 16
(E) 24

$$(1.06)^n = \cancel{\$} 2$$

Solution

$$F = P(F/P, 3\%, n) \qquad 2 = 1(F/P, 3\%, n) \qquad (F/P, 3\%, n) = 2$$

From the 3% compound interest tables we see that n is between 23 and 24 for $(F/P, 3\%, n)$ equal to 2. Thus it will take 24 six-month periods or 12 years for the money to double.

Answer is (B)

PROBLEM 12-2

The uniform annual end-of-year payment to repay a debt in n years, with an interest rate i, is determined by multiplying the capital recovery factor by the

(A) average debt
(B) initial debt plus total interest
(C) average debt plus interest
(D) initial debt plus the first year's interest
(E) initial debt

Solution

The annual payment equals the capital recovery factor multiplied by the initial debt, or $A = P(A/P, i, n)$.

Answer is (E)

PROBLEM 12-3

If $100,000 is invested at 12% interest, compounded monthly, the first year interest, in dollars, is nearest to

(A)   1,000
(B)   12,000
(C)   12,350
(D)   12,700
(E)   144,000

$$(100,000)\left[(1.12)^{\frac{12}{12}} - 1\right]$$

Solution

$$\text{Effective interest rate per year} = (1 + i)^m - 1$$

where i is the interest rate per interest period and m is the number of compoundings per year. In this problem

$$\text{Effective interest rate} = (1 + 0.01)^{12} - 1 = 0.1268 = 12.68\%$$
$$\text{Interest} = 100,000(0.1268) = \$12,680$$

Alternate Solution. At the end of one year the future worth should be

$$F = P(F/P, i, n) = 100,000(F/P, 1\%, 12)$$
$$= 100,000(1.127) = \$112,700$$

The $112,700 consists of the $100,000 investment plus the first year's interest of $12,700.

Answer is (D)

PROBLEM 12-4

A 12% annual interest rate, compounded annually, is equivalent to what annual interest rate (%), compounded quarterly?

(A) 2.9
(B) 3.0
(C) 11.5
(D) 12.0
(E) 12.6

$$1.12 = (1+i)^4$$
$$1.12^{\frac{1}{4}} = 1 + i$$
$$i = 1.12^{\frac{1}{4}} - 1$$

Solution

$$\text{Effective interest rate} = (1+i)^m - 1 = 0.12 = 12\%$$

$$(1+i)^4 = 1.12 \qquad 1.12^{0.25} = 1 + i \qquad 1.0287 = 1 + i$$

$$i \text{ per quarter year} = 0.0287 = 2.87\%$$

$$\text{Annual interest rate} = 4(2.87) = 11.5\%$$

$$\text{Answer is (C)}$$

PROBLEM 12-5

A bank pays 10% nominal annual interest on special three year certificates. Answer the following three questions.

Question 1. If the interest is compounded every three months, the effective annual interest rate (%) is nearest to

(A) 10.000
(B) 10.375
(C) 10.500
(D) 10.750
(E) 30.000

Solution

$$\text{Effective } i = (1+i)^m - 1 = (1 + 0.025)^4 - 1$$

$$= 0.1038 = 10.38\%$$

$$\text{Answer is (B)}$$

Question 2. If interest is compounded daily, the effective annual interest rate (%) is nearest to

(A) 10.000
(B) 10.375
(C) 10.500
(D) 10.750
(E) 30.000

Solution

$$\text{Effective } i = (1 + i)^m - 1 = \left[1 + \frac{0.10}{365}\right]^{365} - 1$$
$$= 0.10516 = 10.516\%$$

Answer is (C)

Question 3. If interest is compounded continuously, the effective annual interest rate (%) is nearest to

(A) 10.000
(B) 10.375
(C) 10.500
(D) 10.750
(E) 30.000

Solution

For continuous compounding, the effective interest rate is $i = e^r - 1$ where r is the nominal interest rate.

$$\text{Hence } i = e^{0.10} - 1 = 0.10517 = 10.517\%$$

Answer is (C)

PROBLEM 12-6

A company is about to purchase two trucks valued at a total of $78,000. The truck dealer offers terms of $5000 downpayment with 12 equal end-of-month payments at an interest rate of 1% per month on the unpaid balance. The monthly payment, in dollars, required by the dealer is closest to

(A) 6000
(B) 6500
(C) 6800
(D) 6900
(E) 9800

Solution

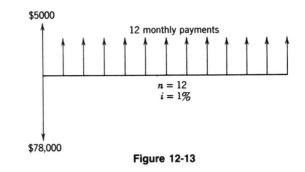

Figure 12-13

$$A = (78{,}000 - 5000)(A/P, 1\%, 12) = 73{,}000(0.0888) = \$6480$$

Answer is (B)

PROBLEM 12-7

An industrial firm is purchasing a machine for $10,000. It estimates there will be a $3000 salvage value at the end of the machine's 12-year useful life. The annual maintenance cost is estimated to be $150. The equivalent uniform annual cost (EUAC) for the machine, in dollars, based on a 10% interest rate, is nearest to

(A) 800
(B) 1000
(C) 1300
(D) 1500
(E) 1600

Solution

$$\text{EUAC} = 10{,}000(A/P, 10\%, 12) + 150 - 3000(A/F, 10\%, 12)$$
$$= 10{,}000(0.1468) + 150 - 3000(0.0468)$$
$$= \$1478$$

Answer is (D)

PROBLEM 12-8

A building costs $500,000. Its estimated life is 20 years. At 12% interest the annual amount, in dollars, that could be spent for extra maintenance, if this would extend the life of the building to 30 years, is closest to

(A) 5,000
(B) 10,000
(C) 30,000
(D) 62,000
(E) 67,000

Solution

One should be willing to spend money on extra annual maintenance until the point is reached where the equivalent uniform annual cost (EUAC) for the building with its life extended to 30 years equals the equivalent uniform annual cost (EUAC) for the building with a 20-year life.

$$\text{EUAC for 30-year life} = \text{EUAC for 20-year life}$$

$500,000(A/P, 12\%, 30) + \text{Extra annual maintenance}$

$$= 500,000(A/P, 12\%, 20)$$

$500,000(0.1241) + \text{Extra annual maintenance} = 500,000(0.1339)$

$\text{Extra annual maintenance} = 500,000(0.1339 - 0.1241) = \$4900$

Answer is (A)

PROBLEM 12-9

A woman wishes to accumulate a total of $10,000 in a savings account at the end of 10 years. If the bank pays 4%, compounded quarterly, the sum that should be deposited now, in dollars, is closest to

(A) 3350
(B) 5000
(C) 6000
(D) 6700
(E) 6760

Solution

$$P = F(P/F, 1\%, 40) = 10,000(P/F, 1\%, 40) = 10,000(0.6717) = \$6717$$

Answer is (D)

PROBLEM 12-10

An engineer has been asked to purchase a mortgage on a home. The holder of the mortgage will receive $8400 at the end of each year for seven years, and a lump sum payment of $120,000 at the end of seven years. How much, in dollars, would the engineer be willing to pay for the mortgage if 10% is considered to be the proper interest rate? The amount is closest to

(A) 80,000
(B) 100,000
(C) 120,000
(D) 140,000
(E) 160,000

Solution

$$P = 8400(P/A, 10\%, 7) + 120,000(P/F, 10\%, 7)$$
$$= 8400(4.868) + 120,000(0.5132) = \$102,500$$

Answer is (B)

PROBLEM 12-11

How much money, in dollars, must be deposited in a bank now at 5% interest to provide $1000 per year for the next 50 years? The amount is closest to

(A) 16,000
(B) 18,000
(C) 20,000
(D) 25,000
(E) 50,000

Solution

$$P = 1000(P/A, 5\%, 50) = 1000(18.256) = \$18,256$$

Answer is (B)

PROBLEM 12-12

An agency of the federal government has loaned $10 million to a city. The loan is at 5% annual interest and is for an infinite period. The annual payment, in dollars, the city must make to the federal agency is nearest to

(A)   0
(B)   50,000
(C) 500,000
(D) 510,000
(E) 600,000

Solution

When n equals infinity, the general equation is $A = Pi$.

Required annual payment $A = Pi = 10,000,000(0.05) = \$500,000$

Answer is (C)

PROBLEM 12-13

A city has been offered land for a park and enough money to pay $10,000 per year maintenance forever. Assuming a 6% interest rate, the amount of money, in dollars, needed to provide the perpetual maintenance is closest to

(A)   6,000
(B)   10,000
(C)   166,667
(D)   600,000
(E) 1,000,000

Solution

When n equals infinity, the capitalized cost P equals A/i.

$$P = \frac{A}{i} = \frac{10,000}{0.06} = \$166,667$$

Answer is (C)

PROBLEM 12-14

A trust fund is to be established to pay $100,000 per year operating costs of a laboratory, and in addition to provide $75,000 of replacement equipment every four years, beginning four years from now. At 8% interest the amount of money, in dollars, now required in the perpetual trust fund is nearest

(A)   115,000
(B)   210,000
(C) 1,250,000
(D) 1,450,000
(E) 1,600,000

Solution

For $100,000 per year operating costs

$$\text{Required amount in the trust fund} = \frac{A}{i} = \frac{100,000}{0.08} = \$1,250,000$$

For $75,000 of replacement equipment every four years the perpetual series is

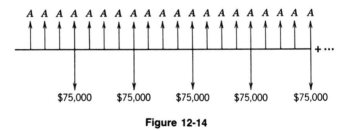

Figure 12-14

If the equivalent amount A is computed, then the $A = Pi$ equation may be used to compute the necessary amount in the trust fund, P. We can solve one portion of the perpetual series for A

Figure 12-15 $75,000

$$A = 75,000(A/F, 8\%, 4) = 75,000(0.2219) = \$16,640$$

This value for A for the four-year period is the same as the value of A for the perpetual series.

$$\text{Required amount in trust fund} = \frac{A}{i} = \frac{16,640}{0.08} = \$208,000$$

For annual maintenance plus periodic equipment replacement the total needed in the trust fund is $1,250,000 + 208,000 = \$1,458,000$

Answer is (D)

PROBLEM 12-15

A subdivider offers lots for sale for $15,000. A downpayment of $1500 is required and $1500 is paid at the end of each year for nine years with no interest charged. Further negotiation reveals that the lots may also be purchased for $10,000 cash. The actual interest rate (%) the buyer will pay if he chooses the installment payment plan is closest to

(A) 4
(B) 6
(C) 8
(D) 10
(E) 12

Solution

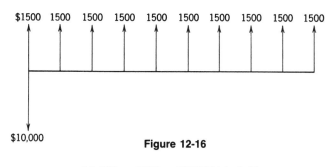

Figure 12-16

$$10,000 = 1500 + 1500(P/A, i, 9)$$

$$(P/A, i, 9) = \frac{8500}{1500} = 5.67$$

From the compound interest tables:

i	$(P/A, i, 9)$
10%	5.759
12%	5.328

By linear interpolation,

$$i = 10\% + (2\%)\left(\frac{5.759 - 5.67}{5.759 - 5.328}\right) = 10\% + (2\%)\left(\frac{0.09}{0.431}\right) = 10.4\%$$

Answer is (D)

PROBLEM 12-16

A machine is to be purchased for $15,500; it has an estimated life of eight years and a salvage value of $600. A sinking fund is to be established so money will be available to purchase a replacement when the first machine wears out at the end of eight years. An amount of $1303 is to be deposited at the end of each year during the lifetime of the first machine into this sinking fund. The interest rate (%) this fund must earn to produce sufficient funds to purchase the replacement machine at the end of eight years is closest to

(A) 6
(B) 7
(C) 8
(D) 10
(E) 12

Solution

The amount of money that must be accumulated in the sinking fund by the end of eight years equals the cost of the new machine minus the salvage value of the original machine, or $15,500 - 600 = $14,900.

$$(A/F, i, 8) = \frac{A}{F} = \frac{1303}{14,900} = 0.0874$$

From the interest tables we see that this is exactly the value of the sinking fund factor when the interest rate i is 10%.

Answer is (D)

PROBLEM 12-17

A manufacturer who produces a single item has a maximum production capacity of 40,000 units per year. The overall and unit costs for different levels of operation are as follows:

Output (units)	Total Cost ($)	Total Cost per Unit ($)
0	60,000	
5,000	85,000	17.00
10,000	109,000	10.90
20,000	155,000	7.75
40,000	243,000	6.08

Early in the year orders are received for 10,000 units at $12 each. Owing to depressed business conditions it is realized that no further domestic orders can be expected in the current year. The manufacturer has an opportunity to sell 10,000 units overseas to a foreign buyer. The lowest selling price per unit, in dollars, the manufacturer is willing to accept on this overseas order is closest to

(A) 4.40
(B) 4.70
(C) 7.75
(D) 10.90
(E) 12.00

Solution

The manufacturer should accept any order where the additional revenue exceeds the cost of supplying the order. The table shows the incremental cost per unit is $4.60.

Output (units)	Total Cost ($)	Incremental Cost ($)	Incremental Cost per Unit ($)
0	60,000		
		25,000	5.00
5,000	85,000		
		24,000	4.80
10,000	109,000		
		46,000	4.60
20,000	155,000		
		88,000	4.40
40,000	243,000		

Answer is (B)

PROBLEM 12-18

Three hundred dollars are deposited in a bank savings account at the beginning of each of 15 years. If the account pays 8% annual interest, the amount, in dollars, in the account at the end of 15 years is nearest to

(A) 2800
(B) 4500
(C) 7600
(D) 8100
(E) 8800

Solution

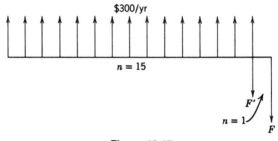

Figure 12-17

$$F' = 300(F/A, 8\%, 15) = 300(27.152) = \$8146$$
$$F = 8146(F/P, 8\%, 1) = 8146(1.080) = \$8798$$

Answer is (E)

PROBLEM 12-19

A firm believes that it will need land for another warehouse 20 years hence. A suitable piece of land can now be bought for $95,000. If purchased, the firm will pay taxes of $3000 per year on the property. A 10% interest rate is to be used. To justify its purchase now, 20 years from now the land must be worth, in dollars, approximately

(A) 155,000
(B) 170,000
(C) 640,000
(D) 700,000
(E) 810,000

Solution

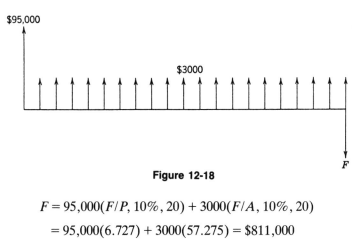

Figure 12-18

$$F = 95{,}000(F/P, 10\%, 20) + 3000(F/A, 10\%, 20)$$
$$= 95{,}000(6.727) + 3000(57.275) = \$811{,}000$$

Answer is (E)

PROBLEM 12-20

A contractor can purchase a dump truck for $28,000. Its estimated salvage value is $8000 at the end of seven years. Daily operating expenses are $250, including maintenance and the cost of the driver. The contractor can hire a similar truck and its driver for $320 a day. At 8% interest the number of days per year the dump truck must be needed to justify its purchase is closest to

(A) 20
(B) 40
(C) 60
(D) 80
(E) 100

Solution

Set the annual cost of hiring a truck equal to the annual cost of truck ownership. Let x = truck utilization in days per year.

$$320x = 28{,}000(A/P, 8\%, 7) + 250x - 8000(A/F, 8\%, 7)$$
$$70x = 28{,}000(0.1921) - 8000(0.1121)$$
$$x = 64 \text{ days per year}$$

Answer is (C)

PROBLEM 12-21

A company which manufactures electric motors has a production capacity of 200 motors per month. The variable costs are $150 per motor. The average selling price of the motors is $275. Fixed costs of the company amount to $20,000 per month, which includes all taxes. The number of motors that must be sold each month to break even is closest to

(A) 40
(B) 80
(C) 120
(D) 160
(E) 200

Solution

Let x = number of motor sales per month to break even.

$$275x = 150x + 20,000$$

$$x = 160 \text{ motors}$$

$$\text{Answer is (D)}$$

PROBLEM 12-22

Given the following data for two capacitors:

	Ferrous	Copper
Installed cost ($)	800	1200
Useful life (years)	10	Unknown
Annual operating cost ($)	63	40

The useful life of the ferrous capacitor is 10 years, but the useful life of the copper alloy capacitor is not known. Either capacitor may be replaced by an identical replacement at the end of its useful life. Neither capacitor has any salvage value. The life, in years, for the copper capacitor to be as economical as the ferrous one, assuming 10% interest, is closest to

(A) 12
(B) 14
(C) 16
(D) 18
(E) 20

Solution

Set the annual cost of the ferrous capacitor equal to the annual cost of the copper capacitor.

$$800(A/P, 10\%, 10) + 63 = 1200(A/P, 10\%, n) + 40$$

$$(A/P, 10\%, n) = \frac{800(0.1627) + 23}{1200} = 0.1276$$

$(A/P, 10\%, n)$ equals 0.1276 when n is very close to 16 years.

Answer is (C)

PROBLEM 12-23

A machine costs $9000 and has an estimated salvage value of $800 at the end of its 10-year life. Depreciation is computed by the straight-line method. When the machine is three years old, its book value, in dollars, will be nearest to

(A) 800
(B) 3260
(C) 6300
(D) 6540
(E) 9000

Solution

$$\text{Straight-line depreciation in any year} = \frac{P - S}{n} = \frac{9000 - 800}{10} = \$820$$

$$\text{Book value} = \text{cost} - \text{depreciation}$$

After three years,

$$\text{Book value} = 9000 - 3(820) = \$6540$$

Answer is (D)

PROBLEM 12-24

A piece of machinery costs $8300 and has a projected $2000 salvage value at the end of its six-year useful life. Sum-of-years digits depreciation, in dollars, for the third year is closest to

(A) 900
(B) 1000
(C) 1100
(D) 1200
(E) 1300

Solution

$$\begin{array}{c}\text{Sum-of-years-digits}\\\text{depreciation in any year}\end{array} = \dfrac{\begin{array}{c}\text{Remaining useful life}\\\text{at beginning of year}\end{array}}{\dfrac{n}{2}(n+1)} (P-S)$$

$$\begin{array}{c}\text{Sum-of-years-digits}\\\text{depreciation for the third year}\end{array} = \dfrac{4}{\frac{6}{2}(6+1)}(8300-2000) = \$1200$$

Answer is (D)

PROBLEM 12-25

A company uses 8000 wheels per year in its manufacture of golf carts. The wheels cost \$15 each and are purchased from an outside supplier. The money invested in the inventory costs 10% per year, and the warehousing cost amounts to an additional 2% per year. It costs \$150 to process each purchase order. The quantity of wheels that should be ordered each time an order is placed is nearest to

(A) 1200
(B) 1600
(C) 2000
(D) 2500
(E) 3000

Solution

The simplest model for the economic order quantity is $EOQ = [2BD/E]^{1/2}$ in which B = ordering cost in \$/order, D = demand per period in units, E = inventory holding cost in \$/unit/period, and EOQ is the economic order quantity in units.
Here

$$EOQ = \left[\frac{2(150)(8000)}{(0.10+0.02)(15)}\right]^{1/2} = 1155 \text{ wheels}$$

Answer is (A)

PROBLEM 12-26

An engineer purchased an automobile for $5000. She expects it to have a $600 resale value at the end of five years. Fixed costs are $180 per year. Fuel and oil are $0.03/km. The engineer drives 15,000 km per year and considers 8% to be a suitable interest rate. Her average annual cost, in dollars, computed by the method of straight-line depreciation plus average interest, is closest to

(A)   900
(B)   1200
(C)   1500
(D)   1800
(E)   1900

Solution

$$\text{Straight-line depreciation} = \frac{P - S}{n} = \frac{5000 - 600}{5} = \$880$$

$$\text{Average interest} = (P - S)\left(\frac{i}{2}\right)\left(\frac{n + 1}{n}\right) + Si$$

$$= (5000 - 600)\left(\frac{0.08}{2}\right)\left(\frac{6}{5}\right) + 600(0.08) = \quad 259$$

$$\text{Fixed costs plus fuel} = 180 + 0.03(15{,}000) = \quad 630$$

$$\text{Average annual cost} = \overline{\$1769}$$

Answer is (D)

PROBLEM 12-27

This is a problem set containing 10 questions. Two machines are being compared.

	Machine *A*	Machine *B*
Initial cost ($)	50,000	120,000
Uniform annual maintenance ($)	4,000	2,000
Useful life (years)	10	20
End of useful life salvage value ($)	10,000	0

Interest rate: 8%

Question 1. The equivalent uniform annual cost, in dollars, for 10 years of machine A is closest to

(A) 4,000
(B) 7,000
(C) 8,900
(D) 10,800
(E) 11,500

Solution

$$\text{EUAC} = 50{,}000(A/P, 8\%, 10) + 4000 - 10{,}000(A/F, 8\%, 10)$$
$$= 50{,}000(0.1490) + 4000 - 10{,}000(0.0690) = \$10{,}760$$
Answer is (D)

Question 2. At the end of 10 years the original machine A will be replaced by another machine A with identical initial cost, annual maintenance, useful life and salvage value. The equivalent uniform annual cost, in dollars, for the 20-year period (10 years of machine A plus 10 years of the replacement machine A) is closest to

(A) 4,000
(B) 7,000
(C) 8,900
(D) 10,800
(E) 11,500

Solution

In Question 1 the equivalent uniform annual cost of 10 years of machine A was computed to be $10,760. During the second 10 years the EUAC will be identical. Therefore the EUAC for the 20-year period is the same as the EUAC for the initial 10-year period.

Answer is (D)

Question 3. The capitalized cost, in dollars, of machine A is nearest

(A) 50,000
(B) 65,000
(C) 80,000
(D) 90,000
(E) 135,000

Solution

Capitalized cost is the present worth of cost for an infinite analysis period. For n equal to infinity, capitalized cost equals A divided by i. From Question 1, the equivalent uniform annual cost for machine A is $10,760.

$$\text{Capitalized cost } P = \frac{A}{i} = \frac{10.760}{0.08} = \$134,500$$

Answer is (E)

Question 4. The present worth of cost, in dollars, for 20 years of machine B is nearest

(A) 8,000
(B) 12,000
(C) 120,000
(D) 140,000
(E) 160,000

Solution

$$\text{Present worth of cost} = 120,000 + 2000(P/A, 8\%, 20)$$
$$= 120,000 + 2000(9.818) = \$139,640$$

Answer is (D)

Question 5. To have sufficient money to replace machine A with an identical machine A at the end of 10 years, annual end-of-year deposits will be made into a sinking fund. If the sinking fund earns 8% interest, the annual deposit, in dollars, is nearest

(A) 2,800
(B) 3,500
(C) 6,800
(D) 12,200
(E) 18,500

Solution

To have sufficient money to replace machine A, one must accumulate $40,000 in the fund. The balance of the needed $50,000 will come from the salvage value of the original machine A.

$$\text{Annual deposit} = 40,000(A/F, 8\%, 10) = 40,000(0.0690) = \$2760$$

Answer is (A)

Question 6. To provide a 10% before-tax rate of return, machine B must have annual benefits, in dollars, nearest to

(A) 8,000
(B) 10,000
(C) 12,000
(D) 14,000
(E) 16,000

Solution

Set the equivalent uniform annual benefits equal to the equivalent uniform annual cost at a 10% interest rate.

$$\text{EAUB} = \text{EUAC}$$
$$= 120{,}000(A/P, 10\%, 20) + 2000$$
$$= 120{,}000(0.1175) + 2000 = \$16{,}100$$

Answer is (E)

Question 7. The uniform annual production benefits from installing machine B are estimated at $12,460 per year. If installed, machine B would have a rate of return (%) nearest

(A) 4
(B) 5
(C) 6
(D) 8
(E) 10

Solution

$$\text{PW of cost} = \text{PW of benefits}$$
$$120{,}000 + 2000(P/A, i, 20) = 12{,}460(P/A, i, 20)$$
$$(P/A, i, 20) = 11.47$$

From compound interest tables, $i = 6\%$.

Answer is (C)

Question 8. The annual straight-line depreciation, in dollars, for machine A is nearest

(A) 4,000
(B) 5,000
(C) 6,000
(D) 8,000
(E) 10,000

Solution

$$\text{Straight-line depreciation in any year} = \frac{P - S}{n} = \frac{50,000 - 10,000}{10} = \$4000$$

Answer is (A)

Question 9. The sum-of-years-digits depreciation, in dollars, for the fourth year of machine A is closest to

(A) 2,000
(B) 4,000
(C) 5,000
(D) 7,000
(E) 28,000

Solution

$$\begin{array}{l} \text{Sum-of-years-digits} \\ \text{depreciation charge for any year} \end{array} = \frac{\begin{array}{c} \text{Remaining useful life} \\ \text{at beginning of year} \end{array}}{\begin{array}{c} \text{Sum-of-years digits} \\ \text{for total useful life} \end{array}} (P - S)$$

where

$$\text{Sum-of-years digits} = \frac{n}{2}(n + 1) = \frac{10}{2}(10 + 1) = 55$$

$$\begin{array}{l} \text{Fourth year sum-of-years-} \\ \text{digits depreciation} \end{array} = \frac{7}{55}(50,000 - 10,000) = \$5090$$

Answer is (C)

Question 10. Based on double declining balance depreciation, the book value of machine A, in dollars, at the end of three years is nearest

(A) 20,000
(B) 26,000
(C) 30,000
(D) 38,000
(E) 40,000

Solution

The important thing to note in double declining balance (DDB) deprecia-tion is that the salvage value is not a component of the calculation.

First year DDB depreciation $= \dfrac{2}{10}$ $(50,000 - 0)$ $= \$10,000$

Second year DDB depreciation $= \dfrac{2}{10}$ $(50,000 - 10,000) =$ $8,000$

Third year DDB depreciation $= \dfrac{2}{10}$ $(50,000 - 18,000) = \underline{\quad 6,400}$

$$\$24,400$$

Book value at the end of three years $= 50,000 - 24,400 = \$25,600$.

Answer is (B)

PROBLEM 12-28

This is a problem set containing 10 questions. A firm is trying to determine which of three different machines it should install in the plant to perform a given task.

	Machine A	Machine B	Machine C
Initial cost ($)	3000	4300	5000
Annual maintenance ($)	1500	1000	1000
Annual operating cost ($)	1500	1500	1100
End of useful life			
salvage value ($)	800	500	0
Useful life (years)	5	5	5

Question 1. Using the method of straight-line depreciation plus average interest, which machine is the best economic choice at a 10% interest rate?

(A) machine A
(B) machine B
(C) machine C
(D) machines A and B are equally good
(E) machines B and C are equally good

Solution

Straight-line depreciation plus average interest

$$= \frac{P - S}{n} + (P - S)\left(\frac{i}{2}\right)\left(\frac{n + 1}{n}\right) + Si$$

Machine A:

$$\text{Annual cost} = \frac{3000 - 800}{5} + (3000 - 800)\left(\frac{0.10}{2}\right)\left(\frac{6}{5}\right) + 800(0.10)$$

$$+ 1500 + 1500 = \$3652$$

Machine B:

$$\text{Annual cost} = \frac{4300 - 500}{5} + (4300 - 500)\left(\frac{0.10}{2}\right)\left(\frac{6}{5}\right) + 500(0.10)$$

$$+ 1000 + 1500 = \$3538$$

Machine C:

$$\text{Annual cost} = \frac{5000 - 0}{5} + (5000 - 0)\left(\frac{0.10}{2}\right)\left(\frac{6}{5}\right) + 1000 + 1100$$

$$= \$3400$$

Answer is (C)

Question 2. If machine B had a useful life of seven years rather than five years, which machine would now be the best economic choice based on straight-line depreciation plus average interest?

(A) machine A
(B) machine B
(C) machine C
(D) machines A and B are equally good
(E) machines B and C are equally good

Solution

Machine B:

$$\text{Annual cost} = \frac{4300 - 500}{7} + (4300 - 500)\left(\frac{0.10}{2}\right)\left(\frac{8}{7}\right) + 500(0.10)$$

$$+ 1000 + 1500 = \$3310$$

Machine B now has the least annual cost, as computed by the method of straight-line depreciation plus average interest.

Answer is (B)

Question 3. The equivalent uniform annual cost, in dollars, of machine A, using a 10% interest rate, is closest to

(A) 3550
(B) 3600
(C) 3650
(D) 3700
(E) 3750

Solution

$$EUAC = (3000 - 800)(A/P, 10\%, 5) + 800(0.10) + 1500 + 1500$$
$$= 2200(0.2638) + 80 + 3000 = \$3660$$

Answer is (C)

Question 4. If machine *C* produces production benefits of $3487 per year, what is its rate of return (%)?

(A) 0
(B) 8
(C) 10
(D) 12
(E) 15

Solution

Set the present worth of cost equal to the present worth of benefits and solve for the unknown rate of return *i*.

PW of cost = PW of benefits

$$5000 + (1000 + 1100)(P/A, i, 5) = 3487(P/A, i, 5)$$

$$(P/A, i, 5) = \frac{5000}{3487 - 2100} = 3.605$$

From compound interest tables, $(P/A, 12\%, 5)$ equals 3.605. The rate of return is exactly 12%.

Answer is (D)

Question 5. If machine *A* is depreciated by the sum-of-years-digits method, its book value, in dollars, at the end of three years is closest to

(A) 440
(B) 1240

(C) 1760
(D) 2020
(E) 2200

Solution

$$\text{Sum} = 1 + 2 + 3 + 4 + 5 = 15$$

$$\begin{array}{l}\text{First year sum-of-years-} \\ \text{digits depreciation}\end{array} = \frac{5}{15}\,(3000 - 800) = \$733.33$$

$$\begin{array}{l}\text{Second year sum-of-years-} \\ \end{array} = \frac{4}{15}\,(3000 - 800) = \quad 586.67$$

$$\begin{array}{l}\text{Third year sum-of-years-} \\ \text{digits depreciation}\end{array} = \frac{3}{15}\,(3000 - 800) = \quad 440.00$$

$$\overline{\hspace{3cm}\$1760.00}$$

$$\text{Book value} = \text{cost} - \text{depreciation} = 3000 - 1760 = \$1240$$

Answer is (B)

Question 6. A sinking fund will be set up to replace machine A if it is the alternative selected. The sinking fund earns 5% annual interest. If the cost of machine A will be 7% higher than the prior year, each year for the five-year period, what uniform annual deposit must be made to the sinking fund to replace machine A at the end of five years? The amount, in dollars, is closest to

(A) 440
(B) 530
(C) 620
(D) 710
(E) 800

Solution

$$\text{Replacement cost of machine } A = 3000(F/P, 7\%, 5)$$

$$= 3000(1.403) = \$4209$$

$$\text{Annual sinking fund deposit} = (4209 - 800)(A/F, 5\%, 5)$$

$$= 3409(0.1810) = \$617$$

Answer is (C)

Question 7. If machine *A* is depreciated by the double declining balance method, the depreciation, in dollars, for year 2 would be nearest to

(A) 600
(B) 720
(C) 880
(D) 1000
(E) 1240

Solution

$$\text{First year DDB depreciation} = \tfrac{2}{5}(3000 - 0) = \$1200$$
$$\text{Second year DDB depreciation} = \tfrac{2}{5}(3000 - 1200) = \$720$$

$$\text{Answer is (B)}$$

Question 8. Assume if machine *B* is installed that the firm could reduce its labor cost by $3403 per year. What rate of return would the firm obtain on this investment? The rate of return (%) is nearest to

(A)   6
(B)   8
(C) 10
(D) 12
(E) 15

Solution

$$\text{EUAC} = \text{EUAB}$$
$$4300(A/P, i, 5) + 1000 + 1500 = 3403 + 500(A/F, i, 5)$$

The equation must be solved by trial and error for the unknown *i*. If $i = 5\%$ is tried, then

$$4300(A/P, 5\%, 5) + 1000 + 1500 = 3403 + 500(A/F, 5\%, 5)$$
$$4300(0.2310) + 2500 = 3403 + 500(0.1810)$$
$$3493 = 3493$$

The rate of return *i* is 5%.

$$\text{Answer is (A)}$$

Question 9. If machine A is installed, the firm would be able to reduce its labor costs by $3800 per year. Based on a 10% interest rate, compute the net present worth of machine A for the five-year period. The net present worth, in dollars, is closest to

(A)   0
(B)   +50
(C)   +500
(D)   +5,000
(E) +10,000

Solution

$$\text{NPW} = \text{PW of benefits} - \text{PW of cost}$$

$$= 3800(P/A, 10\%, 5) + 800(P/F, 10\%, 5)$$

$$- (1500 + 1500)(P/A, 10\%, 5) - 3000$$

$$= 3800(3.791) + 800(0.6209) - 3000(3.791) - 3000 = +530$$

Answer is (C)

Question 10. Some specialized attachments could be purchased for use with one of the machines. The attachments would increase the productivity of the machine. If this increased productivity is valued at $3000 per year, how much could the firm afford to pay for the attachments? Assume a five-year useful life, a $1000 salvage value at the end of five years, and a 10% interest rate. The amount the firm could afford to pay, in dollars, is closest to

(A) 4,000
(B) 10,000
(C) 11,000
(D) 12,000
(E) 15,000

Solution

The firm could pay an amount equal to the present worth of the benefits of having the attachments.

$$\text{PW of benefits} = 3000(P/A, 10\%, 5) + 1000(P/F, 10\%, 5)$$

$$= 3000(3.791) + 1000(0.6209) = \$11,994$$

Answer is (D)

APPENDIX A

SI UNITS AND CONVERSION FACTORS*

The international system of units is a modernized version of the metric system established by international agreement. Officially abbreviated SI, the system is built upon a foundation of seven base units, plus two supplementary units.

Quantity	Unit	Symbol
Length	meter	m
Mass	kilogram	kg
Time	second	s
Electric current	ampere	A
Thermodynamic temperature	kelvin	K
Amount of substance	mole	mol
Luminous intensity	candela	cd
Plane angle	radian	rad
Solid angle	steradian	sr

*This appendix is based on *Standard for Metric Practice*, ANSI/ASTM E 380-76 and is reprinted by permission of the American Society for Testing and Materials.

In addition, there are derived SI units which have special names and symbols.

Quantity	Unit	Symbol	Formula
Frequency (of a periodic phenomenon)	hertz	Hz	$1/s$
Force	newton	N	$kg \cdot m/s^2$
Pressure, stress	pascal	Pa	N/m^2
Energy, work, quantity of heat	joule	J	$N \cdot m$
Power, radiant flux	watt	W	J/s
Quantity of electricity, electric charge	coulomb	C	$A \cdot s$
Electric potential, potential difference, electromotive force	volt	V	W/A
Capacitance	farad	F	C/V
Electric resistance	ohm	Ω	V/A
Conductance	siemens	S	A/V
Magnetic flux	weber	Wb	$V \cdot s$
Magnetic flux density	tesla	T	Wb/m^2
Inductance	henry	H	Wb/A
Luminous flux	lumen	lm	$cd \cdot sr$
Illuminance	lux	lx	lm/m^2
Activity (of radionuclides)	becquerel	Bq	$1/s$

Listed below are some frequently used SI units.

Quantity	Unit	Symbol
Acceleration	meter per second squared	m/s^2
Angular acceleration	radian per second squared	rad/s^2
Angular velocity	radian per second	rad/s
Area	square meter	m^2
Concentration (of amount of substance)	mole per cubic meter	mol/m^3
Current density	ampere per square meter	A/m^2
Density, mass	kilogram per cubic meter	kg/m^3
Electric charge density	coulomb per cubic meter	C/m^3
Electric field strength	volt per meter	V/m
Electric flux density	coulomb per square meter	C/m^2
Energy density	joule per cubic meter	J/m^3
Entropy	joule per kelvin	J/K
Heat capacity	joule per kelvin	J/K
Heat flux density⎫ Irradiance⎭	watt per square meter	W/m^2
Luminance	candela per square meter	cd/m^2
Magnetic field strength	ampere per meter	A/m
Molar energy	joule per mole	J/mol
Molar entropy	joule per mole kelvin	$J/(mol \cdot K)$
Molar heat capacity	joule per mole kelvin	$J/(mol \cdot K)$
Moment of force	newton meter	$N \cdot m$
Permeability	henry per meter	H/m
Permittivity	farad per meter	F/m
Radiance	watt per square meter steradian	$W/(m^2 \cdot sr)$
Radiant intensity	watt per steradian	W/sr
Specific heat capacity	joule per kilogram kelvin	$J/(kg \cdot K)$
Specific energy	joule per kilogram	J/kg
Specific entropy	joule per kilogram kelvin	$J/(kg \cdot K)$
Specific volume	cubic meter per kilogram	m^3/kg
Surface tension	newton per meter	N/m
Thermal conductivity	watt per meter kelvin	$W/(m \cdot K)$
Velocity	meter per second	m/s
Viscosity, dynamic	pascal second	$Pa \cdot s$
Viscosity, kinematic	square meter per second	m^2/s
Volume	cubic meter	m^3

The factors given on the next page allow one to convert easily *from* English *to* SI units; to convert *from* the given SI unit *to* the stated English unit, one divides by the factor.

SI CONVERSION FACTORS

To Convert From	To	Multiply By
	Acceleration	
ft/s^2	meter per second^2 (m/s^2)	0.3048
free fall, standard (g)	meter per second^2 (m/s^2)	9.8066
	Area	
acre (U.S. survey)	meter^2 (m^2)	4.0469×10^3
ft^2	meter^2 (m^2)	9.2903×10^{-2}
in^2	meter^2 (m^2)	6.4516×10^{-4}
yd^2	meter^2 (m^2)	0.8361
	Bending Moment or Torque	
dyne $\cdot$ cm	newton meter (N $\cdot$ m)	1.0000×10^{-7}
kgf $\cdot$ m	newton meter (N $\cdot$ m)	9.8066
lbf $\cdot$ ft	newton meter (N $\cdot$ m)	1.3558
	Energy (includes work)	
Btu (International Table)	joule (J)	1.0551×10^3
calorie (mean)	joule (J)	4.1900
erg	joule (J)	1.0000×10^{-7}
ft $\cdot$ lbf	joule (J)	1.3558
kW $\cdot$ h	joule (J)	3.6000×10^6
	Force	
dyne	newton (N)	1.0000×10^{-5}
kilogram-force	newton (N)	9.8066
kip (1000 lbf)	newton (N)	4.4482×10^3
pound-force (lbf)	newton (N)	4.4482
	Heat	
Btu (International Table)/lb	joule per kilogram (J/kg)	2.3260×10^3

SI CONVERSION FACTORS—continued

To Convert From	To	Multiply By
Length		
inch	meter (m)	2.5400×10^{-2}
foot	meter (m)	0.3048
mile (U.S. survey)	meter (m)	1.6093×10^{3}
Mass		
pound (lbm)	kilogram (kg)	0.4536
Mass per Unit Volume		
g/cm^3	kilogram per meter^3 (kg/m^3)	1.0000×10^{3}
lb/ft^3	kilogram per meter^3 (kg/m^3)	1.6018×10^{1}
Power		
$ft \cdot lbf/min$	watt (W)	2.2597×10^{-2}
horsepower (550 ft · lbf/s)	watt (W)	7.4570×10^{2}
Pressure or Stress		
atmosphere (standard)	pascal (Pa)	1.0132×10^{5}
lbf/ft^2	pascal (Pa)	4.7880×10^{1}
psi	pascal (Pa)	6.8948×10^{3}
Velocity		
ft/s	meter per second (m/s)	0.3048
miles/h	meter per second (m/s)	0.4470
miles/h	kilometers per hour (km/h)	1.6093
Volume		
ft^3	meter^3 (m^3)	2.8317×10^{-2}
gallon (U.S. liquid)	meter^3 (m^3)	3.7854×10^{-3}
yd^3	meter^3 (m^3)	0.7646

APPENDIX B

CENTROIDAL COORDINATES AND MOMENTS OF INERTIA FOR COMMON SHAPES

	Centroid	Moment of Inertia

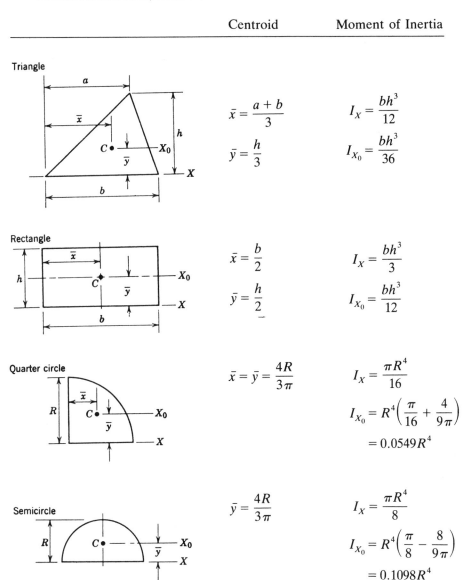

Triangle

$$\bar{x} = \frac{a + b}{3}$$

$$\bar{y} = \frac{h}{3}$$

$$I_X = \frac{bh^3}{12}$$

$$I_{X_0} = \frac{bh^3}{36}$$

Rectangle

$$\bar{x} = \frac{b}{2}$$

$$\bar{y} = \frac{h}{2}$$

$$I_X = \frac{bh^3}{3}$$

$$I_{X_0} = \frac{bh^3}{12}$$

Quarter circle

$$\bar{x} = \bar{y} = \frac{4R}{3\pi}$$

$$I_X = \frac{\pi R^4}{16}$$

$$I_{X_0} = R^4\left(\frac{\pi}{16} + \frac{4}{9\pi}\right)$$

$$= 0.0549 R^4$$

Semicircle

$$\bar{y} = \frac{4R}{3\pi}$$

$$I_X = \frac{\pi R^4}{8}$$

$$I_{X_0} = R^4\left(\frac{\pi}{8} - \frac{8}{9\pi}\right)$$

$$= 0.1098 R^4$$

Circle

$$I_X = \frac{5\pi R^4}{4}$$

$$I_{X_0} = \frac{\pi R^4}{4}$$

APPENDIX C

COMPOUND INTEREST TABLES*

Values of interest factors when $n = \infty$

1. Single payment:

$$(F/P, i, \infty) \ = \infty \qquad (P/F, i, \infty) = 0$$

2. Arithmetic gradient series:

$$(A/G, i, \infty) = 1/i \quad (P/G, i, \infty) = 1/i^2$$

3. Uniform payment series:

$$(A/F, i, \infty) \ = 0 \qquad (A/P, i, \infty) = i$$

$$(F/A, i, \infty) \ = \infty \qquad (P/A, i, \infty) = 1/i$$

* The tables are from *Engineering Economic Analysis* by Donald G. Newnan and are reprinted by permission from Engineering Press, Inc., the copyright owner.

	Single Payment		Uniform Payment Series				Arithmetic Gradient		
	Compound Amount Factor	Present Worth Factor	Sinking Fund Factor	Capital Recovery Factor	Compound Amount Factor	Present Worth Factor	Gradient Uniform Series	Gradient Present Worth	
	Find F Given P F/P	Find P Given F P/F	Find A Given F A/F	Find A Given P A/P	Find F Given A F/A	Find P Given A P/A	Find A Given G A/G	Find P Given G P/G	
n									n
1	1.010	.9901	1.0000	1.0100	1.000	0.990	0	0	1
2	1.020	.9803	.4975	.5075	2.010	1.970	0.498	0.980	2
3	1.030	.9706	.3300	.3400	3.030	2.941	0.993	2.921	3
4	1.041	.9610	.2463	.2563	4.060	3.902	1.488	5.804	4
5	1.051	.9515	.1960	.2060	5.101	4.853	1.980	9.610	5
6	1.062	.9420	.1625	.1725	6.152	5.795	2.471	14.320	6
7	1.072	.9327	.1386	.1486	7.214	6.728	2.960	19.917	7
8	1.083	.9235	.1207	.1307	8.286	7.652	3.448	26.381	8
9	1.094	.9143	.1067	.1167	9.369	8.566	3.934	33.695	9
10	1.105	.9053	.0956	.1056	10.462	9.471	4.418	41.843	10
11	1.116	.8963	.0865	.0965	11.567	10.368	4.900	50.806	11
12	1.127	.8874	.0788	.0888	12.682	11.255	5.381	60.568	12
13	1.138	.8787	.0724	.0824	13.809	12.134	5.861	71.112	13
14	1.149	.8700	.0669	.0769	14.947	13.004	6.338	82.422	14
15	1.161	.8613	.0621	.0721	16.097	13.865	6.814	94.481	15
16	1.173	.8528	.0579	.0679	17.258	14.718	7.289	107.273	16
17	1.184	.8444	.0543	.0643	18.430	15.562	7.761	120.783	17
18	1.196	.8360	.0510	.0610	19.615	16.398	8.232	134.995	18
19	1.208	.8277	.0481	.0581	20.811	17.226	8.702	149.895	19
20	1.220	.8195	.0454	.0554	22.019	18.046	9.169	165.465	20
21	1.232	.8114	.0430	.0530	23.239	18.857	9.635	181.694	21
22	1.245	.8034	.0409	.0509	24.472	19.660	10.100	198.565	22
23	1.257	.7954	.0389	.0489	25.716	20.456	10.563	216.065	23
24	1.270	.7876	.0371	.0471	26.973	21.243	11.024	234.179	24
25	1.282	.7798	.0354	.0454	28.243	22.023	11.483	252.892	25
26	1.295	.7720	.0339	.0439	29.526	22.795	11.941	272.195	26
27	1.308	.7644	.0324	.0424	30.821	23.560	12.397	292.069	27
28	1.321	.7568	.0311	.0411	32.129	24.316	12.852	312.504	28
29	1.335	.7493	.0299	.0399	33.450	25.066	13.304	333.486	29
30	1.348	.7419	.0287	.0387	34.785	25.808	13.756	355.001	30
36	1.431	.6989	.0232	.0332	43.077	30.107	16.428	494.620	36
40	1.489	.6717	.0205	.0305	48.886	32.835	18.178	596.854	40
48	1.612	.6203	.0163	.0263	61.223	37.974	21.598	820.144	48
50	1.645	.6080	.0155	.0255	64.463	39.196	22.436	879.417	50
52	1.678	.5961	.0148	.0248	67.769	40.394	23.269	939.916	52
60	1.817	.5504	.0122	.0222	81.670	44.955	26.533	1 192.80	60
70	2.007	.4983	.00993	.0199	100.676	50.168	30.470	1 528.64	70
72	2.047	.4885	.00955	.0196	104.710	51.150	31.239	1 597.86	72
80	2.217	.4511	.00822	.0182	12i.671	54.888	34.249	1 879.87	80
84	2.307	.4335	.00765	.0177	130.672	56.648	35.717	2 023.31	84
90	2.449	.4084	.00690	.0169	144.863	59.161	37.872	2 240.56	90
96	2.599	.3847	.00625	.0163	159.927	61.528	39.973	2 459.42	96
100	2.705	.3697	.00587	.0159	170.481	63.029	41.343	2 605.77	100
104	2.815	.3553	.00551	.0155	181.464	64.471	42.688	2 752.17	104
120	3.300	.3030	.00435	.0143	230.039	69.701	47.835	3 334.11	120
240	10.893	.0918	.00101	.0110	989.254	90.819	75.739	6 878.59	240
360	35.950	.0278	.00029	.0103	3 495.0	97.218	89.699	8 720.43	360
480	118.648	.00843	.00008	.0101	11 764.8	99.157	95.920	9 511.15	480

1½% Compound Interest Factors 1½%

	Single Payment		Uniform Payment Series				Arithmetic Gradient		
	Compound Amount Factor	Present Worth Factor	Sinking Fund Factor	Capital Recovery Factor	Compound Amount Factor	Present Worth Factor	Gradient Uniform Series	Gradient Present Worth	
	Find F Given P	Find P Given F	Find A Given F	Find A Given P	Find F Given A	Find P Given A	Find A Given G	Find P Given G	
n	F/P	P/F	A/F	A/P	F/A	P/A	A/G	P/G	n
1	1.015	.9852	1.0000	1.0150	1.000	0.985	0	0	1
2	1.030	.9707	.4963	.5113	2.015	1.956	0.496	0.970	2
3	1.046	.9563	.3284	.3434	3.045	2.912	0.990	2.883	3
4	1.061	.9422	.2444	.2594	4.091	3.854	1.481	5.709	4
5	1.077	.9283	.1941	.2091	5.152	4.783	1.970	9.422	5
6	1.093	.9145	.1605	.1755	6.230	5.697	2.456	13.994	6
7	1.110	.9010	.1366	.1516	7.323	6.598	2.940	19.400	7
8	1.126	.8877	.1186	.1336	8.433	7.486	3.422	25.614	8
9	1.143	.8746	.1046	.1196	9.559	8.360	3.901	32.610	9
10	1.161	.8617	.0934	.1084	10.703	9.222	4.377	40.365	10
11	1.178	.8489	.0843	.0993	11.863	10.071	4.851	48.855	11
12	1.196	.8364	.0767	.0917	13.041	10.907	5.322	58.054	12
13	1.214	.8240	.0702	.0852	14.237	11.731	5.791	67.943	13
14	1.232	.8118	.0647	.0797	15.450	12.543	6.258	78.496	14
15	1.250	.7999	.0599	.0749	16.682	13.343	6.722	89.694	15
16	1.269	.7880	.0558	.0708	17.932	14.131	7.184	101.514	16
17	1.288	.7764	.0521	.0671	19.201	14.908	7.643	113.937	17
18	1.307	.7649	.0488	.0638	20.489	15.673	8.100	126.940	18
19	1.327	.7536	.0459	.0609	21.797	16.426	8.554	140.505	19
20	1.347	.7425	.0432	.0582	23.124	17.169	9.005	154.611	20
21	1.367	.7315	.0409	.0559	24.470	17.900	9.455	169.241	21
22	1.388	.7207	.0387	.0537	25.837	18.621	9.902	184.375	22
23	1.408	.7100	.0367	.0517	27.225	19.331	10.346	199.996	23
24	1.430	.6995	.0349	.0499	28.633	20.030	10.788	216.085	24
25	1.451	.6892	.0333	.0483	30.063	20.720	11.227	232.626	25
26	1.473	.6790	.0317	.0467	31.514	21.399	11.664	249.601	26
27	1.495	.6690	.0303	.0453	32.987	22.068	12.099	266.995	27
28	1.517	.6591	.0290	.0440	34.481	22.727	12.531	284.790	28
29	1.540	.6494	.0278	.0428	35.999	23.376	12.961	302.972	29
30	1.563	.6398	.0266	.0416	37.539	24.016	13.388	321.525	30
36	1.709	.5851	.0212	.0362	47.276	27.661	15.901	439.823	36
40	1.814	.5513	.0184	.0334	54.268	29.916	17.528	524.349	40
48	2.043	.4894	.0144	.0294	69.565	34.042	20.666	703.537	48
50	2.105	.4750	.0136	.0286	73.682	35.000	21.428	749.955	50
52	2.169	.4611	.0128	.0278	77.925	35.929	22.179	796.868	52
60	2.443	.4093	.0104	.0254	96.214	39.380	25.093	988.157	60
70	2.835	.3527	.00817	.0232	122.363	43.155	28.529	1 231.15	70
72	2.921	.3423	.00781	.0228	128.076	43.845	29.189	1 279.78	72
80	3.291	.3039	.00655	.0215	152.710	46.407	31.742	1 473.06	80
84	3.493	.2863	.00602	.0210	166.172	47.579	32.967	1 568.50	84
90	3.819	.2619	.00532	.0203	187.929	49.210	34.740	1 709.53	90
96	4.176	.2395	.00472	.0197	211.719	50.702	36.438	1 847.46	96
100	4.432	.2256	.00437	.0194	228.802	51.625	37.529	1 937.43	100
104	4.704	.2126	.00405	.0190	246.932	52.494	38.589	2 025.69	104
120	5.969	.1675	.00302	.0180	331.286	55.498	42.518	2 359.69	120
240	35.632	.0281	.00043	.0154	2 308.8	64.796	59.737	3 870.68	240
360	212.700	.00470	.00007	.0151	14 113.3	66.353	64.966	4 310.71	360
480	1 269.7	.00079	.00001	.0150	84 577.8	66.614	66.288	4 415.74	480

	Single Payment		Uniform Payment Series				Arithmetic Gradient		
	Compound Amount Factor	Present Worth Factor	Sinking Fund Factor	Capital Recovery Factor	Compound Amount Factor	Present Worth Factor	Gradient Uniform Series	Gradient Present Worth	
	Find F Given P F/P	Find P Given F P/F	Find A Given F A/F	Find A Given P A/P	Find F Given A F/A	Find P Given A P/A	Find A Given G A/G	Find P Given G P/G	
n									n
1	1.030	.9709	1.0000	1.0300	1.000	0.971	0	0	1
2	1.061	.9426	.4926	.5226	2.030	1.913	0.493	0.943	2
3	1.093	.9151	.3235	.3535	3.091	2.829	0.980	2.773	3
4	1.126	.8885	.2390	.2690	4.184	3.717	1.463	5.438	4
5	1.159	.8626	.1884	.2184	5.309	4.580	1.941	8.889	5
6	1.194	.8375	.1546	.1846	6.468	5.417	2.414	13.076	6
7	1.230	.8131	.1305	.1605	7.662	6.230	2.882	17.955	7
8	1.267	.7894	.1125	.1425	8.892	7.020	3.345	23.481	8
9	1.305	.7664	.0984	.1284	10.159	7.786	3.803	29.612	9
10	1.344	.7441	.0872	.1172	11.464	8.530	4.256	36.309	10
11	1.384	.7224	.0781	.1081	12.808	9.253	4.705	43.533	11
12	1.426	.7014	.0705	.1005	14.192	9.954	5.148	51.248	12
13	1.469	.6810	.0640	.0940	15.618	10.635	5.587	59.419	13
14	1.513	.6611	.0585	.0885	17.086	11.296	6.021	68.014	14
15	1.558	.6419	.0538	.0838	18.599	11.938	6.450	77.000	15
16	1.605	.6232	.0496	.0796	20.157	12.561	6.874	86.348	16
17	1.653	.6050	.0460	.0760	21.762	13.166	7.294	96.028	17
18	1.702	.5874	.0427	.0727	23.414	13.754	7.708	106.014	18
19	1.754	.5703	.0398	.0698	25.117	14.324	8.118	116.279	19
20	1.806	.5537	.0372	.0672	26.870	14.877	8.523	126.799	20
21	1.860	.5375	.0349	.0649	28.676	15.415	8.923	137.549	21
22	1.916	.5219	.0327	.0627	30.537	15.937	9.319	148.509	22
23	1.974	.5067	.0308	.0608	32.453	16.444	9.709	159.656	23
24	2.033	.4919	.0290	.0590	34.426	16.936	10.095	170.971	24
25	2.094	.4776	.0274	.0574	36.459	17.413	10.477	182.433	25
26	2.157	.4637	.0259	.0559	38.553	17.877	10.853	194.026	26
27	2.221	.4502	.0246	.0546	40.710	18.327	11.226	205.731	27
28	2.288	.4371	.0233	.0533	42.931	18.764	11.593	217.532	28
29	2.357	.4243	.0221	.0521	45.219	19.188	11.956	229.413	29
30	2.427	.4120	.0210	.0510	47.575	19.600	12.314	241.361	30
31	2.500	.4000	.0200	.0500	50.003	20.000	12.668	253.361	31
32	2.575	.3883	.0190	.0490	52.503	20.389	13.017	265.399	32
33	2.652	.3770	.0182	.0482	55.078	20.766	13.362	277.464	33
34	2.732	.3660	.0173	.0473	57.730	21.132	13.702	289.544	34
35	2.814	.3554	.0165	.0465	60.462	21.487	14.037	301.627	35
40	3.262	.3066	.0133	.0433	75.401	23.115	15.650	361.750	40
45	3.782	.2644	.0108	.0408	92.720	24.519	17.156	420.632	45
50	4.384	.2281	.00887	.0389	112.797	25.730	18.558	477.480	50
55	5.082	.1968	.00735	.0373	136.072	26.774	19.860	531.741	55
60	5.892	.1697	.00613	.0361	163.053	27.676	21.067	583.052	60
65	6.830	.1464	.00515	.0351	194.333	28.453	22.184	631.201	65
70	7.918	.1263	.00434	.0343	230.594	29.123	23.215	676.087	70
75	9.179	.1089	.00367	.0337	272.631	29.702	24.163	717.698	75
80	10.641	.0940	.00311	.0331	321.363	30.201	25.035	756.086	80
85	12.336	.0811	.00265	.0326	377.857	30.631	25.835	791.353	85
90	14.300	.0699	.00226	.0323	443.349	31.002	26.567	823.630	90
95	16.578	.0603	.00193	.0319	519.272	31.323	27.235	853.074	95
100	19.219	.0520	.00165	.0316	607.287	31.599	27.844	879.854	100

	Single Payment		Uniform Payment Series				Arithmetic Gradient		
	Compound Amount Factor	Present Worth Factor	Sinking Fund Factor	Capital Recovery Factor	Compound Amount Factor	Present Worth Factor	Gradient Uniform Series	Gradient Present Worth	
	Find F Given P F/P	Find P Given F P/F	Find A Given F A/F	Find A Given P A/P	Find F Given A F/A	Find P Given A P/A	Find A Given G A/G	Find P Given G P/G	
n									n
1	1.050	.9524	1.0000	1.0500	1.000	0.952	0	0	1
2	1.102	.9070	.4878	.5378	2.050	1.859	0.488	0.907	2
3	1.158	.8638	.3172	.3672	3.152	2.723	0.967	2.635	3
4	1.216	.8227	.2320	.2820	4.310	3.546	1.439	5.103	4
5	1.276	.7835	.1810	.2310	5.526	4.329	1.902	8.237	5
6	1.340	.7462	.1470	.1970	6.802	5.076	2.358	11.968	6
7	1.407	.7107	.1228	.1728	8.142	5.786	2.805	16.232	7
8	1.477	.6768	.1047	.1547	9.549	6.463	3.244	20.970	8
9	1.551	.6446	.0907	.1407	11.027	7.108	3.676	26.127	9
10	1.629	.6139	.0795	.1295	12.578	7.722	4.099	31.652	10
11	1.710	.5847	.0704	.1204	14.207	8.306	4.514	37.499	11
12	1.796	.5568	.0628	.1128	15.917	8.863	4.922	43.624	12
13	1.886	.5303	.0565	.1065	17.713	9.394	5.321	49.988	13
14	1.980	.5051	.0510	.1010	19.599	9.899	5.713	56.553	14
15	2.079	.4810	.0463	.0963	21.579	10.380	6.097	63.288	15
16	2.183	.4581	.0423	.0923	23.657	10.838	6.474	70.159	16
17	2.292	.4363	.0387	.0887	25.840	11.274	6.842	77.140	17
18	2.407	.4155	.0355	.0855	28.132	11.690	7.203	84.204	18
19	2.527	.3957	.0327	.0827	30.539	12.085	7.557	91.327	19
20	2.653	.3769	.0302	.0802	33.066	12.462	7.903	98.488	20
21	2.786	.3589	.0280	.0780	35.719	12.821	8.242	105.667	21
22	2.925	.3419	.0260	.0760	38.505	13.163	8.573	112.846	22
23	3.072	.3256	.0241	.0741	41.430	13.489	8.897	120.008	23
24	3.225	.3101	.0225	.0725	44.502	13.799	9.214	127.140	24
25	3.386	.2953	.0210	.0710	47.727	14.094	9.524	134.227	25
26	3.556	.2812	.0196	.0696	51.113	14.375	9.827	141.258	26
27	3.733	.2678	.0183	.0683	54.669	14.643	10.122	148.222	27
28	3.920	.2551	.0171	.0671	58.402	14.898	10.411	155.110	28
29	4.116	.2429	.0160	.0660	62.323	15.141	10.694	161.912	29
30	4.322	.2314	.0151	.0651	66.439	15.372	10.969	168.622	30
31	4.538	.2204	.0141	.0641	70.761	15.593	11.238	175.233	31
32	4.765	.2099	.0133	.0633	75.299	15.803	11.501	181.739	32
33	5.003	.1999	.0125	.0625	80.063	16.003	11.757	188.135	33
34	5.253	.1904	.0118	.0618	85.067	16.193	12.006	194.416	34
35	5.516	.1813	.0111	.0611	90.320	16.374	12.250	200.580	35
40	7.040	.1420	.00828	.0583	120.799	17.159	13.377	229.545	40
45	8.985	.1113	.00626	.0563	159.699	17.774	14.364	255.314	45
50	11.467	.0872	.00478	.0548	209.347	18.256	15.223	277.914	50
55	14.636	.0683	.00367	.0537	272.711	18.633	15.966	297.510	55
60	18.679	.0535	.00283	.0528	353.582	18.929	16.606	314.343	60
65	23.840	.0419	.00219	.0522	456.795	19.161	17.154	328.691	65
70	30.426	.0329	.00170	.0517	588.525	19.343	17.621	340.841	70
75	38.832	.0258	.00132	.0513	756.649	19.485	18.018	351.072	75
80	49.561	.0202	.00103	.0510	971.222	19.596	18.353	359.646	80
85	63.254	.0158	.00080	.0508	1 245.1	19.684	18.635	366.800	85
90	80.730	.0124	.00063	.0506	1 594.6	19.752	18.871	372.749	90
95	103.034	.00971	.00049	.0505	2 040.7	19.806	19.069	377.677	95
100	131.500	.00760	.00038	.0504	2 610.0	19.848	19.234	381.749	100

6%   Compound Interest Factors   6%

	Single Payment		Uniform Payment Series				Arithmetic Gradient		
	Compound Amount Factor	Present Worth Factor	Sinking Fund Factor	Capital Recovery Factor	Compound Amount Factor	Present Worth Factor	Gradient Uniform Series	Gradient Present Worth	
	Find F Given P F/P	Find P Given F P/F	Find A Given F A/F	Find A Given P A/P	Find F Given A F/A	Find P Given A P/A	Find A Given G A/G	Find P Given G P/G	
n									n
1	1.060	.9434	1.0000	1.0600	1.000	0.943	0	0	1
2	1.124	.8900	.4854	.5454	2.060	1.833	0.485	0.890	2
3	1.191	.8396	.3141	.3741	3.184	2.673	0.961	2.569	3
4	1.262	.7921	.2286	.2886	4.375	3.465	1.427	4.945	4
5	1.338	.7473	.1774	.2374	5.637	4.212	1.884	7.934	5
6	1.419	.7050	.1434	.2034	6.975	4.917	2.330	11.459	6
7	1.504	.6651	.1191	.1791	8.394	5.582	2.768	15.450	7
8	1.594	.6274	.1010	.1610	9.897	6.210	3.195	19.841	8
9	1.689	.5919	.0870	.1470	11.491	6.802	3.613	24.577	9
10	1.791	.5584	.0759	.1359	13.181	7.360	4.022	29.602	10
11	1.898	.5268	.0668	.1268	14.972	7.887	4.421	34.870	11
12	2.012	.4970	.0593	.1193	16.870	8.384	4.811	40.337	12
13	2.133	.4688	.0530	.1130	18.882	8.853	5.192	45.963	13
14	2.261	.4423	.0476	.1076	21.015	9.295	5.564	51.713	14
15	2.397	.4173	.0430	.1030	23.276	9.712	5.926	57.554	15
16	2.540	.3936	.0390	.0990	25.672	10.106	6.279	63.459	16
17	2.693	.3714	.0354	.0954	28.213	10.477	6.624	69.401	17
18	2.854	.3503	.0324	.0924	30.906	10.828	6.960	75.357	18
19	3.026	.3305	.0296	.0896	33.760	11.158	7.287	81.306	19
20	3.207	.3118	.0272	.0872	36.786	11.470	7.605	87.230	20
21	3.400	.2942	.0250	.0850	39.993	11.764	7.915	93.113	21
22	3.604	.2775	.0230	.0830	43.392	12.042	8.217	98.941	22
23	3.820	.2618	.0213	.0813	46.996	12.303	8.510	104.700	23
24	4.049	.2470	.0197	.0797	50.815	12.550	8.795	110.381	24
25	4.292	.2330	.0182	.0782	54.864	12.783	9.072	115.973	25
26	4.549	.2198	.0169	.0769	59.156	13.003	9.341	121.468	26
27	4.822	.2074	.0157	.0757	63.706	13.211	9.603	126.860	27
28	5.112	.1956	.0146	.0746	68.528	13.406	9.857	132.142	28
29	5.418	.1846	.0136	.0736	73.640	13.591	10.103	137.309	29
30	5.743	.1741	.0126	.0726	79.058	13.765	10.342	142.359	30
31	6.088	.1643	.0118	.0718	84.801	13.929	10.574	147.286	31
32	6.453	.1550	.0110	.0710	90.890	14.084	10.799	152.090	32
33	6.841	.1462	.0103	.0703	97.343	14.230	11.017	156.768	33
34	7.251	.1379	.00960	.0696	104.184	14.368	11.228	161.319	34
35	7.686	.1301	.00897	.0690	111.435	14.498	11.432	165.743	35
40	10.286	.0972	.00646	.0665	154.762	15.046	12.359	185.957	40
45	13.765	.0727	.00470	.0647	212.743	15.456	13.141	203.109	45
50	18.420	.0543	.00344	.0634	290.335	15.762	13.796	217.457	50
55	24.650	.0406	.00254	.0625	394.171	15.991	14.341	229.322	55
60	32.988	.0303	.00188	.0619	533.126	16.161	14.791	239.043	60
65	44.145	.0227	.00139	.0614	719.080	16.289	15.160	246.945	65
70	59.076	.0169	.00103	.0610	967.928	16.385	15.461	253.327	70
75	79.057	.0126	.00077	.0608	1 300.9	16.456	15.706	258.453	75
80	105.796	.00945	.00057	.0606	1 746.6	16.509	15.903	262.549	80
85	141.578	.00706	.00043	.0604	2 343.0	16.549	16.062	265.810	85
90	189.464	.00528	.00032	.0603	3 141.1	16.579	16.189	268.395	90
95	253.545	.00394	.00024	.0602	4 209.1	16.601	16.290	270.437	95
100	339.300	.00295	.00018	.0602	5 638.3	16.618	16.371	272.047	100

	Single Payment		Uniform Payment Series				Arithmetic Gradient		
	Compound Amount Factor	Present Worth Factor	Sinking Fund Factor	Capital Recovery Factor	Compound Amount Factor	Present Worth Factor	Gradient Uniform Series	Gradient Present Worth	
	Find F Given P	Find P Given F	Find A Given F	Find A Given P	Find F Given A	Find P Given A	Find A Given G	Find P Given G	
n	F/P	P/F	A/F	A/P	F/A	P/A	A/G	P/G	n
1	1.070	.9346	1.0000	1.0700	1.000	0.935	0	0	1
2	1.145	.8734	.4831	.5531	2.070	1.808	0.483	0.873	2
3	1.225	.8163	.3111	.3811	3.215	2.624	0.955	2.506	3
4	1.311	.7629	.2252	.2952	4.440	3.387	1.416	4.795	4
5	1.403	.7130	.1739	.2439	5.751	4.100	1.865	7.647	5
6	1.501	.6663	.1398	.2098	7.153	4.767	2.303	10.978	6
7	1.606	.6227	.1156	.1856	8.654	5.389	2.730	14.715	7
8	1.718	.5820	.0975	.1675	10.260	5.971	3.147	18.789	8
9	1.838	.5439	.0835	.1535	11.978	6.515	3.552	23.140	9
10	1.967	.5083	.0724	.1424	13.816	7.024	3.946	27.716	10
11	2.105	.4751	.0634	.1334	15.784	7.499	4.330	32.467	11
12	2.252	.4440	.0559	.1259	17.888	7.943	4.703	37.351	12
13	2.410	.4150	.0497	.1197	20.141	8.358	5.065	42.330	13
14	2.579	.3878	.0443	.1143	22.551	8.745	5.417	47.372	14
15	2.759	.3624	.0398	.1098	25.129	9.108	5.758	52.446	15
16	2.952	.3387	.0359	.1059	27.888	9.447	6.090	57.527	16
17	3.159	.3166	.0324	.1024	30.840	9.763	6.411	62.592	17
18	3.380	.2959	.0294	.0994	33.999	10.059	6.722	67.622	18
19	3.617	.2765	.0268	.0968	37.379	10.336	7.024	72.599	19
20	3.870	.2584	.0244	.0944	40.996	10.594	7.316	77.509	20
21	4.141	.2415	.0223	.0923	44.865	10.836	7.599	82.339	21
22	4.430	.2257	.0204	.0904	49.006	11.061	7.872	87.079	22
23	4.741	.2109	.0187	.0887	53.436	11.272	8.137	91.720	23
24	5.072	.1971	.0172	.0872	58.177	11.469	8.392	96.255	24
25	5.427	.1842	.0158	.0858	63.249	11.654	8.639	100.677	25
26	5.807	.1722	.0146	.0846	68.677	11.826	8.877	104.981	26
27	6.214	.1609	.0134	.0834	74.484	11.987	9.107	109.166	27
28	6.649	.1504	.0124	.0824	80.698	12.137	9.329	113.227	28
29	7.114	.1406	.0114	.0814	87.347	12.278	9.543	117.162	29
30	7.612	.1314	.0106	.0806	94.461	12.409	9.749	120.972	30
31	8.145	.1228	.00980	.0798	102.073	12.532	9.947	124.655	31
32	8.715	.1147	.00907	.0791	110.218	12.647	10.138	128.212	32
33	9.325	.1072	.00841	.0784	118.934	12.754	10.322	131.644	33
34	9.978	.1002	.00780	.0778	128.259	12.854	10.499	134.951	34
35	10.677	.0937	.00723	.0772	138.237	12.948	10.669	138.135	35
40	14.974	.0668	.00501	.0750	199.636	13.332	11.423	152.293	40
45	21.002	.0476	.00350	.0735	285.750	13.606	12.036	163.756	45
50	29.457	.0339	.00246	.0725	406.530	13.801	12.529	172.905	50
55	41.315	.0242	.00174	.0717	575.930	13.940	12.921	180.124	55
60	57.947	.0173	.00123	.0712	813.523	14.039	13.232	185.768	60
65	81.273	.0123	.00087	.0709	1 146.8	14.110	13.476	190.145	65
70	113.990	.00877	.00062	.0706	1 614.1	14.160	13.666	193.519	70
75	159.877	.00625	.00044	.0704	2 269.7	14.196	13.814	196.104	75
80	224.235	.00446	.00031	.0703	3 189.1	14.222	13.927	198.075	80
85	314.502	.00318	.00022	.0702	4 478.6	14.240	14.015	199.572	85
90	441.105	.00227	.00016	.0702	6 287.2	14.253	14.081	200.704	90
95	618.673	.00162	.00011	.0701	8 823.9	14.263	14.132	201.558	95
100	867.720	.00115	.00008	.0701	12 381.7	14.269	14.170	202.200	100

8% Compound Interest Factors 8%

	Single Payment		Uniform Payment Series				Arithmetic Gradient		
	Compound Amount Factor	Present Worth Factor	Sinking Fund Factor	Capital Recovery Factor	Compound Amount Factor	Present Worth Factor	Gradient Uniform Series	Gradient Present Worth	
	Find F Given P F/P	Find P Given F P/F	Find A Given F A/F	Find A Given P A/P	Find·F Given A F/A	Find P Given A P/A	Find A Given G A/G	Find P Given G P/G	
n									n
1	1.080	.9259	1.0000	1.0800	1.000	0.926	0	0	1
2	1.166	.8573	.4808	.5608	2.080	1.783	0.481	0.857	2
3	1.260	.7938	.3080	.3880	3.246	2.577	0.949	2.445	3
4	1.360	.7350	.2219	.3019	4.506	3.312	1.404	4.650	4
5	1.469	.6806	.1705	.2505	5.867	3.993	1.846	7.372	5
6	1.587	.6302	.1363	.2163	7.336	4.623	2.276	10.523	6
7	1.714	.5835	.1121	.1921	8.923	5.206	2.694	14.024	7
8	1.851	.5403	.0940	.1740	10.637	5.747	3.099	17.806	8
9	1.999	.5002	.0801	.1601	12.488	6.247	3.491	21.808	9
10	2.159	.4632	.0690	.1490	14.487	6.710	3.871	25.977	10
11	2.332	.4289	.0601	.1401	16.645	7.139	4.240	30.266	11
12	2.518	.3971	.0527	.1327	18.977	7.536	4.596	34.634	12
13	2.720	.3677	.0465	.1265	21.495	7.904	4.940	39.046	13
14	2.937	.3405	.0413	.1213	24.215	8.244	5.273	43.472	14
15	3.172	.3152	.0368	.1168	27.152	8.559	5.594	47.886	15
16	3.426	.2919	.0330	.1130	30.324	8.851	5.905	52.264	16
17	3.700	.2703	.0296	.1096	33.750	9.122	6.204	56.588	17
18	3.996	.2502	.0267	.1067	37.450	9.372	6.492	60.843	18
19	4.316	.2317	.0241	.1041	41.446	9.604	6.770	65.013	19
20	4.661	.2145	.0219	.1019	45.762	9.818	7.037	69.090	20
21	5.034	.1987	.0198	.0998	50.423	10.017	7.294	73.063	21
22	5.437	.1839	.0180	.0980	55.457	10.201	7.541	76.926	22
23	5.871	.1703	.0164	.0964	60.893	10.371	7.779	80.673	23
24	6.341	.1577	.0150	.0950	66.765	10.529	8.007	84.300	24
25	6.848	.1460	.0137	.0937	73.106	10.675	8.225	87.804	25
26	7.396	.1352	.0125	.0925	79.954	10.810	8.435	91.184	26
27	7.988	.1252	.0114	.0914	87.351	10.935	8.636	94.439	27
28	8.627	.1159	.0105	.0905	95.339	11.051	8.829	97.569	28
29	9.317	.1073	.00962	.0896	103.966	11.158	9.013	100.574	29
30	10.063	.0994	.00883	.0888	113.283	11.258	9.190	103.456	30
31	10.868	.0920	.00811	.0881	123.346	11.350	9.358	106.216	31
32	11.737	.0852	.00745	.0875	134.214	11.435	9.520	108.858	32
33	12.676	.0789	.00685	.0869	145.951	11.514	9.674	111.382	33
34	13.690	.0730	.00630	.0863	158.627	11.587	9.821	113.792	34
35	14.785	.0676	.00580	.0858	172.317	11.655	9.961	116.092	35
40	21.725	.0460	.00386	.0839	259.057	11.925	10.570	126.042	40
45	31.920	.0313	.00259	.0826	386.506	12.108	11.045	133.733	45
50	46.902	.0213	.00174	.0817	573.771	12.233	11.411	139.593	50
55	68.914	.0145	.00118	.0812	848.925	12.319	11.690	144.006	55
60	101.257	.00988	.00080	.0808	1 253.2	12.377	11.902	147.300	60
65	148.780	.00672	.00054	.0805	1 847.3	12.416	12.060	149.739	65
70	218.607	.00457	.00037	.0804	2 720.1	12.443	12.178	151.533	70
75	321.205	.00311	.00025	.0802	4 002.6	12.461	12.266	152.845	75
80	471.956	.00212	.00017	.0802	5 887.0	12.474	12.330	153.800	80
85	693.458	.00144	.00012	.0801	8 655.7	12.482	12.377	154.492	85
90	1 018.9	.00098	.00008	.0801	12 724.0	12.488	12.412	154.993	90
95	1 497.1	.00067	.00005	.0801	18 701.6	12.492	12.437	155.352	95
100	2 199.8	.00045	.00004	.0800	27 484.6	12.494	12.455	155.611	100

10%   Compound Interest Factors   10%

	Single Payment		Uniform Payment Series				Arithmetic Gradient		
	Compound Amount Factor	Present Worth Factor	Sinking Fund Factor	Capital Recovery Factor	Compound Amount Factor	Present Worth Factor	Gradient Uniform Series	Gradient Present Worth	
n	Find F Given P F/P	Find P Given F P/F	Find A Given F A/F	Find A Given P A/P	Find F Given A F/A	Find P Given A P/A	Find A Given G A/G	Find P Given G P/G	n
1	1.100	.9091	1.0000	1.1000	1.000	0.909	0	0	1
2	1.210	.8264	.4762	.5762	2.100	1.736	0.476	0.826	2
3	1.331	.7513	.3021	.4021	3.310	2.487	0.937	2.329	3
4	1.464	.6830	.2155	.3155	4.641	3.170	1.381	4.378	4
5	1.611	.6209	.1638	.2638	6.105	3.791	1.810	6.862	5
6	1.772	.5645	.1296	.2296	7.716	4.355	2.224	9.684	6
7	1.949	.5132	.1054	.2054	9.487	4.868	2.622	12.763	7
8	2.144	.4665	.0874	.1874	11.436	5.335	3.004	16.029	8
9	2.358	.4241	.0736	.1736	13.579	5.759	3.372	19.421	9
10	2.594	.3855	.0627	.1627	15.937	6.145	3.725	22.891	10
11	2.853	.3505	.0540	.1540	18.531	6.495	4.064	26.396	11
12	3.138	.3186	.0468	.1468	21.384	6.814	4.388	29.901	12
13	3.452	.2897	.0408	.1408	24.523	7.103	4.699	33.377	13
14	3.797	.2633	.0357	.1357	27.975	7.367	4.996	36.801	14
15	4.177	.2394	.0315	.1315	31.772	7.606	5.279	40.152	15
16	4.595	.2176	.0278	.1278	35.950	7.824	5.549	43.416	16
17	5.054	.1978	.0247	.1247	40.545	8.022	5.807	46.582	17
18	5.560	.1799	.0219	.1219	45.599	8.201	6.053	49.640	18
19	6.116	.1635	.0195	.1195	51.159	8.365	6.286	52.583	19
20	6.728	.1486	.0175	.1175	57.275	8.514	6.508	55.407	20
21	7.400	.1351	.0156	.1156	64.003	8.649	6.719	58.110	21
22	8.140	.1228	.0140	.1140	71.403	8.772	6.919	60.689	22
23	8.954	.1117	.0126	.1126	79.543	8.883	7.108	63.146	23
24	9.850	.1015	.0113	.1113	88.497	8.985	7.288	65.481	24
25	10.835	.0923	.0102	.1102	98.347	9.077	7.458	67.696	25
26	11.918	.0839	.00916	.1092	109.182	9.161	7.619	69.794	26
27	13.110	.0763	.00826	.1083	121.100	9.237	7.770	71.777	27
28	14.421	.0693	.00745	.1075	134.210	9.307	7.914	73.650	28
29	15.863	.0630	.00673	.1067	148.631	9.370	8.049	75.415	29
30	17.449	.0573	.00608	.1061	164.494	9.427	8.176	77.077	30
31	19.194	.0521	.00550	.1055	181.944	9.479	8.296	78.640	31
32	21.114	.0474	.00497	.1050	201.138	9.526	8.409	80.108	32
33	23.225	.0431	.00450	.1045	222.252	9.569	8.515	81.486	33
34	25.548	.0391	.00407	.1041	245.477	9.609	8.615	82.777	34
35	28.102	.0356	.00369	.1037	271.025	9.644	8.709	83.987	35
40	45.259	.0221	.00226	.1023	442.593	9.779	9.096	88.953	40
45	72.891	.0137	.00139	.1014	718.905	9.863	9.374	92.454	45
50	117.391	.00852	.00086	.1009	1 163.9	9.915	9.570	94.889	50
55	189.059	.00529	.00053	.1005	1 880.6	9.947	9.708	96.562	55
60	304.482	.00328	.00033	.1003	3 034.8	9.967	9.802	97.701	60
65	490.371	.00204	.00020	.1002	4 893.7	9.980	9.867	98.471	65
70	789.748	.00127	.00013	.1001	7 887.5	9.987	9.911	98.987	70
75	1 271.9	.00079	.00008	.1001	12 709.0	9.992	9.941	99.332	75
80	2 048.4	.00049	.00005	.1000	20 474.0	9.995	9.961	99.561	80
85	3 299.0	.00030	.00003	.1000	32 979.7	9.997	9.974	99.712	85
90	5 313.0	.00019	.00002	.1000	53 120.3	9.998	9.983	99.812	90
95	8 556.7	.00012	.00001	.1000	85 556.9	9.999	9.989	99.877	95
100	13 780.6	.00007	.00001	.1000	137 796.3	9.999	9.993	99.920	100

12% Compound Interest Factors 12%

	Single Payment		Uniform Payment Series				Arithmetic Gradient		
	Compound Amount Factor	Present Worth Factor	Sinking Fund Factor	Capital Recovery Factor	Compound Amount Factor	Present Worth Factor	Gradient Uniform Series	Gradient Present Worth	
n	Find F Given P F/P	Find P Given F P/F	Find A Given F A/F	Find A Given P A/P	Find F Given A F/A	Find P Given A P/A	Find A Given G A/G	Find P Given G P/G	n
1	1.120	.8929	1.0000	1.1200	1.000	0.893	0	0	1
2	1.254	.7972	.4717	.5917	2.120	1.690	0.472	0.797	2
3	1.405	.7118	.2963	.4163	3.374	2.402	0.925	2.221	3
4	1.574	.6355	.2092	.3292	4.779	3.037	1.359	4.127	4
5	1.762	.5674	.1574	.2774	6.353	3.605	1.775	6.397	5
6	1.974	.5066	.1232	.2432	8.115	4.111	2.172	8.930	6
7	2.211	.4523	.0991	.2191	10.089	4.564	2.551	11.644	7
8	2.476	.4039	.0813	.2013	12.300	4.968	2.913	14.471	8
9	2.773	.3606	.0677	.1877	14.776	5.328	3.257	17.356	9
10	3.106	.3220	.0570	.1770	17.549	5.650	3.585	20.254	10
11	3.479	.2875	.0484	.1684	20.655	5.938	3.895	23.129	11
12	3.896	.2567	.0414	.1614	24.133	6.194	4.190	25.952	12
13	4.363	.2292	.0357	.1557	28.029	6.424	4.468	28.702	13
14	4.887	.2046	.0309	.1509	32.393	6.628	4.732	31.362	14
15	5.474	.1827	.0268	.1468	37.280	6.811	4.980	33.920	15
16	6.130	.1631	.0234	.1434	42.753	6.974	5.215	36.367	16
17	6.866	.1456	.0205	.1405	48.884	7.120	5.435	38.697	17
18	7.690	.1300	.0179	.1379	55.750	7.250	5.643	40.908	18
19	8.613	.1161	.0158	.1358	63.440	7.366	5.838	42.998	19
20	9.646	.1037	.0139	.1339	72.052	7.469	6.020	44.968	20
21	10.804	.0926	.0122	.1322	81.699	7.562	6.191	46.819	21
22	12.100	.0826	.0108	.1308	92.503	7.645	6.351	48.554	22
23	13.552	.0738	.00956	.1296	104.603	7.718	6.501	50.178	23
24	15.179	.0659	.00846	.1285	118.155	7.784	6.641	51.693	24
25	17.000	.0588	.00750	.1275	133.334	7.843	6.771	53.105	25
26	19.040	.0525	.00665	.1267	150.334	7.896	6.892	54.418	26
27	21.325	.0469	.00590	.1259	169.374	7.943	7.005	55.637	27
28	23.884	.0419	.00524	.1252	190.699	7.984	7.110	56.767	28
29	26.750	.0374	.00466	.1247	214.583	8.022	7.207	57.814	29
30	29.960	.0334	.00414	.1241	241.333	8.055	7.297	58.782	30
31	33.555	.0298	.00369	.1237	271.293	8.085	7.381	59.676	31
32	37.582	.0266	.00328	.1233	304.848	8.112	7.459	60.501	32
33	42.092	.0238	.00292	.1229	342.429	8.135	7.530	61.261	33
34	47.143	.0212	.00260	.1226	384.521	8.157	7.596	61.961	34
35	52.800	.0189	.00232	.1223	431.663	8.176	7.658	62.605	35
40	93.051	.0107	.00130	.1213	767.091	8.244	7.899	65.116	40
45	163.988	.00610	.00074	.1207	1 358.2	8.283	8.057	66.734	45
50	289.002	.00346	.00042	.1204	2 400.0	8.304	8.160	67.762	50
55	509.321	.00196	.00024	.1202	4 236.0	8.317	8.225	68.408	55
60	897.597	.00111	.00013	.1201	7 471.6	8.324	8.266	68.810	60
65	1 581.9	.00063	.00008	.1201	13 173.9	8.328	8.292	69.058	65
70	2 787.8	.00036	.00004	.1200	23 223.3	8.330	8.308	69.210	70
75	4 913.1	.00020	.00002	.1200	40 933.8	8.332	8.318	69.303	75
80	8 658.5	.00012	.00001	.1200	72 145.7	8.332	8.324	69.359	80
85	15 259.2	.00007	.00001	.1200	127 151.7	8.333	8.328	69.393	85
90	26 891.9	.00004		.1200	224 091.1	8.333	8.330	69.414	90
95	47 392.8	.00002		.1200	394 931.4	8.333	8.331	69.426	95
100	83 522.3	.00001		.1200	696 010.5	8.333	8.332	69.434	100

APPENDIX D

STATE BOARDS
OF REGISTRATION

Alabama	State Board of Registration for Professional Engineers and Surveyors 300 Interstate Park Drive, Suite 301 Montgomery 36130
Alaska	State Board of Registration for Architects, Engineers and Land Surveyors State Office Building, 9th Foor P.O. Box D-LIC Juneau 99811
Arizona	State Board of Technical Registration 1951 W. Camelback Road, Suite 250 Phoenix 85015
Arkansas	State Board of Registration for Professional Engineers and Land Surveyors Twin City Bank Building, Suite 700 P.O. Box 2541 Little Rock 72203
California	Board of Registration for Professional Engineers and Land Surveyors 1428 Howe Avenue, Suite 56 Sacramento 95825-3298

Colorado

State Board of Registration for
Professional Engineers and Professional
Land Surveyors
600-B State Services Building
1525 Sherman Street
Denver 80203

Connecticut

State Board of Examiners for Professional
Engineers and Land Surveyors
State Office Building, Room G-3A
165 Capitol Avenue
Hartford 06106

Delaware

Delaware Association of Professional
Engineers
2005 Concord Pike
Wilmington 19803

District of Columbia

Board of Registration for Professional
Engineers
614 H Street, N.W., Room 923
Washington 20001

Florida

Board of Professional Engineers
130 North Monroe Street
Tallahassee 32399-0755

Georgia

State Board of Registration for
Professional Engineers and Land Surveyors
166 Pryor Street, S.W.
Atlanta 30303

Guam

Territorial Board of Registration for
Professional Engineers, Architects and
Land Surveyors
Guam Department of Public Works
P.O. Box 2950
Agaña 96910

Hawaii

Board of Registration of Professional
Engineers, Architects, Surveyors and
Landscape Architects
1010 Richards Street
P.O. Box 3469
Honolulu 96801

Idaho	Board of Professional Engineers and Land Surveyors 842 La Cassia Drive Boise 83705
Illinois	Department of Professional Regulation Professional Engineers' Examining Committee 320 West Washington Street, 3rd Floor Springfield 62786
Indiana	State Board of Registration for Professional Engineers and Land Surveyors 1021 State Office Building 100 North Senate Avenue Indianapolis 46204
Iowa	Engineering & Land Surveying Examining Board 1918 S.E. Hulsizer Ankeny 50021
Kansas	State Board of Technical Professions Landon State Office Building 900 Jackson, Suite 507 Topeka 66612-1214
Kentucky	State Board of Registration for Professional Engineers and Land Surveyors Kentucky Engineering Center 160 Democrat Drive Frankfort 40601
Louisiana	State Board of Registration for Professional Engineers and Land Surveyors 1055 St. Charles Avenue, Suite 415 New Orleans 70130-3997
Maine	State Board of Registration for Professional Engineers State House, Station 92 Augusta 04333
Maryland	State Board of Registration for Professional Engineers 501 St. Paul Place, Room 902 Baltimore 21202

Massachusetts	Board of Registration of Professional Engineers and Land Surveyors Saltonstall Building, Room 1512 100 Cambridge Street Boston 02202
Michigan	Board of Professional Engineers 611 West Ottawa P.O. Box 30018 Lansing 48909
Minnesota	State Board of Registration for Architects, Engineers, Land Surveyors and Landscape Architects 402 Metro Square 121 East 7th Street St. Paul 55101
Mississippi	State Board of Registration for Professional Engineers and Land Surveyors 239 N. Lamar, Suite 501 P.O. Box 3 Jackson 39205
Missouri	Board for Architects, Professional Engineers and Land Surveyors 3523 North Ten Mile Drive P.O. Box 184 Jefferson City 65102
Montana	Board of Professional Engineers and Land Surveyors 1424 9th Avenue Helena 59620-0407
Nebraska	State Board of Examiners for Professional Engineers and Architects 301 Centennial Mall, South P.O. Box 94751 Lincoln 68509
Nevada	State Board of Registered Professional Engineers and Land Surveyors 1755 East Plumb Lane, Suite 135 Reno 89502

New Hampshire	Board of Professional Engineers Storrs Street Concord 03301
New Jersey	State Board of Professional Engineers and Land Surveyors 1100 Raymond Blvd, Room 317 Newark 07102
New Mexico	Board of Registration for Professional Engineers and Surveyors 440 Cerrillos Road, Suite A Santa Fe 87501
New York	State Board for Engineering and Land Surveying State Education Dept. Cultural Education Center Madison Avenue Albany 12230
North Carolina	Board of Registration for Professional Engineers and Land Surveyors 3620 Six Forks Road Raleigh 27609
North Dakota	State Board of Registration for Professional Engineers and Land Surveyors 420 Avenue B East P.O. Box 1357 Bismarck 58502
Northern Mariana Islands	Board of Professional Licensing P.O. Box 449 CHRB Saipan, MP 96950
Ohio	State Board of Registration for Professional Engineers and Surveyors 77 South High Street, 16th Floor Columbus 43266-0314
Oklahoma	State Board of Registration for Professional Engineers and Land Surveyors Oklahoma Engineering Center, Room 120 201 N.E. 27th Street Oklahoma City 73105

Oregon	Board of Engineering Examiners 750 Front Street, N.E., Suite 240 Salem 97310
Pennsylvania	State Registration Board for Professional Engineers 613 Transportation & Safety Building P.O. Box 2649 Harrisburg 17105-2649
Puerto Rico	Board of Examiners of Engineers, Architects and Surveyors Ciputacion Provencial Building, Room 201 P.O. Box 3271 San Juan 00904
Rhode Island	State Board of Registration for Professional Engineers and Land Surveyors 233 Richmond Street, Suite 230 Providence 02903
South Carolina	State Board of Registration for Professional Engineers and Land Surveyors 2221 Devine Street, Suite 404 P.O. Drawer 50408 Columbia 29250
South Dakota	State Commission of Engineering and Architectural Examiners 2040 West Main Street, Suite 304 Rapid City 57702-2497
Tennessee	State Board of Architectural and Engineering Examiners 500 James Robertson Parkway, 3rd Floor Nashville 37219
Texas	State Board of Registration for Professional Engineers 1917 IH 35 South P.O. Drawer 18329 Austin 78760

Utah	Representative Committee for Professional Engineers and Land Surveyors 160 East 300 South P.O. Box 45802 Salt Lake City 84145
Vermont	Board of Registration for Professional Engineers Pavilion Building Montpelier 05602
Virginia	Board for Architects, Professional Engineers, Land Surveyors and Landscape Architects 3600 West Broad Street Richmond 23230-4917
Virgin Islands	Board for Architects, Engineers and Land Surveyors Property and Procurement Building No. 1 Sub Base, Room 205 St. Thomas 00801
Washington	State Board of Registration for Professional Engineers and Land Surveyors 1300 S. Quince Street, 2nd Floor P.O. Box 9649 Olympia 98504
West Virginia	State Board of Registration for Professional Engineers 608 Union Building Charleston 25301
Wisconsin	Examining Board of Architects, Professional Engineers, Designers and Land Surveyors 1400 East Washington Avenue P.O. Box 8935 Madison 53708-8935
Wyoming	State Board of Registration for Professional Engineers and Professional Land Surveyors Herschler Building, Room 4135E Cheyenne 82002

INDEX